S276 Geology
Science: Level 2

Book 3
Fossils and Sedimentary Rocks

Written by Peter Sheldon (Book Chair), Angela Coe and Fiona Hyden, based in part on earlier contributions by Evelyn Brown, Peter Skelton and Chris Wilson.

This publication forms part of the Open University course S276. Details of this and other Open University courses can be obtained from the Student Registration and Enquiry Service, The Open University, PO Box 197, Milton Keynes MK7 6BJ, United Kingdom (tel. +44 (0)845 300 60 90; email general-enquiries@open.ac.uk).

Alternatively, you may visit the Open University website at www.open.ac.uk where you can learn more about the wide range of courses and packs offered at all levels by The Open University.

To purchase a selection of Open University course materials visit www.ouw.co.uk, or contact Open University Worldwide, Walton Hall, Milton Keynes MK7 6AA, United Kingdom for a brochure (tel. +44 (0)1908 858793; fax +44 (0)1908 858787; email ouw-customer-services@open.ac.uk).

The Open University
Walton Hall, Milton Keynes
MK7 6AA

First published 2010.

Copyright © 2010 The Open University

All rights reserved. No part of this publication may be reproduced, stored in a retrieval system, transmitted or utilised in any form or by any means, electronic, mechanical, photocopying, recording or otherwise, without written permission from the publisher or a licence from the Copyright Licensing Agency Ltd. Details of such licences (for reprographic reproduction) may be obtained from the Copyright Licensing Agency Ltd, Saffron House, 6–10 Kirby Street, London EC1N 8TS (website www.cla.co.uk).

Open University course materials may also be made available in electronic formats for use by students of the University. All rights, including copyright and related rights and database rights, in electronic course materials and their contents are owned by or licensed to The Open University, or otherwise used by The Open University as permitted by applicable law.

In using electronic course materials and their contents you agree that your use will be solely for the purposes of following an Open University course of study or otherwise as licensed by The Open University or its assigns.

Except as permitted above you undertake not to copy, store in any medium (including electronic storage or use in a website), distribute, transmit or retransmit, broadcast, modify or show in public such electronic materials in whole or in part without the prior written consent of The Open University or in accordance with the Copyright, Designs and Patents Act 1988.

Edited and designed by The Open University.

Typeset by The Open University.

Printed and bound in the United Kingdom by Halstan Printing Group, Amersham.

ISBN 978 1 84873204 9

1.1

The paper used in this publication contains pulp sourced from forests independently certified to the Forest Stewardship Council (FSC) principles and criteria. Chain of custody certification allows the pulp from these forests to be tracked to the end use (see www.fsc-uk.org).

The S276 Course Team

Course Team Chair
Peter Sheldon

Course Manager
Glynda Easterbrook

Main Authors
Tom Argles
Stephen Blake (Book 2 Chair)
Angela Coe
Nigel Harris
Fiona Hyden
Simon Kelley
Peter Sheldon (Book 3 Chair)
Peter Webb (Book 1 Chair)

External Course Assessor
Dr Alan Boyle
(University of Liverpool)

Other Course Team Members
Kevin Church (Consultant Reader)
Roger Courthold (Graphic Artist)
Sue Cozens (Warehouse Production Manager, Home Kit)
Sarah Davies (eLearning Advisor)
Ruth Drage (Media Project Manager)
Linda Fowler (Exam Board Chair)
Michael Francis (Media Developer, Sound and Vision)
Sara Hack (Graphic Artist)
Chris Hough (Graphic Designer)
Richard Howes (Lead Media Assistant)
Martin Keeling (Picture Researcher)
Jane MacDougall (Consultant Reader)
Clive Mitchell (Consultant Reader)
Corinne Owen (Media Assistant)
Andrew Rix (Digital Kit video filming)
Colin Scrutton (Consultant Reader)
Bob Spicer (Reader)
Andy Sutton (Software Developer)
Andy Tindle (Digital Kit and Virtual Microscope photography)
Margaret Tindle (Kit Technician)
Pamela Wardell (Editor)

Course Secretary
Ashea Tambe

Other contributors are acknowledged in specific book, video and multimedia credits. The Course Team would also like to thank all authors and others who contributed to the previous Level 2 *Geology* course S260, from which parts of S276 are derived.

The cover photograph shows Rackwick Bay on the island of Hoy, Orkney. The boulders are mostly Devonian sandstones and conglomerates of the Old Red Sandstone, interspersed with dark-grey, generally smaller, boulders and cobbles of basaltic lava, also of Devonian age. Copyright © Andy Sutton.

Contents

1	**Introduction**	**1**
1.1	The observation → process → environment concept	5
1.2	Summary of Chapter 1	8
1.3	Objectives for Chapter 1	8
2	**Introduction to fossils**	**9**
2.1	The scientific value of fossils	11
2.2	Interpreting morphological features in fossils	13
2.2	Summary of Chapter 2	16
2.3	Objectives for Chapter 2	16
3	**Fossil classification and palaeobiology**	**19**
3.1	Classifying body fossils	19
3.2	Making sense of body fossils	25
3.3	Interpreting fossils as living organisms	29
3.4	Relationships and classification	63
3.5	Body fossils of other organisms	66
3.6	Trace fossils	80
3.7	A brief perspective on the history of life	91
3.8	The use of fossils in biostratigraphy	95
3.9	Summary of Chapter 3	98
3.10	Objectives for Chapter 3	102
4	**Fossilisation**	**103**
4.1	From death to final burial	104
4.2	Diagenetic processes following final burial	111
4.3	Metamorphism of fossils	120
4.4	Summary of Chapter 4	121
4.5	Objectives for Chapter 4	122
5	**Fossils as evidence of past environments**	**123**
5.1	Clues to physical and chemical conditions	125
5.2	Palaeoecological relationships between fossil organisms	129
5.3	Summary of Chapter 5	133
5.4	Objectives for Chapter 5	133
6	**From rocks to sediments**	**135**
6.1	The weathering process	135
6.2	Physical weathering	135
6.3	Biological weathering	136
6.4	Chemical weathering	136
6.5	Summary of Chapter 6	139
6.6	Objectives for Chapter 6	140
7	**Sediments on the move**	**143**
7.1	Sediment transport by fluids: water and wind	143
7.2	Erosional structures	150

7.3	Sediment deposition	151
7.4	Aqueous bedforms	155
7.5	Aeolian bedforms	166
7.6	Transport and deposition by sediment gravity flows	170
7.7	Post-depositional structures	176
7.8	Glacial processes	177
7.9	Summary of Chapter 7	178
7.10	Objectives for Chapter 7	180
8	**Sediments from solution**	**183**
8.1	Carbonate sediments	183
8.2	Siliceous sediments	188
8.3	Carbonaceous sediments	189
8.4	Evaporites	190
8.5	Summary of Chapter 8	190
8.6	Objectives for Chapter 8	191
9	**From sediments to rocks**	**193**
9.1	Compaction	193
9.2	Cementation in sandstones	194
9.3	Cementation in limestones	197
9.4	Cementation in mudstones	198
9.5	Nodules	198
9.6	Classifying sedimentary rocks	199
9.7	Summary of Chapter 9	204
9.8	Objectives for Chapter 9	206
10	**Introduction to sedimentary environments**	**207**
10.1	Factors that control sediment deposition	208
10.2	Facies	210
10.3	Vertical successions of facies and graphic logs	214
10.4	Facies and environmental models	218
10.5	Summary of Chapter 10	218
10.6	Objectives for Chapter 10	219
11	**Alluvial environments**	**221**
11.1	River flow	221
11.2	The river valley	222
11.3	Fluvial channel styles and their typical sedimentary successions	223
11.4	Alluvial fans	233
11.5	Summary of Chapter 11	234
11.6	Objectives for Chapter 11	235
12	**Deserts**	**237**
12.1	Wind in the desert	237
12.2	Water in the desert	240
12.3	Summary of Chapter 12	243
12.4	Objectives for Chapter 12	243

13	**Siliciclastic coastal and continental shelf environments**	**245**
13.1	Subdivision of the coastal and continental shelf environment	246
13.2	Interaction of fluvial, tidal and wave processes	247
13.3	Strandplains	252
13.4	Barrier islands	257
13.5	Deltas	260
13.6	Summary of Chapter 13	265
13.7	Objectives for Chapter 13	266
14	**Shallow-marine carbonate environments**	**267**
14.1	Controls and processes affecting carbonate deposition	268
14.2	Similarities and differences between shallow-marine carbonates and siliciclastics	271
14.3	Carbonate platforms	272
14.4	Peritidal facies	273
14.5	Carbonate buildups (including reefs) and biostromes	277
14.6	Carbonate platform facies models	280
14.7	Summary of Chapter 14	286
14.8	Objectives for Chapter 14	287
15	**Deep-sea environments**	**289**
15.1	Introduction	289
15.2	Processes and sedimentation in the deep sea	290
15.3	Sediment gravity flows in the deep sea	291
15.4	Hemipelagic and pelagic sediments	298
15.5	Summary of Chapter 15	300
15.6	Objectives for Chapter 15	301
16	**Glacial environments**	**303**
16.1	Introduction	303
16.2	Erosive glacial landforms	304
16.3	Depositional glacial landforms	306
16.4	Glacial lake sediments	308
16.5	Wind-blown loess	308
16.6	Glaciomarine environments	308
16.7	Glacial successions	310
16.8	Summary of Chapter 16	312
16.9	Objectives for Chapter 16	312
17	**A matter of time**	**313**
Answers to questions		**317**
Acknowledgements		**338**
Index		**341**

Chapter 1 Introduction

This book is about fossils, sediments and sedimentary rocks. As in Book 2, there is a lot about geological processes – in this case the sedimentary processes that act on the Earth's surface. An understanding of processes happening today is essential for interpreting events and environments that occurred in the past and which are recorded in the geological record. The idea that 'the present is the key to the past' has been an important concept since the early days of geology, and it still is. Today, with concerns about global warming and other effects of climate change, *understanding the past is now seen as a key to the future.*

Earth's surface processes are extremely complex, involving a web of interconnected and open systems linking the planet's interior and land surface with its atmosphere, oceans and biosphere. Understanding how these processes and systems interact over time requires an interdisciplinary approach. Crucial to this endeavour is an understanding of sedimentary rocks. Strata are the diaries of Earth history and the evolution of life, containing within them a record of the past laid down one layer upon another, like successive pages in a journal. Of course, many such pages are missing from the diaries, and even where the pages survive, there may be little or nothing about life written on them, all evidence of life having long vanished. Returning to a theme introduced in Book 1, deciphering the history of Earth and its life is a piece of detective work that, more than any other science, routinely involves changing scales of time and space. Bear in mind as you study this book that the inorganic and organic worlds interact with each other over many scales in highly complex ways.

To give you an overall flavour of Book 3, and to illustrate some principles and methods for studying sedimentary rocks and their fossils, the first activity involves watching a video sequence about features seen along a modern coast. The features of this modern coast are compared with ancient sedimentary deposits that are also interpreted as having been formed along a coast, in this case over 300 Ma ago in the Carboniferous Period. In both ancient and modern examples, the animals and plants (or their remains), which together may be termed the **biota**, are considered too.

On Ordnance Survey maps of Britain, the coastline is defined by the mean high-tide level. However, the position of the coastline changes rapidly. About 30% of the entire coastline of England is estimated to be eroding at rates in excess of 10 cm y^{-1}. Several parts of the coastline of eastern England are eroding at a long-term average rate of about 2 m y^{-1}. For example, since Roman times the Holderness coast of East Yorkshire has retreated about 4–5 km, with the loss of at least 30 villages. Locally, rates of retreat may be even higher, especially where the cliffs are made of glacially deposited soft sands and clays (Figure 1.1). At Covehithe in Suffolk, on the East Anglian coast, the rate of retreat has averaged about 6 m y^{-1} for several decades, and a storm surge in 1990 caused overnight recession of 35 m locally. Conversely, deposition may extend the coastline; what is removed in one place may be deposited in another. For example, the rate of growth of new land at Dungeness in Kent averaged 5 m y^{-1} between 1600 and 1800. More recently, growth averaged less, about 1–2 m y^{-1}, and in the past few decades human movement of shingle to prevent erosion near two power stations has altered the pattern of sediment accumulation in the area. Overall, climate change is expected to increase rates of erosion around the coast through accelerated rates of sea-level rise and changes in wave conditions.

(a)

(b)

Figure 1.1 Rapid erosion on the East Anglian coast. Spot the difference in these views of houses at the end of Beach Road, Happisburgh, Norfolk, England, taken in (a) May 1998 and (b) December 2009. Note in (b) the large igneous boulders brought by boat from Norway to inhibit erosion of the glacial sands and clays forming the cliffs.

Chapter 1 Introduction

Coastlines are dynamic environments that are constantly changing in response to the interaction of various tectonic, climatic and sedimentary processes, and the activities of organisms. Over geological timescales, coastlines may change so dramatically that the outline of many countries, including Britain, would be unrecognisable further back than about 10 000 years ago. Sea level has fluctuated widely before and during the Quaternary Ice Age. For example, until about 9000 years ago, when rapidly melting ice on land caused sea level to rise, it was possible for human ancestors to walk to and fro between southeast England and France, Holland or Belgium, only getting their feet wet in rivers. For much of geological time, though, Britain, has been entirely, or at least largely, covered by sea.

As is often the case in geology, rare, short-lived events can have significant, often long-lasting consequences. For example, a massive subduction-related earthquake off Sumatra, Indonesia on 26 December 2004 triggered a tsunami that affected the coasts of most landmasses bordering the Indian Ocean, killing nearly 230 000 people in eleven countries. The total length of the earthquake rupture was about 1600 km. In Aceh Province, northern Sumatra, (Figure 1.2), the wave reached a height of 24 m when coming ashore along large stretches of the coastline, rising to 30 m in some areas when travelling inland. In this area, the tsunami removed almost the entire suite of depositional landforms of beaches, sand dunes and wetlands, and eroded the coast back for about 500 m except at the rocky headlands. In places, though, a new coast started to form within weeks, closely resembling the pre-tsunami version. In the Aceh region, the level of the land had changed relatively little after the earthquake, with local subsidence of 10–20 cm. Elsewhere, the effects were more marked and will last much longer. The earthquake released stresses that had been building up for many years, and the warped crust sprang back to its pre-stressed profile. Depending on their proximity to the rupture zone, the result was that some islands (e.g. the Andaman Islands and Simeulue) rose in places by up to about 2 m; conversely, parts of some islands further away from the rupture (e.g. Great Nicobar) subsided by over 2 m.

Figure 1.2 View from a helicopter of the west coast of Aceh Province, Sumatra, Indonesia after the tsunami on 26 December 2004. Note the near-total devastation of the coastal plain and signs of run-up far inland in the distant valley.

3

Figure 1.3 provides some global context within which to view the accumulation of sedimentary rocks and the evolution of life during the course of the Phanerozoic.

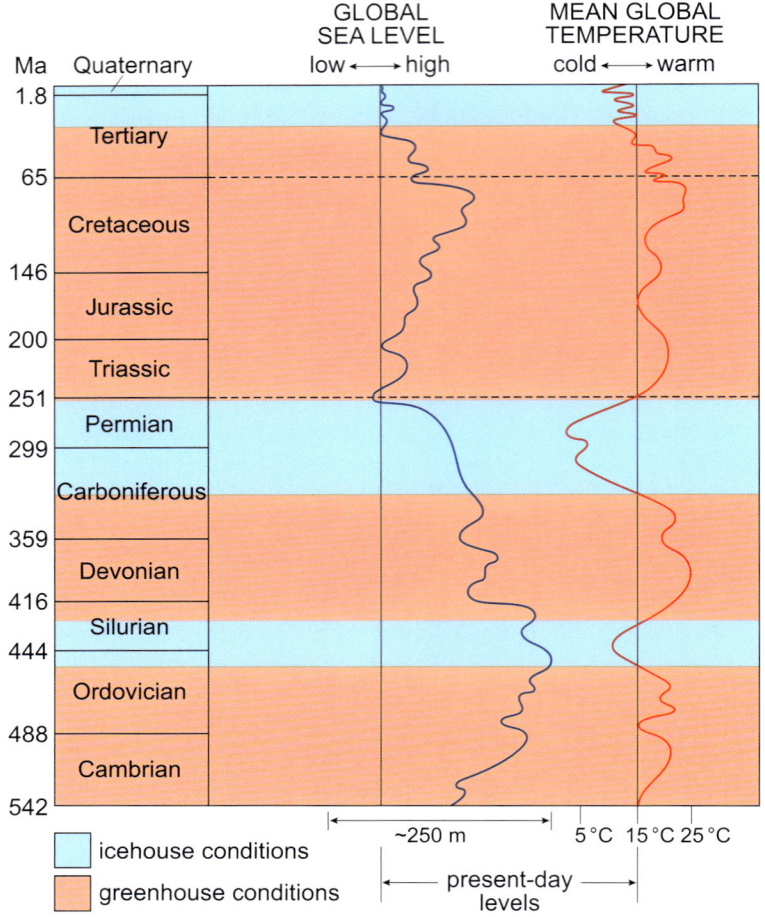

Figure 1.3 Changes in global sea level and global temperature during the Phanerozoic.

- During the Phanerozoic, intervals of warm 'greenhouse' conditions have been interspersed with cold 'icehouse' conditions. Which of these two types of global climate has tended to last longer?

  ☐ Greenhouse conditions.

The curves for changes in global sea level and mean global temperature depicted in Figure 1.3 are, of course, highly generalised and do not show the high-frequency fluctuations revealed by high-resolution analysis of the sedimentary record. Changes in sea level and temperature, both globally and locally, are caused by many interacting processes, discussion of which is beyond the scope of this course.

Question 1.1

According to Figure 1.3, during the Phanerozoic, when has (a) global sea level been (i) highest and (ii) lowest; (b) global temperature been (i) highest and (ii) lowest?

Chapter 1 Introduction

Activity 1.1 Coastal processes

This activity involves watching the video sequence *Coastal processes* on DVD 4 and answering some questions.

Activity 1.1 and the brief account of the 2004 Indian Ocean tsunami are reminders that beaches are much more geologically active than the idyllic scenes depicted in holiday postcards suggest. Complex interactions between moving water, shifting sediments and the activities of living organisms are taking place continuously. As you saw in the video sequence, these same processes leave behind clues that may be preserved in sedimentary rocks.

1.1 The observation → process → environment concept

Once evidence has been gathered from rocks by making careful observations, these observations can be used to deduce the processes involved in their formation. Often, the overall objective is to reconstruct the environments in which the rocks were formed. This is the useful three-stage conceptual framework of making *observations*, deducing *processes* and reconstructing *environments*, i.e. observation → process → environment.

A geologist studying sedimentary rocks and their fossils hunts for clues, like a detective, decides what the evidence reveals about the processes that shaped those rocks, and reconstructs the 'scene of the crime'. In essence, geologists use the principle of **consilience** (i.e. corroboration of a hypothesis by independent testimony, as in a legal trial), a principle frequently demonstrated by Sherlock Holmes, the fictional detective. Geologists are, however, perhaps more fortunate than most detectives because much of what is found in strata at the site of deposition is relevant evidence. An important skill, though, is learning to recognise what counts, and what does not.

When studying sedimentary rocks and their fossils, many useful and diagnostic observations can be made on hand specimens and on exposures in the field without the need for sophisticated equipment. The processes of formation and deposition of sedimentary rocks can be readily deduced from putting together these observations; for instance, questions such as 'was the sedimentary rock deposited by wind or water?' can in many cases be answered by looking at the texture of the grains in the rock, any sedimentary structures that formed at the time of deposition, and any fossils the rock contains. Similarly, a question such as 'was the direction of transport always the same or was it changing?' can often be answered by making observations on the type of sedimentary structure and deducing how the structures were formed. The correct answers to these kinds of questions allow us to hypothesise about the different possible environments in which the rocks may have been deposited. However, to determine which specific environment is represented, many observations about the rock and its field relationships are often required, though it is surprising how much one can sometimes determine from just a single hand specimen.

An example of how the observation → process → environment concept can be used is illustrated in the following question.

Question 1.2

For all the parts of this exercise, you will need to use RS 10 from the Home Kit and the small card showing the grain-size scale and comparison charts for rock texture. Part (a) involves completing Table 1.1; parts (b) and (c) involve answering a number of short questions.

(a) Observations

Observations on sedimentary rocks can be grouped into the useful checklist given in Table 1.1, so that you know you have not missed anything. Remember that even 'negative' observations, i.e. recording that something is *not* present, are often as important as 'positive' ones. Do not forget to use your grain-size scale. Do not be daunted if you cannot see all the features (which are difficult to ascertain in fine-grained sedimentary deposits, for example). Just record as much as you can on the right-hand side of Table 1.1. Remember that basic features such as colour and hardness give an indication of the composition.

Table 1.1 Checklist for the description of sedimentary rocks. For use with Question 1.2.

Property			Questions	Observations for RS 10
Composition	Grains		What is the composition of the most abundant grains?	
	Matrix		Is there any finer-grained fragmentary material infilling the spaces between larger grains? If so, what is it?	
	Cement		Is there any crystalline material precipitated around the edges of grains, or in the spaces between grains? If so, what is it?	
Texture	Crystalline or fragmentary		Is the rock crystalline or fragmentary?	
	Grain size		What is the most abundant grain size present (use your grain-size scale)?	
	Grain sorting		Are the grains all more or less of the same size (i.e. well sorted) or different sizes (i.e. poorly sorted)?	
	Grain **morphology**:			
		shape or form	Are the grains long and thin or equidimensional?	
		roundness	Do the grains have rounded or angular corners?	
		sphericity	Are the grains like spheres?	
	Grain surface texture		Are any quartz grains present smooth and glassy, or are they frosted?	
	Grain fabric (packing)		Are the grains orientated in any preferred direction?	
			Are the grains closely packed together? Are the grains matrix-supported or grain-supported?	
Fossils			Can you see the remains of any biota or the results of their activities?	
Sedimentary structures			Are there any obvious layers in the rocks?	

(b) Processes

From the observations in part (a), you can deduce a lot of information about the processes that were responsible for the deposition of the original sediment. Answering the following questions will guide you through this stage.

(i) Was RS 10 deposited where the energy of the transporting medium was relatively high or low, or where there was a mixture of high and low energy?

(*Hint*: In siliciclastic rocks, grain size generally indicates the energy conditions: clay and silt indicate fairly low energy, whereas larger grains indicate increasing energy because higher current speeds are required to move them and the clay and silt grains have been winnowed away. In carbonate rocks, the size of the clasts is often dependent on the organisms present rather than the energy conditions, because some produce large shells and skeletons and others small ones.)

(ii) Do the grains look as though they have been moved around and sorted or been derived from a rock with well-sorted grains, or are they poorly sorted because they have not been transported far or because they have been derived from a rock with a mixture of grain sizes?

(*Hint*: The sorting and grain morphology of RS 10 should form the basis of your answer to this question.)

(iii) Was the sediment deposited in water or in air?

(*Hint*: Remember from Book 1 that the surface texture of the grains tells you this. If the grains are smooth and glassy, the sediment has been deposited in water because the grains are cushioned; if it was deposited in air, the grains are constantly colliding and knocking tiny chips off each other, creating surface pits that give the grains a frosted appearance.)

(iv) What other evidence (or lack of it) is there to support your answer to part (iii)?

(v) Does the presence (or absence) of any sedimentary structures tell you anything about the processes?

(c) Environment

In what possible environments could this rock have been deposited?

Now think back to the video sequence you watched in Activity 1.1. There was evidence for several different processes having operated, including waves, tides, wind and the action of animals and plants, and the features they produce along a modern coastline. This knowledge then allowed observations to be made on the Carboniferous rocks, and hence the processes responsible for depositing these rocks could be deduced.

In this instance, it was relatively easy because the presenters were making a direct comparison between the modern-day coast and a Carboniferous example to teach you the concepts. However, usually when geologists look at a series of rock exposures and specimens, it takes time to decide the environment of deposition: they initially have to make observations and then determine the processes from first principles. With experience, this can be done quite quickly – but beware, even professional geologists can make mistakes through lack of careful observations.

The rest of this book is in three parts. Chapters 2–5 look at various aspects of fossils – how to observe, classify and interpret them, how fossils form, and how they can provide evidence about ancient environments (i.e. **palaeoenvironments**). Chapters 6–9 consider the processes that form sediments and sedimentary rocks, and how to make relevant observations to reveal which processes have been operating. Chapters 10–16 demonstrate how observation of rocks and deduction of processes enable geologists to reconstruct palaeoenvironments (e.g. a deep sea, desert or ice sheet), one of the main goals of geology. Chapter 17, at the end of the book, gives some perspective on timescales.

Finally, bear in mind that an understanding of fossils, sedimentary rocks and palaeoenvironments is essential for geologists searching for fossil fuels – oil, natural gas and coal. Detailed discussion of the origin, occurrence and exploration for fossil fuels is beyond the scope of this course, but much of the content in this book is good grounding for geologists involved in the extraction of these and other natural resources found in sedimentary rocks.

1.2   Summary of Chapter 1

1. Given concerns about global warming and climate change, understanding the past is now considered a key to the future. Investigating how Earth's highly complex surface processes interact now and did in the past requires an interdisciplinary approach.

2. Sedimentary rocks are the diaries of Earth history and the evolution of life. Reconstruction of ancient sedimentary environments and the organisms that lived in them is a frequent objective when studying sedimentary rocks. A three-stage conceptual framework of making *observations*, deducing *processes* and reconstructing *environments* is often appropriate.

3. Coastlines exemplify dynamic environments that are constantly changing in response to the interaction of various tectonic, climatic and sedimentary processes, and the activities of organisms.

4. Observations of sedimentary rocks can be grouped into four major categories: composition, texture and grain morphology, fossils, and sedimentary structures.

1.3   Objectives for Chapter 1

Now you have completed this chapter, you should be able to:

1.1 Describe, in very general terms, the relationships between sediments and sedimentary rocks; living biota and fossils; and sedimentary structures and bedforms in modern and ancient coastal environments.

1.2 List the key observations that can be made about sedimentary rocks that reveal the processes, and hence environments, that contributed to their formation.

1.3 Give examples from the video sequence *Coastal processes* or RS 10 of how observations can be used to interpret sedimentary processes, from which an environmental setting can then be deduced.

1.4 Give examples of rapid rates of coastal processes, such as erosion and uplift from earthquakes, based on modern observations.

Chapter 2   Introduction to fossils

A **fossil** is simply any evidence of ancient life, naturally preserved within the materials that make up the Earth. Usually, such evidence is found within a sedimentary rock, but other possibilities for entombment include natural tars and resins, ice or even lava flows. There is no strict dividing line in terms of age between recent organic remains and fossils. As a rough guide, most palaeontologists would probably consider any evidence of life over about 10 000 years old to be a fossil; the question of definition is not often an issue, however, as most fossils are millions of years old.

Body fossils preserve something of the *bodily remains* of animals or plants, such as shells, bones (Figure 2.1) and leaves, or their impression in the enclosing sediment. It does not matter whether or not the parts of the body have been altered in chemical composition or physical structure. **Trace fossils** preserve evidence of the *activity* of animals, such as their tracks (Figures 2.2 and 2.3), trails, burrows, or borings. Trace fossils are often the only evidence we have of extinct organisms whose bodies lacked any hard parts. Unlike body fossils, in which the body may have been transported after death a long way away from where the original organism lived (by currents, for example), most trace fossils are direct, *in situ* evidence of the environment at the time and place the organism was living (Section 3.6). The wide range of fossilisation processes, particularly those involved in the formation of body fossils, are discussed in Chapter 4.

Figure 2.1   Complete skeleton (2.8 m long) of a female Jurassic ichthyosaur *Stenopterygius* from Holzmaden, Germany. Near the rear of the rib cage of the mother, with its snout in her pelvic area, lies a baby, which has been interpreted either as dying along with the mother during childbirth, or as being born spontaneously, a little sooner than otherwise, at the moment the mother died of another cause. Other babies lie within her, although they are hard to see and are disarticulated, probably due to decay within a void (i.e. the body cavity). Ichthyosaurs gave birth to live young (rather than hatching eggs), as in some modern marine reptiles (sea snakes). In a study of 45 *Stenopterygius* females found with young inside the body cavity, the majority (64%) had just one or two embryos, but litter size ranged up to a maximum of eleven embryos.

Figure 2.2 Spectacular parallel tracks of two Cretaceous dinosaurs, which are interpreted as walking together, i.e. showing social behaviour. Note the person wearing a red top, bottom left, for scale. These huge quadrupedal herbivores (titanosaurs) were walking on four feet across the floor of a shallow, mostly dried-up lake. Their tracks are preserved here on a steeply dipping limestone bed at a cement factory in the Bolivian Andes.

Figure 2.3 Close-up view of some of the Cretaceous dinosaur tracks in Figure 2.2. A single titanosaur footprint, including the rim, is almost a metre across. In addition to the tracks of the titanosaurs, a series of tridactyl (three-digit) footprints of a single theropod dinosaur (a bipedal carnivore) can be seen crossing from right to left in the middle of the photograph. This can be traced for over 550 m, and is the longest known dinosaur trackway in the world. Note that the limestone bed displays many parallel tensional fractures.

A common practical distinction used in the study of body fossils is based on size: a **macrofossil** is one large enough to be studied with the naked eye or a hand lens, while a **microfossil** is one best studied under a microscope, although microfossils are often visible with a hand lens, too. Microfossils (Section 3.5.3) are derived from many types of organism, both large and small.

In some rocks, the only evidence of life may be chemical residues known to have been produced solely by life processes. For example, certain diagnostic breakdown products (biomarkers) from cell membranes allow organic matter derived from higher plants to be distinguished from residues of algal or bacterial origin. Such **chemical fossils** can only be detected with specialist analytical equipment and, although of growing value, especially in studies of Precambrian life, are much less important overall than body fossils and trace fossils.

Pseudofossils are misleading structures produced by inorganic processes that by chance look as if they are evidence of ancient life; they are therefore not fossils.

2.1 The scientific value of fossils

The scientific study of fossils is known as **palaeontology** (which means 'the study of ancient life'). What we can infer from fossils may be outlined under various headings:

- *The biology of ancient organisms*. The reconstruction of the anatomy and life habits of ancient organisms is called **palaeobiology**. The study of how ancient organisms interacted with their environment and with each other is called **palaeoecology**.

- *Environmental reconstruction*. Information about the palaeobiology and palaeoecology of organisms can be combined with sedimentological observations to interpret the original depositional environments of the rocks in which the fossils are found. Investigation of the manner of death and disintegration of ancient organisms (as well as their biology when living), often helps in the reconstruction of palaeoenvironments. Fossils have provided a great deal of data on past environmental and climatic changes, at all scales from local to global.

- *Evolution and extinction*. Fossils provide the only direct evidence of the existence of long-extinct species and the course of life's evolution through time. Without fossils, we would not know, for example, that dinosaurs ever existed or when trees first appeared. Palaeontology has been contributing to the development of evolutionary theory since Darwin's time. The fossil record has, for example, yielded much information on the transition between major groups of organisms, such as between fish and amphibians, and between reptiles and mammals. High-resolution studies, in which large numbers of specimens (usually invertebrates) can be collected from many thin, successive stratigraphic intervals representing relatively short amounts of time, have been able to demonstrate patterns of evolution within single species and the formation of new species. New and sometimes dramatic fossils are continually being found. For example, in the 1990s, dinosaurs with feathers preserved were discovered in the Liaoning Province of northeast

China (Figure 2.4). Initially, all such specimens were found in rocks of early Cretaceous age, and thus were *younger* than the earliest known bird, *Archaeopteryx*, from the late Jurassic of Germany. Then, in 2009, again in the Liaoning Province, the first dinosaurs with feathers were discovered in Jurassic rocks – rocks 30 million years older than those where the first feathered dinosaurs were discovered. These specimens *pre-dated Archaeopteryx* by some 5–10 million years. The newly found, crow-sized form, called *Anchiornis*, whilst not the direct ancestor of birds, showed that bird-like feathered dinosaurs were around early enough for related forms to be the ancestors of birds. The temporal paradox had been removed.

- *Biodiversity*. Fossils provide data on long-term changes in biodiversity – data that are highly relevant today, given recently elevated rates of extinction resulting from habitat destruction, introduction of non-native species, pollution, human population growth and overharvesting.
- *Stratigraphy*. The distribution of fossils within sedimentary successions, both vertically and laterally, can be used for the relative dating and correlation of strata (Section 3.8). This scientific information is of great economic value, particularly in exploration for fossil fuels (coal, oil and natural gas) but also for other resources found in rocks, such as deposits containing metals and industrial minerals.

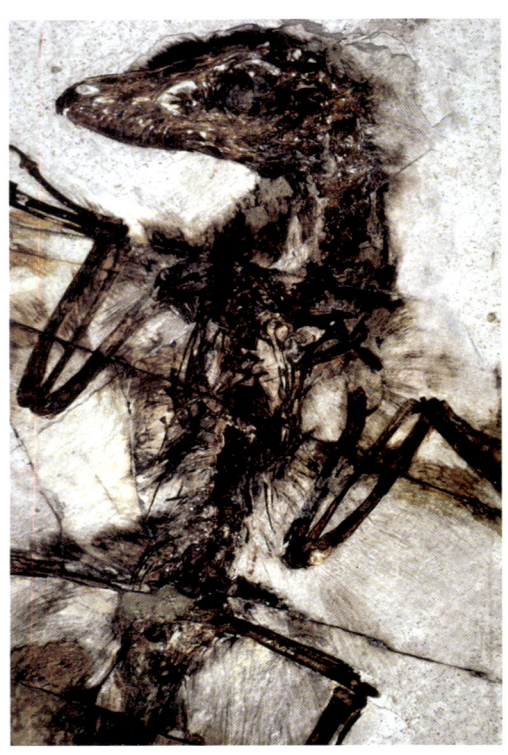
(a)

(b)

Figure 2.4 Feathered dinosaurs from the early Cretaceous of China. (a) *Sinornithosaurus*. This is probably a juvenile specimen, covered in finely branched feathers. Length (with tail) 60 cm. (b) *Microraptor*. With its long tail (only part of which is shown), this dinosaur was about 75 cm in length. It had feathers on all four limbs, as did the even earlier Jurassic form *Anchiornis* that predated the earliest known bird, *Archaeopteryx*.

- *Reconstruction of ancient geography and changing plate configurations.* The geographic distribution of fossils, combined with other information mentioned above, can assist in reconstructing the arrangements of ancient oceans, landmasses, continental shelves, seaways and other major features of the Earth's surface that change with time. In addition, the analysis of strain in distorted fossils found in mountain belts can reveal the stresses to which rocks have been subjected during plate movements (Book 2, Section 8.3).

This book concentrates on the first two topics listed above – palaeobiology and environmental reconstruction.

2.2 Interpreting morphological features in fossils

Organisms are the products of evolution. The morphology of any species combines a fundamental organisation, inherited from very distant ancestors, with more recently evolved adaptations that enhance the survival and reproduction of individuals in the species' own particular circumstances. Any modification that has evolved as a consequence of natural selection is called an **adaptation**. All organisms have numerous adaptations with particular functions. The likely functional significance of adaptations in fossil organisms can sometimes be recognised by comparison with living organisms (or even similarly constructed working models), and this information can then be used to interpret the conditions in which the fossil organisms lived.

It is important to distinguish between features that are similar in two species due to *inheritance from a common ancestor*, and those features whose similarity has arisen *independently* in two species as a result of evolution under similar conditions.

To illustrate the distinction between these types of similarity, consider the extinct woolly mammoth. Suppose there were no specimens of mammoths with hair or soft tissues preserved. The skeletons alone would indicate merely some kind of elephant, identifiable as such from its massive, ridged molar teeth, for example (Figure 2.5). These teeth were inherited by the mammoth as the evolutionary legacy from the ancestral elephant's adaptation to a rough vegetarian diet. Nothing in the mammoth's skeleton specifically reveals adaptation to cold conditions.

Figure 2.5 A molar tooth (20 cm long) of a woolly mammoth, found in a gravel pit in East Anglia, England. The biting surface of the tooth, which slopes down from top right to bottom left, consists of hard enamel ridges separated by softer dentine and cement.

Today's two surviving types of elephant (the African elephant and the Asian elephant) happen to live in hot climates. With only the skeletons of mammoths available, we might be tempted to infer that Siberia, for example, must have been hot, too, when this extinct elephant lived there.

- ■ What would be wrong with such a line of reasoning?
- □ As noted above, the mammoth is recognised as a kind of elephant because of features such as its skeleton and the massive, ridged molar teeth inherited from the common ancestor of all elephants. Though reflecting certain ancestral traits (such as diet in the case of its molar teeth), these features offer no reason why the mammoth should not have been adapted to cold conditions. Just because it was a type of elephant, it need not have shared the same particular adaptations of living elephants.

Contrary to popular belief, and old reconstructions showing mammoths roaming around in nothing but a white landscape, mammoths did not live in a habitat dominated by ice and snow. Rather, they lived in a mostly grassy steppe–tundra environment, beyond the ice sheets. The mammoth's vegetarian diet was mainly grasses, sedges and occasional herbs, although hundreds of plant types have been identified in the stomachs of frozen individuals.

Similarity of organisation inherited from a common ancestor is termed **homology**. As the example above shows, homology is an unreliable guide to the specific conditions experienced by past organisms. Homologous features may indeed become adapted to very diverse functions, as illustrated by vertebrate forelimbs: the wings of birds, the front legs of horses, the flippers of seals, the front legs of moles and our own arms, are all homologous. All these forelimbs retain the basic pattern of bones inherited from early amphibians, i.e. the humerus, ulna, radius, carpals, and so on – and have been subsequently modified for functions as diverse as flying, running, swimming and burrowing. At best, we may sometimes be able to make some generalisations from homologous features, where little modification is evident, such as inferring the retention of a rough vegetarian diet from the teeth of mammoths. On the other hand, homology is the essential key to recognising evolutionary relationships, and thus classification.

Figure 2.6a shows a baby female mammoth carcass discovered in 2007 by a nomadic reindeer herdsman in northern Siberia. It had remained deep-frozen in permafrost since its death about 42 000 years ago, when it became trapped in, and suffocated by, thick mud. In the baby's stomach are residues of its mother's milk. This fossil, like some other exceptional specimens, has a certain amount of preserved hair. The coarse, wiry outer hairs of a woolly mammoth (Figure 2.6b) were about six times thicker than a typical human hair. Their orange–brown colour today is probably not natural, but the result of pigment change during long burial. Below this coarse hair was a dense layer of short, fine hairs. Many unrelated mammals living today – such as the musk ox – have evolved long, shaggy coats in very cold climates, so it is reasonable to infer that mammoths with similarly long coats might also have been adapted to such conditions. This idea could be tested by seeking other evidence from, say, association of the fossils with glacial sediments (i.e. by consilience).

Chapter 2 Introduction to fossils

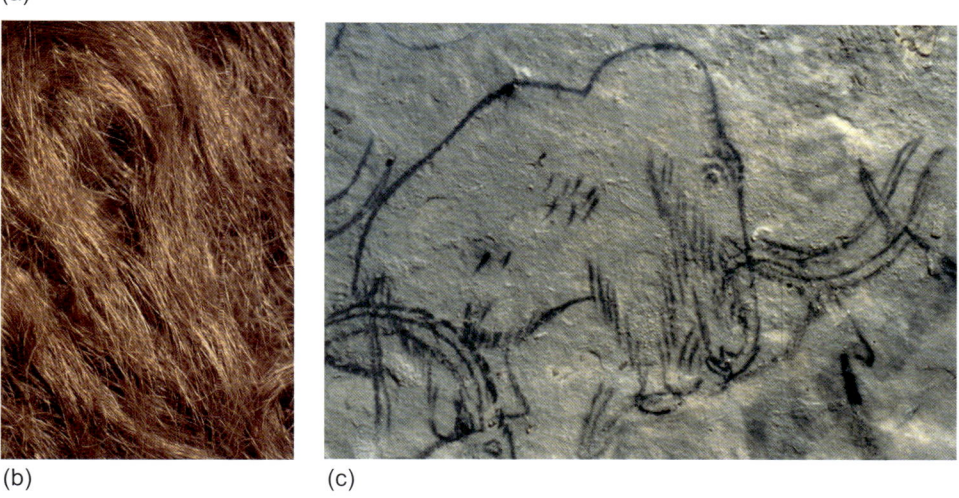

Figure 2.6 (a) A baby female mammoth, found in northern Siberia in 2007, that died aged 1–2 months about 42 000 years ago. It is about 1.3 m long, and retains soft parts and the body's full three-dimensional shape. Much of the fur has been lost, but some tufts of hair are preserved. (b) The outer hair of a woolly mammoth. See text for discussion. (c) Cave painting of a woolly mammoth, from Rouffignac Cave, France. The artist has accurately drawn two projections at the end of the trunk, matching the trunks of frozen carcasses, and has portrayed how the foot swelled when weight was placed on it, just as it does in living elephants.

Similarity that arises as a result of independent adaptation to the same conditions by different organisms is termed **analogy**. The analogous features are produced during convergent evolution, when the morphology of independently evolving groups of organisms converges, so that they come to look more similar. Though superficially similar, reflecting their common function, analogous features may differ markedly in basic structure, especially when they have evolved in organisms that are only very distantly related. Such is the case, for example, with the wings of birds and insects; the common ancestor of birds and insects (probably in the late Precambrian) did not have wings, so their wings are analogous. However, analogous features in more closely related organisms may be more similar in detail if based on the same (i.e. homologous) original components. Thus, in the example of the different woolly mammals discussed above, it is the exaggerated shagginess of their coats that is analogous among them; the hair itself (which they inherited from the ancestral mammal) is homologous. So, analogy can offer precise clues as to how fossil organisms lived.

To interpret fossils correctly then, we need to know about the inherited common structural organisation of each group of organisms – its **bodyplan** – in order to distinguish the signals of ancient history from more recently acquired adaptations to prevailing circumstances. That involves classification of the major groups often found as fossils, a topic of the next chapter.

2.3 Summary of Chapter 2

1. A fossil is any evidence of ancient life, naturally preserved within the materials that make up the Earth (usually a sedimentary rock). Body fossils preserve something of the bodily remains of animals or plants, or their impression in the enclosing sediment. A macrofossil is one large enough to be studied with the naked eye or a hand lens, while a microfossil is best studied under a microscope. Trace fossils preserve evidence of the activity of animals. Chemical fossils are biochemical residues detectable only with specialist equipment.

2. Fossils provide scientific information on: the biology of ancient organisms (palaeobiology) and how such organisms interacted with their environment and with each other (palaeoecology); the reconstruction of palaeoenvironments; the existence of extinct species and the course of evolution through time; evolutionary theory, including the origin and extinction of species; changes in biodiversity; the relative dating and the correlation of strata; the reconstruction of ancient geography and changing plate configurations.

3. The morphology of any species combines a fundamental organisation, inherited from very distant ancestors, with more recently evolved adaptations that enhance the survival and reproduction of individuals in that species' own particular circumstances. It is important to distinguish between features that are similar in two species *due to inheritance from a common ancestor* (homology), and those features whose similarity has arisen *independently* in two species as a result of evolution in similar conditions (analogy).

2.4 Objectives for Chapter 2

When you have completed this chapter, you should be able to:

2.1 Define, and distinguish between, body fossils (macrofossils and microfossils), trace fossils, chemical fossils and pseudofossils.

2.2 Outline the scientific value of fossils.

2.3 Explain the difference between homologous and analogous features of organisms and recognise the value of each to the interpretation of fossils.

Now try the following questions to test your understanding of Chapter 2.

Question 2.1

Place the following items in the appropriate general category of fossil (e.g. a trace fossil): (a) the hair of a woolly mammoth (Figure 2.6b); (b) a cave painting of a woolly mammoth (Figure 2.6c); (c) petroleum compounds (hydrocarbons) found in Jurassic shales; (d) crystals of manganese dioxide resembling a fern (Figure 2.7a); (e) the moulted shell of a Palaeogene crab; (f) the bite marks of a carnivorous dinosaur on the skull of a herbivorous dinosaur; (g) the tooth of a Cretaceous shark (Figure 2.7b); (h) the impression of the external surface of a Jurassic snail shell preserved in limestone (the shell itself having been dissolved away); (i) the footprint of a dinosaur (Figure 2.7c).

(a)

(b)

(c)

Figure 2.7 (a) Crystals of manganese dioxide resembling a fern, precipitated from solutions passing along a crack in limestone. Field of view 6 cm across. (b) Tooth of a Cretaceous shark, *Cretalamna* (length 2.5 cm). (c) Left footprint of a Cretaceous theropod dinosaur, 30 cm across. The impressions of four digits are visible, including the backward pointing hallux (or 'big toe'), bottom right. For use with Question 2.1.

Question 2.2

Ammonites (discussed later in Section 3.3.4) are an extinct group of marine invertebrates that had a long tubular shell coiled up over the animal's body. Figure 2.8a shows a typical ammonite shell. Though only distantly related to ammonites, the living pearly nautilus (Figure 2.8b and c) has a similar shell. In life, much of the pearly nautilus shell is filled with gas, providing buoyancy and allowing the animal to swim around. The position of the coiled, gas-filled part of the shell in the water, situated above the denser, soft parts of the body, also maintains the animal in a stable vertical orientation as it swims. Ammonites are interpreted to have swum in a similar fashion. Although ammonites and nautiluses apparently inherited a gas-filled shell from a distant common ancestor, the coiling of the shell above the body apparently evolved independently in the two groups. On the basis of this information, decide which of the following features should be considered homologous, and which analogous, between the two groups: (a) the ability to swim; (b) coiling of the shell; (c) buoyancy; (d) the maintenance of stability by the shell during swimming.

Figure 2.8 (a) A fossil ammonite shell, *Asteroceras*, Jurassic (15 cm); (b) the shell of a modern pearly nautilus (*Nautilus*) (15 cm); (c) a living pearly nautilus. For use with Question 2.2.

Chapter 3 Fossil classification and palaeobiology

This chapter is closely linked to your study of the fossil replicas, labelled A–P in the Home Kit (and from which you will also need the hand lens). You will need to look at the replicas both as you read this chapter and when you undertake the activities, so you should ensure that you can study in a place where you can conveniently lay them out. It would also be useful to have ready access to your computer as, after covering a particular group, you are usually directed to explore the fossils of that group in the Digital Kit; some activities also use the Digital Kit. The Digital Kit includes images of all the replicas in the Home Kit, as well as many other fossils spanning a wider range of groups. Sometimes, you may choose to check some feature of a replica in the Digital Kit as you study the replica, rather than at the end. Sometimes, in activities, you are directed to the Book 3 video clips of modern biota on the Book 3 Resources *DVD, though these can be watched as you encounter the relevant group in the text.*

In the Digital Kit, a replica is indicated by an 'R' after a name in the menu for a fossil group. Replicas are shown in black and white to distinguish them from real specimens, which are shown in colour. Different views of the same specimen (whether a replica or not) are indicated by 'a', 'b', 'c', and so on, after the name. In this book (and the workbook) we refer to the fossils in the Home Kit by the abbreviation FS.

3.1 Classifying body fossils

To start with, you will be classifying the Home Kit specimens using a simple key system that is based on a few readily observable features. This is to help you become familiar with the specimens themselves before progressing to other members in the groups they represent and looking at their biology in greater detail, in Section 3.2. Before proceeding, however, you should be aware of the main advantage and disadvantage of key systems. They make classification relatively simple because you have to consider only one feature at a time. So, by going through a sequence of choices, you can eventually classify a specimen into one particular pigeonhole. The process is quick and easy, but it has one major drawback. As a consequence of evolution, one or more of the diagnostic features for a group may be modified or even lost in a given example, so an attempt to classify it using the key too rigidly would lead to problems. It is therefore necessary in biological key systems to allow for frequent exceptions. As you become more familiar with fossils, you should become used to recognising combinations of features in a given specimen, and be able to cope with such exceptions, and so arrive at a correct classification.

The replica fossils (FS) A–M all represent the shelly hard parts of marine invertebrate animals. These specimens have been chosen because they are the sorts of macrofossil you are most likely to encounter in the field.

- Why are the hard parts of shelly marine invertebrates the most commonly encountered macrofossils?

- Environments of deposition are commonplace in the sea, but are scattered and often short-lived on land. The remains of marine organisms thus stand a greater chance of burial, and hence preservation, than do those of terrestrial organisms. Organisms with hard skeletal parts, especially the most abundant shelly forms, obviously stand the greatest chance of leaving durable remains that might eventually be buried. Invertebrates are much more abundant than vertebrates.

Activity 3.1 Numbers of skeletal elements

The starting point for the fossil identification key is to consider the number of discrete skeletal elements in each specimen. In Activity 3.1, you will examine some of the fossil replicas in the Home Kit to determine whether the original skeletons were composed of one, two or more elements.

3.1.1 Specimens with one skeletal element

From Activity 3.1, you may have noticed that FS A, D and E can be distinguished further on their external shape and internal structure. Each shell was originally grown as an expanding tube, but while FS E is approximately straight, FS A and D are both coiled. The interior of FS A was filled with sedimentary matrix, smoothly plugging the shell opening, while FS D is an **internal mould** (i.e. the lithified filling of a fossil shell that records internal surface features), so the site of its shell opening (the widest end of the coiled tube) is also solid. In all three cases, the soft parts of the animal that built the shell were originally housed inside the tube and projected from its open end in life.

- Inspect the sites of the original shell openings and classify each of FS A, D and E according to whether it is: (a) subdivided by numerous thin projections running inwards from the shell wall, or (b) unrestricted (other than by the sedimentary infill).

- You should have classified the specimens as follows: (a) FS E; (b) FS A and D. The opening to the body space of FS E is subdivided by thin, projecting radial walls that extend towards the centre from the rim of the cone. By contrast, the sediment-filled areas corresponding to the shell openings of the other two specimens are smooth, indicating a lack of any such radial walls.

The radial walls visible in FS E are known as **septa** (singular: **septum**). They are a characteristic feature of the skeleton of most **corals**. FS E is from a **solitary** coral, the skeleton having been built by one individual. Some other corals are **colonial**, with a skeleton consisting of either openly branching or closely joined **corallites**, each of which is like the skeletal tube of a solitary coral, but is joined to many other such tubes.

Now turn to FS D.

■ What features on the internal mould of FS D might indicate that the body space occupied only part of the shell interior?

☐ The specimen shows a series of intricately folded fine grooves encircling the coiled internal mould and spaced out at intervals along it, up to nearly half a whorl back from the site of the original shell opening. (A whorl is a single turn in a spiral shell.) Since the specimen is an internal mould, these grooves must indicate where thin shelly partitions within the shell met the inner surface of the originally enclosing shell wall. The body could thus only have occupied the final half whorl of the shell, where these partitions stopped.

Hence most of the shell cavity in FS D was divided into a series of chambers by transverse partitions, also called **septa** (to avoid confusion, it is worth remembering that the septa of corals are *radial*, while those seen here are *transverse*). The complex pattern noted in the question above, where each septum joined the inside wall of the coiled shell, is termed a **suture**. The outer margin of each septum was thrown into a series of tiny folds, rather like the edge of a leaf of curly kale. Possession of such intricately folded septa is diagnostic of the shells of an extinct group of organisms known as **ammonites**. Notice also that FS D is coiled in a single plane; this is common, though not universal, in ammonites.

It is not possible to tell whether FS A contained internal septa like those in ammonites because only the outer surface of the shell is visible. In fact, it did not, and it is not an ammonite. However, its asymmetrical coiling (coiled not in one plane, but as in a spiral staircase), is typical of the shells of most, but by no means all, **gastropods**, a huge group that includes not only snails, whelks and winkles, but also slugs – in which the shell is reduced or lost.

3.1.2 Specimens with two skeletal elements

In FS B and F, the shell consists of two parts (known as **valves**) that fit closely together. In life, the valves were hinged along one margin and the shell could be opened and shut: when shut, the shell entirely enclosed the animal's soft parts.

Both FS B and F show **bilateral symmetry** (i.e. the whole shell can be divided by an imaginary plane into two halves that are mirror images of each other). However, the two specimens can readily be distinguished by the different orientations of the symmetry plane with respect to the shell. Figure 3.1 illustrates the two kinds of symmetry.

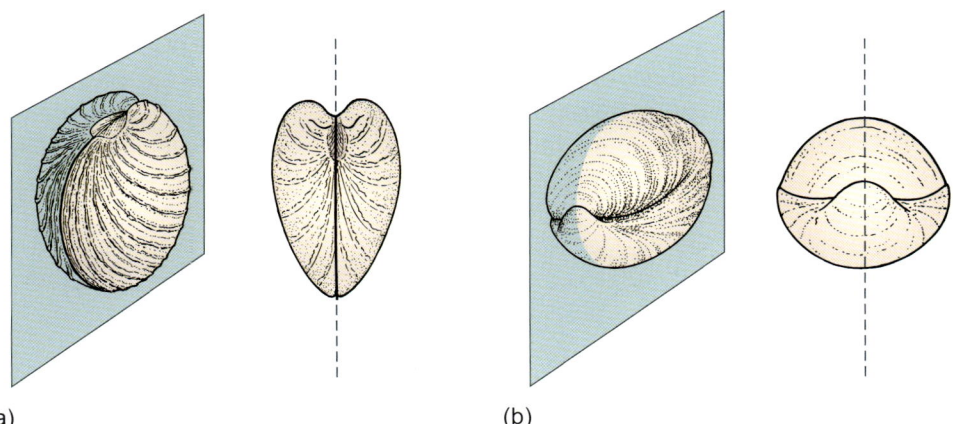

Figure 3.1 Contrast between the orientation of the plane of bilateral symmetry in (a) bivalves, and (b) brachiopods.

(a)   (b)

- Classify FS B and F according to whether they show a plane of symmetry that (i) lies between the two valves so that each valve is a mirror image of the other (Figure 3.1a), or (ii) cuts through both valves so that, although the two valves are differently shaped, each can be divided into two halves that are mirror images of each other (Figure 3.1b).

□ (i) FS B; (ii) FS F.

The first kind of symmetry (Figure 3.1a) is typical of a group known as **bivalves** while the second kind (Figure 3.1b) characterises **brachiopods**. There are, however, several exceptions. Oysters, for example, are bivalves that have distinctly asymmetrical valves. Nevertheless, those bivalves that lack the typical symmetry of the group never show perfect brachiopod symmetry (though scallops are bivalves that sometimes come close to that). A few brachiopods also deviate from the usual symmetry of their group.

3.1.3 Specimens with many skeletal elements

Symmetry again provides the next step in the key for FS C, G and H.

- Classify FS C, G and H according to whether they show (approximately) (a) bilateral symmetry of a row of transverse skeletal elements, or (b) radial symmetry around a central axis, with the shell composed of several sectors of similar elements.

□ (a) FS G; (b) FS C and H. The symmetries of FS G and FS C are obvious. That of FS H is less clear because only part of the specimen is visible. Nevertheless, if you look at it from the end where the stem emerges, you can see that it shows a number of similar radial sectors of plates.

Now inspect FS G. Notice that there is a broad, mask-like **head-shield** (or **cephalon**) at one end, bearing two prominent knobs that are eyes. Behind this is a central trunk region (the **thorax**) consisting of a number of transverse segments that were articulated in life. The **tailpiece** (or **pygidium**) behind the trunk region is of similar construction, though here the segments were fused together.

- How can you tell that the tailpiece segments were fused together?

□ The smooth form of the rim running around the outside of the tailpiece segments shows that they were fused in one piece.

The central part of the body in FS G, from head to tail, known as the **axis**, is slightly raised and the two flanks on either side are mirror images of each other. This threefold (trilobed) longitudinal division of the body gives the name of the group to which the specimen belongs – the **trilobites**. This specimen represents the shelly segments of hard 'skin' that covered the upper side of the animal. An external shelly skeleton of this kind is termed an **exoskeleton**. The skin of the lower surface and of the limbs lacked the mineral component (calcite) responsible for the 'shelliness' of the upper skin and is, therefore, very rarely fossilised.

Now look at FS C and H.

- ■ How many similar radial sectors can you detect in FS C and H?

- ☐ In FS C, you should be able to count five sectors with large plates bearing two rows of large knobs, alternating with another five sectors with smaller plates bearing two rows of smaller knobs flanked by rows of small pits. If you look closely at FS H, particularly around where the stem attaches, you will see that it could (and in fact it does) have five radially arranged sectors of plates, though one is hidden on the unexposed side of the specimen.

Hence both specimens effectively show approximate **fivefold symmetry** around a central axis. Both specimens belong to a major animal group known as the **echinoderms**, most members of which have approximate fivefold symmetry. Important exceptions exist, however, including many forms in which the shell has been modified during evolution so as to become bilaterally symmetrical, with a distinct front and back. In such forms the fivefold arrangement of plates remains, though the sectors are no longer radially symmetrical. This is a good example of the modification of homologous structures to suit new functions, as discussed in Chapter 2.

FS C and H are nevertheless clearly different. FS C is an **echinoid** (sea urchin), one of a group of echinoderms that can move around. FS H, with its distinctive cluster of arms, is a **crinoid**.

- ■ What feature might suggest that FS H was not capable of moving around?

- ☐ The stem leading off from the base of the specimen suggests that the crinoid was attached to the sea floor, in contrast to the echinoid which was free-living (i.e. not attached to anything).

Now that you have sorted FS A–H into their respective groups, summarise your conclusions by answering Question 3.1.

Question 3.1

Working your way through the key in Figure 3.2 (overleaf) to remind yourself of the principles of classification, allocate FS A–H to their appropriate groups, filling in the boxes shown at the foot of the key.

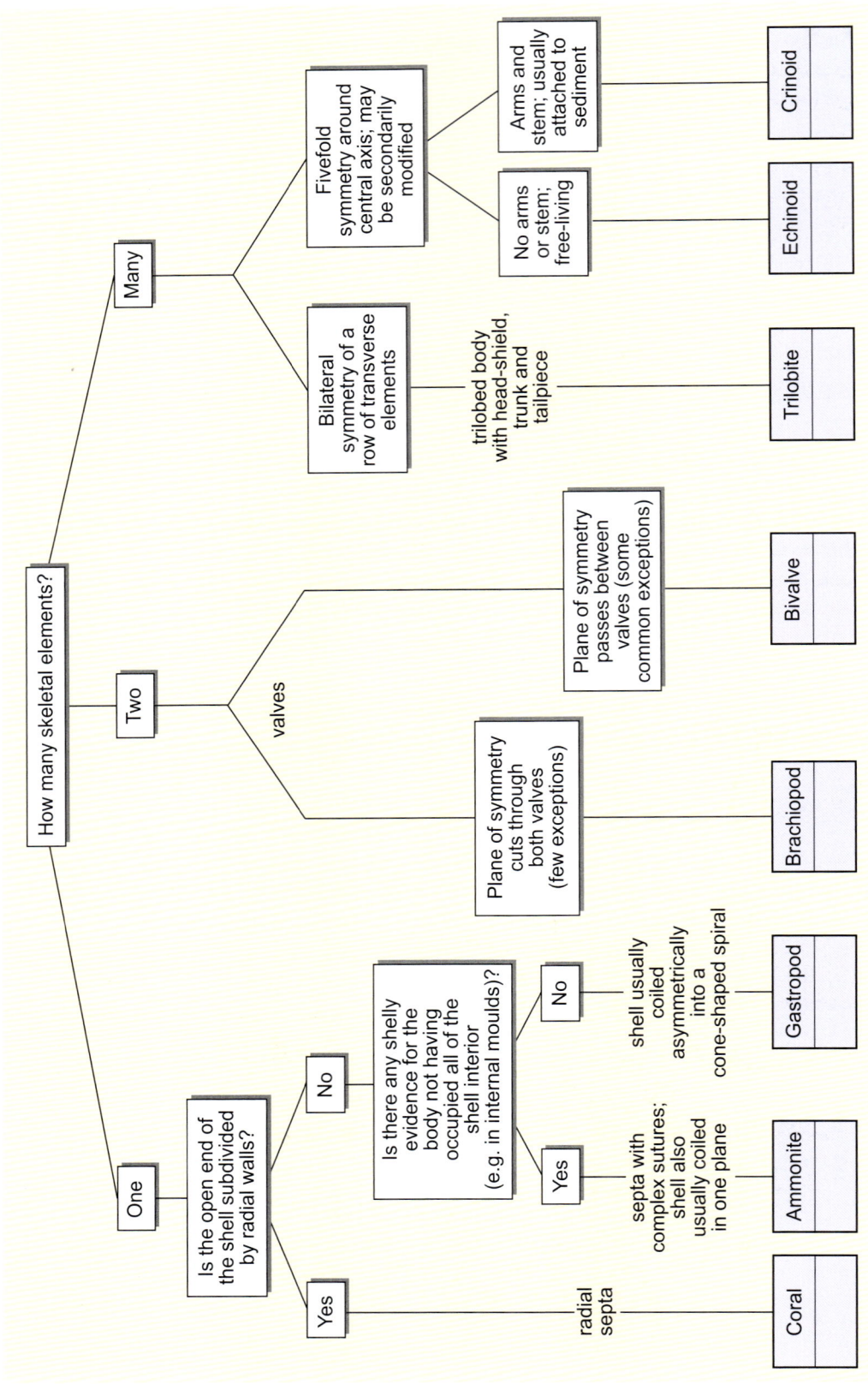

Figure 3.2 Key for the classification of some major marine invertebrate fossil groups discussed in Chapter 3.

3.2 Making sense of body fossils

Assigning a fossil to a particular group is only the first step in understanding it. The following two aspects of body fossils, studied in this section, are widely relevant:

- the ecology of marine invertebrates and their fossil counterparts
- the **morphology** (i.e. shape and structure) of fossils, especially the relationship between the surviving hard parts and the now vanished soft parts.

3.2.1 Ecology of marine animals and their fossils

In studying FS A–M, three questions should be asked: Where did these organisms live in relation to the seabed? Did they move around (were they, for example, burrowers through the sediment, attached to rocks, or active swimmers)? How did they feed? Before tackling these questions, however, we need some idea of the range of possible answers derived from living organisms.

One way of describing the habitat of aquatic organisms is by reference to the sea floor (or, in non-marine settings, the bottom of a lake). Various possible living positions and habits, which are often used as a way of classifying these organisms, are summarised in Figure 3.3. The term **epifauna** applies to animals that live on

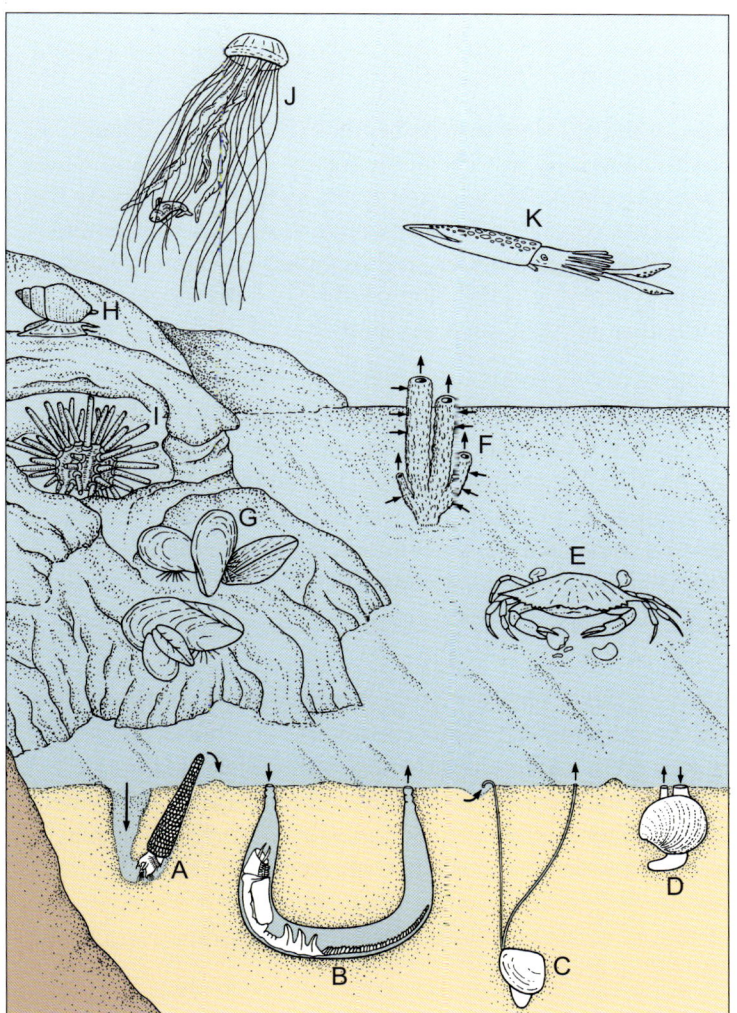

Figure 3.3 The life habits of some adult marine invertebrates. The infauna includes vagrant worms (A) digging through the sediment, more sessile worms (B), and sessile to intermittently vagrant bivalves (C and D), all maintaining connections with the overlying seawater. The epifauna includes vagrant and sessile animals on both soft sediment and hard rocky surfaces. On soft sediment, a vagrant crab (E) and sessile sponges (F) are depicted, while on the rocks there are sessile bivalves (G), a vagrant gastropod (H) and an echinoid (I). In the overlying seawater, the pelagic animals include planktonic jellyfish (J) and nektonic squid (K). Most of these invertebrates have planktonic larval forms. Small arrows indicate feeding currents. The worm (A) selects sand grains of similar size and glues them together to form an agglutinated tube open at both ends.

the sea floor, while animals that live entirely buried within the sediment are the **infauna** – although, in many cases, a connection is maintained with the overlying seawater. In either case, the animals may be **sessile**, habitually staying in one place, or they may be **vagrant**, frequently moving around. Sessile epifauna may just lie on, encrust, or otherwise attach themselves to the surface, while vagrant forms move over it. Many infaunal animals are sessile, living in more or less permanent burrows, but other, vagrant, forms move through the sediment in search of food. Infaunal and epifaunal animals are classed together as bottom dwellers or **benthos**; they are said to be **benthic**.

Not all kinds of sea floor are equally accommodating. Solid rock impedes the establishment of an infauna, though some animals have evolved the ability to bore into solid rock. Epifaunal animals might be expected to be rare in water-rich soupy mud as they would sink into it. Some members of the epifauna, however, are adapted to live on soft sediment with the aid of stilts, spines, broad thin shells, or soft parts that spread the load. Both infauna and epifauna usually have little problem invading muddy sand, which can be penetrated by burrowers and also support surface dwellers.

In discussing the infauna and epifauna, we distinguished between sessile and vagrant animals in both cases, but this needs qualifying: infaunal animals that are sessile (dwelling in permanent burrows) are usually able to move in an emergency, and, in some cases, no strict division can be drawn between vagrant and sessile infauna.

Pelagic animals live above the sea floor and its benthos (Figure 3.3). Pelagic animals that float or swim so weakly as to be at the mercy of currents and winds are collectively known as the **plankton**, whereas active swimmers comprise the **nekton**. Much of the plankton consists of minute animals, such as the floating larvae (juvenile forms) of benthic animals, as well as algae. The larger plankton consists mostly of various types of jellyfish, some with gas-filled floats. The nekton includes fish and other animals, such as squid.

Like most classifications involving living organisms, the one given for their ecology here is not neat and decisive, as many animals do not invariably fall into one category or another. For example, the common prawn buries itself (i.e. is infaunal) during the day, but at night emerges to join the epifauna while it feeds. Other epifaunal animals will bury themselves during low tide. Despite the difficulties in pigeonholing some species, this classification by habitats is widely used, especially as some groups have almost constant habitats. As you will see, adult brachiopods (FS F and L), for example, are practically always epifaunal and sessile.

Benthic species often have a rather patchy distribution across the sea floor. This is due to factors such as variation in sediment type and the amount of suspended sediment in the water, as well as clustered settlement of their larvae. Pelagic animals, by contrast, may be independent of bottom conditions and have wider geographical ranges. Furthermore, while the dead remains of the benthos may be transported by bottom currents, distances of transport are usually fairly limited. However, the carcasses of pelagic animals may float considerable distances (especially when buoyed up by gases produced during decay), sometimes far beyond the range of the living animals.

In addition to divisions based on habitats, animals can be classified by their feeding methods (Table 3.1), though again there is much variation. Many **predators** will actively pursue vagrant prey. Others feed on sessile prey animals, which have to rely on such defences as protective shells or unpalatable taste as a deterrent. **Scavengers** will attack corpses and other decaying animal material, but some scavengers will turn predator if the opportunity arises. Similarly, at times some predators will scavenge.

Table 3.1 Principal feeding types and habits in marine invertebrates.

Feeding type	Habits
predators	hunt and consume live animal prey
scavengers	consume dead and rotting animal material
deposit feeders	select particulate grains (collectors) or eat sediment (swallowers)
suspension feeders	strain food particles from seawater
grazers	scrape food, usually algae, off sediment or rock surfaces, or graze on large plants
parasites	have a long-term attachment to a host, which serves as a food source: the host is usually much larger than the parasite

Certain sediments are rich in residual organic fragments, or **detritus**, including faecal matter and associated bacteria that coat detritus and sediment grains alike. The detritus, and particularly the bacteria, are the food source for **deposit feeders**. Because the total surface area of the grains is much larger in fine-grained sediments, deposit feeders prefer muddy sediments to coarse-grained sands. Two sorts of deposit feeder are distinguished by the degree of selectivity shown during feeding. **Collectors** select a limited size or type of particle, whereas **swallowers** are less discriminate and ingest sediment wholesale, passing out processed sediment in large quantities, as in the abundant worm casts you may have seen on tidal flats.

Seawater contains microscopic suspended food particles that may be strained off, or trapped on an array of tentacles, by **suspension feeders**. In open water, microscopic plankton, especially unicellular algae, make up the bulk of their food material, although close to the sea floor there may be sizeable quantities of detritus resuspended by currents. Commonly, the feeding organ of suspension feeders consists of a grid or fan of tentacles with abundant **cilia** (singular: cilium; minute, regularly beating, hair-like structures projecting from cell surfaces) that both generate a water current and help to trap food particles from it with the aid of secreted mucus. The mucus is then drawn into the mouth. Many suspension feeders have sorting mechanisms to bypass or reject unwanted particles, but in excessive turbidity the animal will stop feeding.

- ■ Suspension feeders are often rare in fine-grained sediments. Why might this be so?
- □ Fine-grained sediments (muds) would tend to clog the feeding organ.

Grazers are animals that rasp or scrape organic mats, especially those composed of algae, from sediment or rock surfaces. They may also feed on larger plants.

The final feeding type to be considered is parasitism, where one animal directly uses another organism as a more or less permanent food source. **Parasites** differ from other feeding types in that it is important that the host remains alive. Death of the host may also kill the parasite, though in some cases the host finally dies only after the parasite has completed its own growth stage. Parasites often have parts of their anatomy simplified (e.g. the gut may be absent), but may have complicated life cycles. Although parasites are extremely common and outnumber free-living species, rather few examples have been documented from the fossil record.

As with the classification based on habitats, the feeding classification is not rigid because of many borderline cases. The overlap between predators and scavengers has already been noted. Another borderline case is that between suspension feeders and animals that trap very small live prey, as is the case with many coral polyps, for example.

Although these two classifications based on habitats and feeding methods are not foolproof, they are simple and particularly applicable to fossils. They can be amalgamated to give a matrix showing all possible combinations (Table 3.2). You will be returning to Table 3.2 as you work through the next section, so you may find it useful to insert a marker to aid quick reference. Note that because the 'habitat' for a parasite is usually its host, parasites are not included in this table.

Table 3.2 Combined feeding and habitat classification system for marine animals.

Habitat		predator	scavenger	deposit feeder collector	deposit feeder swallower	suspension feeder	grazer
pelagic	nekton						
pelagic	plankton						
epifaunal	vagrant						
epifaunal	sessile						
infaunal	vagrant						
infaunal	sessile						

Some combinations in Table 3.2 are, of course, rather unlikely. Thus a wholly pelagic animal could not be a deposit feeder, although deposit feeders might occasionally swim from one feeding site to another. The same argument applies to grazers other than those forms that drift around with floating seaweed. An infaunal suspension feeder might also seem unlikely, but if you refer to Figure 3.3 again, you can see that many infaunal animals maintain connections with the overlying water and so can draw in suspended food.

3.3 Interpreting fossils as living organisms

Several questions should occur to you when looking at FS A–M. The fossils, of course, formed from the hard skeletons, but what were the soft parts like? How did the animals move and on what did they feed? In this section, we will concentrate on building up a picture of what the soft parts of the animals looked like, and then link the functions of both the specific organs and the entire animals to their original ecology.

You will study two groups in detail – the bivalves and the echinoids. One advantage of looking at these groups, as with most of those represented in the Home Kit, is that they have living relatives. Much of the original anatomy of the fossil specimens can thus be inferred by homology. This advantage will be apparent when we try to reconstruct the life and habits of the extinct trilobites.

With the bivalves and echinoids, the extent to which the former soft parts leave marks on the available hard parts will be stressed. However, many soft organs have no direct effect on the hard parts and their reconstruction must be based entirely on living relatives, or, in exceptional cases, preservation of the soft parts.

3.3.1 The bivalve *Circomphalus* (fossil specimen I)

It is important to satisfy yourself that you can see the features of FS I that are described here (again, you will be able to check your observations against the Digital Kit). The hand lens will be an essential aid, and Activity 3.2 will also help your observations by getting you to sketch certain features on an incomplete photograph.

FS I is from a Neogene species of the genus *Circomphalus*, of which some other species are still living. We start by introducing some standard terms that are used to orientate animals. These will help guide your exploration of the specimen, rather like the bearings on a map, as well as assisting comparison with other animals. *Note that these terms are for anatomical comparison alone and do not necessarily indicate how the animal is orientated in life.* Some animals, for example, live with their anatomical 'backs' facing downwards (though this was not, in fact, the case with FS I).

In Section 3.1.2, you looked at FS B (a genus called *Crassatella*), and saw how in bivalves the plane of bilateral symmetry, giving two mirror images, runs *between* the two valves. This means that one valve is placed on each side of the body and the valves are said to be **lateral**. How is the rest of the animal orientated? In bivalves, the back, or **dorsal**, surface of the animal is where the valves are hinged, so that the opposite anatomical underside is **ventral** (Figure 3.4).

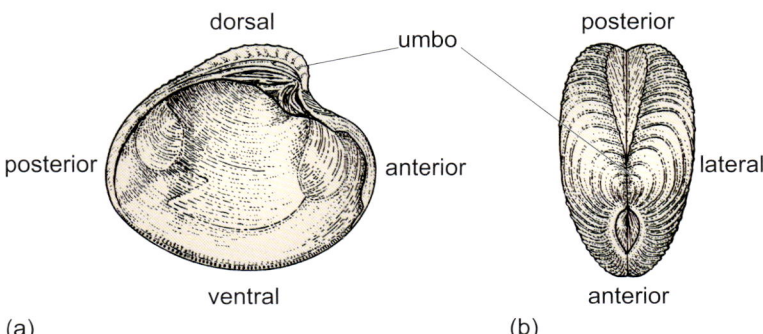

Figure 3.4 Orientation of a bivalve shell: (a) interior of left valve (i.e. the valve on the left side looking towards the anterior, with the dorsal side uppermost); (b) dorsal view of both valves.

In the majority of living animals, the front end, or **anterior**, is marked by the head and mouth, while the anus opens towards, or at, the rear end, or **posterior**. The soft body of bivalves, bearing the mouth and anus (there being no distinct head), rots away after death. This means that in fossil bivalves various shell features must be used to identify the anterior and posterior ends (Figure 3.4). These features are introduced below.

FS I is just one of the valves of *Circomphalus* – the one from the animal's left side (compare with Figure 3.4a). Single, detached valves are common as fossils in bivalves (unlike in brachiopods). Bivalve shells may be made of aragonite (which often dissolves away) or calcite, or a mixture of both. The original shell of *Circomphalus* was composed of aragonite (a polymorph of calcium carbonate), and it had only a thin outer layer of non-living organic material, though, as with the soft parts, this organic layer usually does not fossilise. The shell is thus an exoskeleton (Section 3.1.3). The interior of the shell is relatively smooth, with markings that, as you will see, are directly related to some of the soft parts.

The exterior of each valve is broadly convex, terminating dorsally in a blunt protrusion known as the **umbo** (plural: **umbones**). In most bivalve genera, including *Circomphalus*, the umbo is curved towards the anterior. The exterior has many sharp, broadly spaced ridges arranged concentrically around the umbo.

- What finer surface features can you see between the ridges of FS I? You will need the hand lens to look.

- The ridges are separated by much finer lines, in this case running parallel with the ridges.

The shell was grown by progressive addition of material around its margins and over its inner surface. The fine lines you have just been studying mark the successive increments of this accretionary growth, and are called **growth lines**. Accretionary growth means that the entire history of an individual is recorded from the moment it starts producing the shell. In the case of *Circomphalus*, imagine subtracting successive growth increments from the margin, to leave progressively smaller growth stages, ending in the juvenile shell at the top of the umbo. In doing this imaginary exercise, you will see that the overall shape of the shell remained relatively unchanged as it increased in size. However, in a few bivalves, as well as other groups, there may be a marked change in morphology between the juvenile and adult shells.

Look now at the inner surface of the shell. The dorsal area beneath the umbo consists of a platform that carries grooves and raised areas. This area marks the zone of hinging, or articulation, between the two valves of the original shell. In life, the dorsal margins of the valves were connected by a strip of horny organic material known as the **ligament** (Figure 3.5). Although the ligament itself is not preserved here, its site of attachment to the valve can be detected as a shallow crescent-shaped groove – the **ligament groove** – curving back from the tip of the umbo, alongside the dorsal margin. When the valves were closed (Figure 3.5a), the elastic ligament was deformed like a C-shaped spring, with its inner part compressed and its outer part stretched. So, when the muscles (the **adductor muscles**) that closed the valves relaxed, the stressed ligament opened the valves (Figure 3.5b). The economy of this arrangement is that the bivalve avoided expending energy whilst keeping its valves open – a necessary posture for feeding.

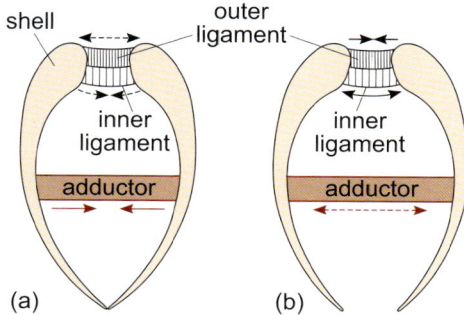

Figure 3.5 Counteraction of the ligament and adductor muscles in the closure (a), and opening (b), of the bivalve shell, seen in diagrammatic section across the valves. Solid arrows show the forces exerted, and dashed arrows show the forces imposed.

Activity 3.2 Bivalve shell morphology

In this activity, you will draw and label part of the shell of *Circomphalus*.

Teeth and **sockets** are integral parts of a bivalve's articulation. They guide opening and tight closing, and so prevent the two valves shearing past one another. Bivalve ligaments are well adapted to cope with the stresses of normal shell opening and closing, but are more susceptible to the twisting stresses arising from shear. The particular arrangement of teeth and sockets seen in FS I is one of several types found in bivalves.

In life, the inner surface of a bivalve shell is lined with a sheet of tissue, known as the **mantle**, which secretes the shell. The mantle margins are complex and bear various sensory organs, as well as a muscular fold or flap in each valve (Figure 3.6), running around the inside of the valve margin. The folds act as 'curtains' and are used to control the size of the gap between the open valves. In many bivalves, the opposing folds are fused together. The points and extent of this fusion vary in different species, but their chief effect is to separate areas of water inflow and outflow.

The muscles that run into the mantle margins, especially the folds mentioned above, are attached to the inside of each valve, leaving an attachment scar known as the **pallial line**.

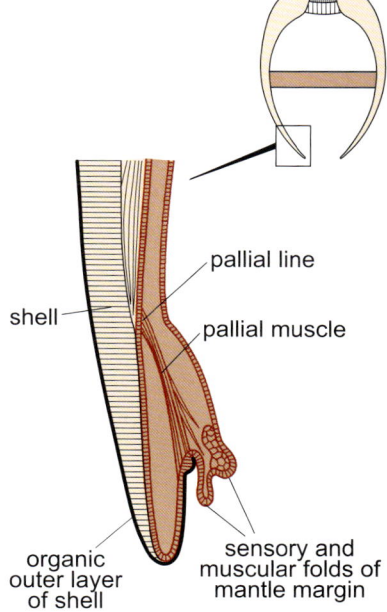

Figure 3.6 Section through the outer edge of one valve, and its mantle lobe, of a bivalve.

Question 3.2

Find the pallial line in FS I. Is it entirely parallel to the ventral margin of the shell, or is any deviation visible?

The posterior kink in the pallial line referred to in the answer to Question 3.2 is called the **pallial sinus**, and is an invaluable indicator of the original presence of **siphons**, which are fleshy tube-like extensions of the cross-fused mantle

(Figure 3.7a). The pallial sinus is invariably posterior and so is another useful guide for orientating the shell. Siphons are typical of infaunal bivalves and serve to connect the animal with the overlying water. If threatened by predators or inclement conditions, the bivalve withdraws its siphons and closes its shell. The space to accommodate the retracted siphons is marked by the pallial sinus. Thus, the absence of a pallial sinus indicates very short siphons, or no siphons, with the bivalve living either on, or just within, the sediment.

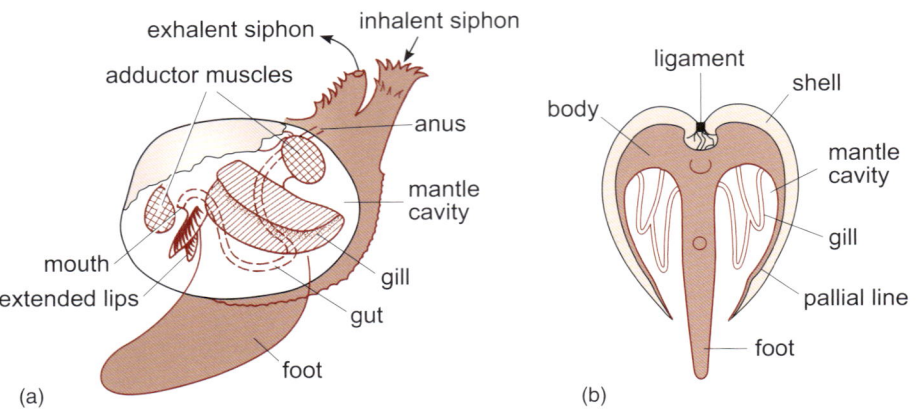

Figure 3.7 Bivalve anatomy: major soft parts in relation to the shell: (a) side view; (b) in cross-section.

Typically, the siphons form an inhalent and exhalent pair (Figures 3.7a and 3.8a). The inhalent siphon serves either to draw in water with suspended food particles or, in some other bivalve species, to search across the sediment surface picking up detrital particles like a vacuum cleaner. The exhalent siphon removes waste material. The siphons also take in oxygenated, and expel deoxygenated, water.

Figure 3.8 Living bivalves: (a) edible cockle, *Cerastoderma edule*, with extended siphons clearly visible; (b) spiny cockle, *Acanthocardia aculeata*, with muscular foot extended, about to burrow.

While studying the pallial line and sinus, you may have noticed two roughly oval markings at the anterior and posterior ends of the shell interior. These are **adductor muscle scars**; they mark the sites of attachment of the adductor muscles that ran between the valves (Figure 3.5). As noted earlier, the adductor muscles shut the valves and so they act in opposition to the elastic ligament.

The remainder of the soft parts in FS I would have consisted of the body, with laterally suspended **gills**, extended lips and a **foot**, which is a muscular ventral extension of the body (Figures 3.7 and 3.8b), attached to the inside of the shell by **protractor** and **retractor muscles** (Figure 3.9). These muscles, respectively, extend and retract the foot. The bivalve body contains the gut and reproductive organs and is suspended from the dorsal part of the shell beneath the hinge area (Figure 3.7b).

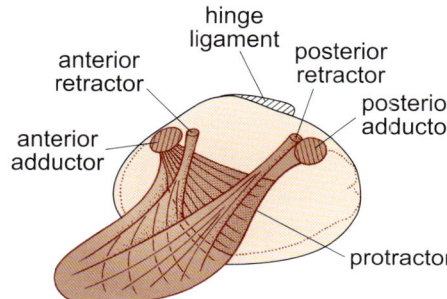

Figure 3.9 The muscles associated with the bivalve foot (and the adductor muscles).

- By reference to Figure 3.9, try to detect any markings corresponding to where the anterior muscles from the foot and body wall might have attached to the inside of the shell in FS I. (You will need to peer obliquely into the valve interior with the posterior rim towards you.)

- By looking at the inside of the shell beneath the hinge area, you may be able to see a small scar immediately behind the anterior adductor scar, where the anterior retractor muscle from the foot was inserted, followed by a row of tiny pits that show where small muscles from the body were attached. The scars for the protractor and posterior retractor muscles cannot be so clearly distinguished.

Arising from both sides of the body would have been the large folded gills that hung like porous curtains in the space surrounding the body, or **mantle cavity** (Figure 3.7). At the anterior end of the gills four fleshy elongated lips would have surrounded the mouth, two of which are visible on Figure 3.7a. These and the gills together formed the feeding organ. We know from living bivalves that the beating of cilia on the gills produces a water current that draws water and food particles in through the inhalent siphon and across the gills, where food particles are trapped and then transported in a stream of mucus to the lips and thence to the mouth. Waste products from digestion are expelled with the water that has passed through the gills and leave via the exhalent siphon. Movements of the overlying seawater ensure a constant renewal of food, and in most circumstances living *Circomphalus* can remain stationary.

- ■ Suggest a physical event that could have disturbed the living position of *Circomphalus*.

- □ During storms, either extensive deposition or erosion of sediment would have forced *Circomphalus* to move upwards or downwards.

When necessary, *Circomphalus* would have burrowed through the sediment – a process involving complex interactions of the shell, contraction of the adductor muscles, dilation of the foot with blood, and movements of the muscles that extend and retract the foot.

- ■ Study the overall shape and roughness of your specimen of *Circomphalus*, and try to decide whether it was a rapid burrower.

- □ The relatively swollen shape of the valves combined with the prominent concentric ribs suggest that rapid movement would have been prevented by high friction, so this species was probably a sluggish burrower.

In some bivalves, the external ornamentation is so arranged as to help burrowing, but some of the most rapid burrowers have smooth compressed shells that can slice through the sediment.

Question 3.3

Thinking back to the discussion of life habits (Section 3.2.1), how would you now classify *Circomphalus* in Table 3.2 (p. 28)?

Bivalves are a very large group, and although the majority of species are (or were) burrowing suspension feeders, such as *Circomphalus*, other modes of life are also well represented. Attachment to the sea floor, for example, is quite common, either by means of anchoring threads of organic material secreted by the foot (e.g. mussels), or by direct cementation of one valve (e.g. oysters). Scallops mostly attach in the first way, though some lie freely on the sediment surface and can actually swim for short distances by clapping their valves, if menaced by predators such as starfish. The characteristic bilateral symmetry of burrowing bivalves is frequently lost in such epifaunal types. Indeed, in the case of the swimming scallops (Figure 3.10), secondary adaptation of the shell for symmetrical streamlining of the clapping valves has superficially invested it with a brachiopod style of bilateral symmetry (Figure 3.1b), though internal features, such as the ligament and an adductor muscle scar, unmask it as a bivalve.

(a)

(b)

Figure 3.10 (a) External and (b) internal view of one valve of a Recent scallop (6 cm). Note the single adductor muscle scar visible on the inner surface in (b).

Some oysters such as *Gryphaea* (Figure 3.11) were cemented only very early in life, after which they were free-living. In adults, the larger and heavier valve sat on the sea floor; the animal could open its lid-like, convex upper valve and close it with a single, centrally placed muscle (conspicuous as a muscle scar in fossils; Figure 3.11b). Such oyster species may appear to have bilateral symmetry, but closer inspection reveals they never do; for example, the umbo always points to one side, away from the midline. Other oysters, such as *Lopha* (Figure 3.12), are much less regular, and shape varies a lot from individual to individual.

(a)

(b)

Figure 3.11 Two species of the Jurassic oyster *Gryphaea*. (a) *Gryphaea arcuata* (5 cm). (b) *Gryphaea dilatata*. Left: complete shell of this Jurassic oyster. Note the lack of symmetry (compare with brachiopods). Right: internal view of a detached upper valve showing the single, central muscle scar (8 cm).

Burrowing has also been replaced in some instances by boring into hard materials, such as rock, stiff cohesive clay or wood. Not all bivalves are suspension feeders: there are several kinds of deposit feeders (collectors), especially among primitive forms, and even a few predators that suck small worms and other such prey into the mantle cavity with highly modified gills. In addition, there are non-marine bivalves, such as freshwater mussels. Thus, a wide variety of habitats and feeding types is encompassed, though careful consideration of the morphology of fossil forms usually means adaptations to specific environmental conditions can be identified.

Bivalves are sometimes useful for global correlation, especially in parts of the Mesozoic and Cenozoic.

Figure 3.12 Two views of a single shell of the oyster *Lopha* (9 cm). Shell shape varies much from individual to individual. Drawing of a Jurassic specimen.

> Bivalves are entirely aquatic. Most are marine, living on shallow sea floors, though some inhabit freshwater. They range in age from early Cambrian to Recent.

Now explore the bivalves in the Digital Kit.

3.3.2 The echinoids *Pseudodiadema* (fossil specimen C) and *Micraster* (fossil specimen J)

The lower surface of FS C is somewhat obscured where it has been encrusted by tubes made by worms. Otherwise, it shows many well-preserved details, although you may need the hand lens to help you see them clearly. For Activity 3.3, which follows on immediately from your study of this specimen, you will also need FS J.

As an introduction to echinoids, the Jurassic genus *Pseudodiadema*, of which FS C is an example, is suitable because it shows fairly generalised features, and relatives of this Jurassic sea urchin are alive today, so that a reconstruction of the form of the soft parts can be made fairly accurately.

The key you completed in Question 3.1 included two important features characteristic of practically all echinoids. These are:

- fivefold symmetry (modified to bilateral symmetry in some forms)
- a skeleton composed of individual plates that are closely joined, giving it the superficial appearance of a single unit.

A hollow skeleton such as this, composed of many interlocking pieces, is known as a **test**. As with *Circomphalus*, the skeleton is composed of calcium carbonate, but in this case, as in all echinoids, it is composed of calcite rather than aragonite. The mode of secretion is also rather different. Instead of the calcium carbonate being deposited *above* a layer of mantle tissue, plates are secreted *within* a thin layer of soft tissue. This means that the echinoid skeleton is strictly internal and so is an **endoskeleton**, not an exoskeleton. Each echinoid plate is not solid, but is riddled with fine pores so that under the microscope the skeleton resembles a sponge-like three-dimensional meshwork (Figure 3.13). In life, this meshwork is filled mostly with a fibrous protein known as **collagen**. It is the collagen fibres that bind the calcite plates together in life. This type of meshwork skeleton is unique to echinoderms. Its texture is usually preserved in fossil material, which means that even small fragments can be distinguished from the skeletal debris of other groups.

A curious characteristic of echinoderm plates, which is very convenient for identification in thin section, is that each plate is a single crystal. Thus, when seen between crossed polars, the whole plate goes into extinction at the same time, despite the pores running through it. Echinoderm plates consist of single, optically continuous crystals, which are often murky-looking in thin section because they are full of pores. The pores may become secondarily filled with calcite, in which case the plates break along cleavage planes, giving flat, reflective surfaces.

The radial appearance of *Pseudodiadema*, whereby all sides of the test look rather similar, means that the terms used to orientate the animal have to be somewhat different from those used for bilaterally symmetrical animals such as *Circomphalus*. The flatter surface, with the mouth, is called the **oral surface**, and opposite that lies the domed **aboral surface**. In life, the oral surface (Figure 3.14b) rested on the sea floor.

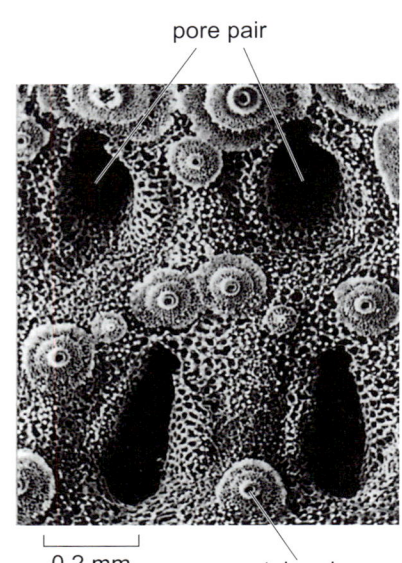

Figure 3.13 The sponge-like structure of the echinoid endoskeleton revealed by an electron microscope.

Chapter 3 Fossil classification and palaeobiology

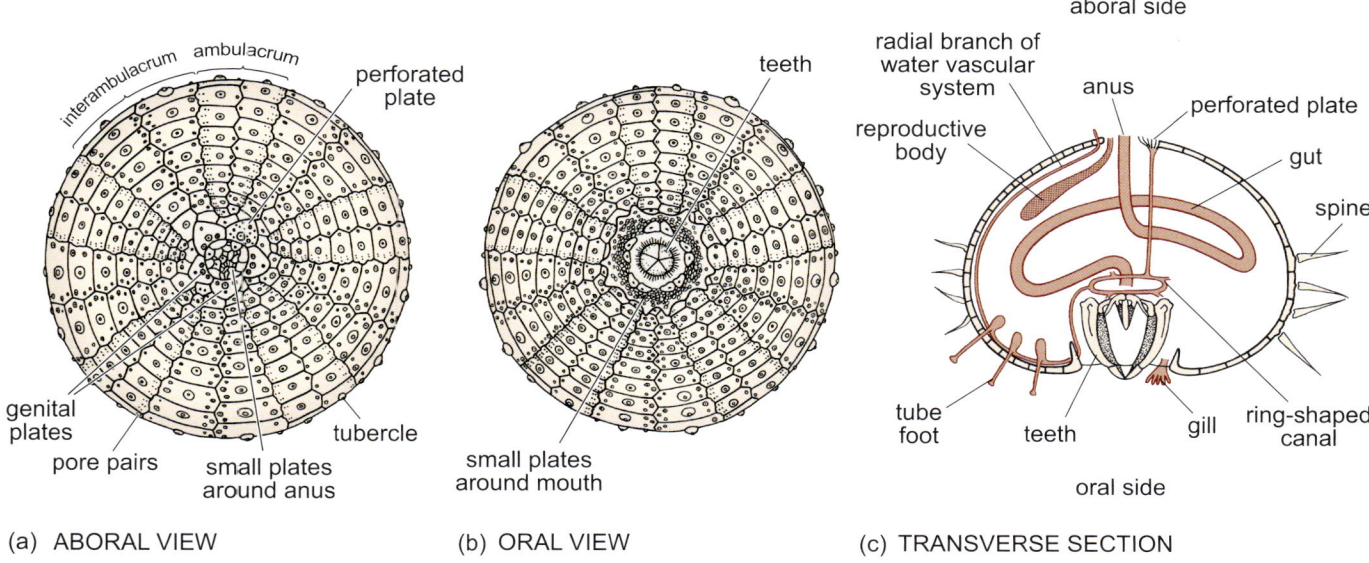

Figure 3.14 Echinoid anatomy: (a) aboral view; (b) oral view; (c) cross-section.

Question 3.4

Look down onto the domed aboral surface of FS C (corresponding to Figure 3.14a). How many *types* of radial zone can you make out on the basis of the ornamentation, and how are they characterised?

The wide zones referred to in the answer to Question 3.4 are termed **interambulacra** (singular: **interambulacrum**), and the narrower and more subdued zones are termed **ambulacra** (singular: **ambulacrum**) (Figure 3.14a). The knobs (of all sizes) are known as **tubercles**. Both ambulacra and interambulacra are arranged with fivefold symmetry, and we shall discuss their significance shortly.

In *Circomphalus*, the valve you studied was effectively intact. The specimen of *Pseudodiadema*, however, has suffered some loss of hard parts. The centre of the oral surface is occupied by an expanse of featureless material that represents infilling sediment. In life, this area was covered by a flexible membrane studded with plates, with the mouth opening in the middle (Figure 3.14b). The mouth contained a complex structure that included a circle of five teeth operated by a series of muscles (Figure 3.14c). Figure 3.15 shows the oral surface of an echinoid in which the teeth are still in place. Such a feeding apparatus was probably used to rasp algae from rocky surfaces, and/or sessile prey such as coral polyps. The smaller hole on the aboral surface was also originally occupied by a group of plates that surrounded the anus (Figure 3.14a). In both oral and aboral regions, the rotting of the membranes on death released the embedded plates and, with the loss of teeth, sediment could enter the skeleton.

Figure 3.15 Oral surface of a Jurassic echinoid, *Paracidaris smithii*, now lacking spines, but with the teeth of the jaw apparatus still in place (7 cm across).

37

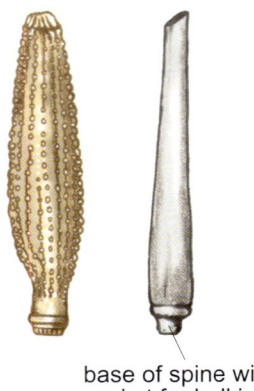

Figure 3.16 Examples of large spines from two fossil echinoid species. The broken tip of the right-hand one is flat because it has broken across a calcite cleavage plane. Length of each 3 cm.

The other major change in the skeleton of FS C has been the loss of all the **spines** that formerly were attached to each tubercle and so covered the animal (Figure 3.14c). In living echinoids, the spines act in defence, and in some species they are poisonous. They are also used in locomotion to lever the animal across the sea floor. The largest spines were on the most prominent tubercles (on the interambulacra). By looking closely at the skeleton, you will see numerous other smaller tubercles that mark the former position of small spines.

The spines (Figure 3.16) are often found isolated in the sediment and it is only in exceptional circumstances that they remain attached (Figure 3.17). Inspect one of the large tubercles on FS C together with Figure 3.18. Between the spine and its base is a ball and socket joint, with the spherical tubercle acting as the ball. The central hole in the tubercle marks where a strand of tissue ran to the spine. The spine was moved by sets of muscles that ran from its base to the outer margin of the tubercle. Look closely at the outer margin of a tubercle and you should see that it forms a flattened ring. The spine muscles were attached onto this ring and consisted of an inner and outer layer (Figure 3.18). If the spine itself was touched, the inner muscle clamped the spine rigidly to its base. If, however, the animal was touched elsewhere, the outer muscles would swivel the spine towards the disturbance. After death, and rotting of the muscles, the spines would tend to fall off the test. Notice how the size of the tubercles, and hence the spines, decreased towards the oral region so that *Pseudodiadema* was not kept away from the sea floor and its food.

Figure 3.17 Underside of the Cretaceous echinoid *Tylocidaris clavigera* (6.5 cm across, including spines). The large spines were defensive. The jaw apparatus has fallen out.

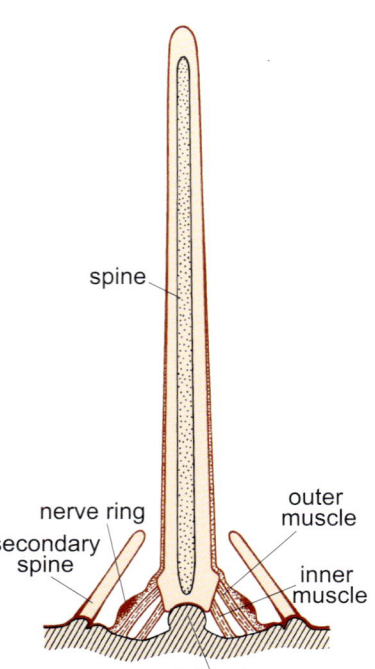

Figure 3.18 Longitudinal section of an echinoid spine and tubercle.

Echinoids and their relatives are unique in possessing arrays of hydraulically operated tentacles connected to a system of fluid-filled canals inside the body known as the **water vascular system** (Figures 3.14c and 3.19). This consists of a ring-shaped canal around the gut, which gives rise to five radial branches, each one running up the inside of an ambulacrum. The ring canal is connected to the exterior of the animal by a tube leading to a single perforated skeletal plate (the **perforated plate**). The perforations allow the pressure between the water vascular system and the seawater to remain in equilibrium.

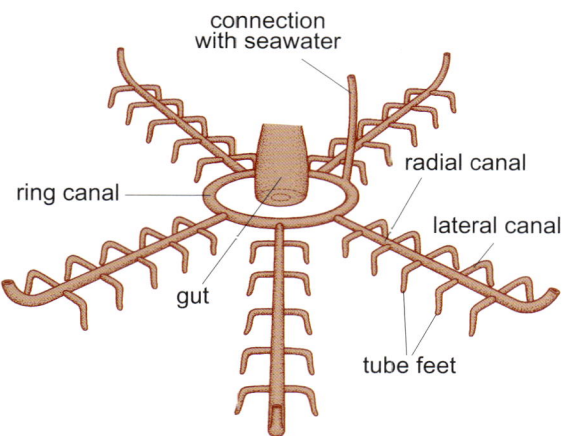

Figure 3.19 Water vascular system of a generalised echinoderm. The mouth is shown facing downwards, as in echinoids (in crinoids, it faces upwards).

■ Look down on the aboral surface of FS C again and, using the hand lens, try to identify this perforated plate amongst the ring of plates immediately surrounding the central hole.

☐ Surrounding the central hole are several small plates. You should be able to see that there are five larger plates with another five smaller plates wedged between them. One of the larger plates is slightly bigger, more convex, and has a porous appearance. This is the perforated plate.

Each of the radial branches of the water vascular system bears numerous soft tentacles called **tube feet** that project through the skeleton into the surrounding water (Figures 3.14c and 3.20).

Figure 3.20 The edible sea urchin, *Echinus esculentus*, with long tube feet extending beyond its spines.

■ Try to detect the holes through which the tube feet emerged in the plates of the skeleton in FS C. Whereabouts are they?

☐ By looking at the edges of one of the ambulacra, you may see a line of small paired pores running down each side.

Each *pair* of pores, called a **pore pair**, led to *one* tube foot outside the test, connecting it with the water vascular system inside. The pore pair represents the divided base (not shown on Figure 3.19) of a single tube foot. The tube feet could be inflated by hydrostatic pressure produced by the action of muscles around the bulb-shaped bases of the tube feet, and flexed by small muscles around them. You can see from the distribution of pore pairs that each ambulacrum must have had two rows of tube feet (Figure 3.19), one on either side of its entire length. Tube feet have a wide variety of functions in echinoids, including feeding, gas exchange (respiration), locomotion and, in some species, constructing burrows.

As you have already seen, the mouth and anus were opposite one another on the skeleton. The intervening gut was a coiled structure (Figure 3.14c). The interior of the skeleton of echinoids is divided by thin membranes into a series of fluid-filled cavities. The bulk of the interior is formed by one large cavity, but smaller cavities surround the anus and mouth region. Suspended within the large cavity are five reproductive bodies (one of which is shown in Figure 3.14c). Look again at the aboral surface of FS C and the ring of ten plates around the anus. You should see that the perforated plate and three of the four larger plates have a small pore; the other plate (almost opposite the perforated plate) seems to be abnormal in that it has two pores. Well, none of us is perfect. The reproductive bodies released eggs or sperm via these pores (the sexes are separate in echinoids).

Question 3.5

From the above description, classify *Pseudodiadema* in terms of habitat and feeding as discussed in Section 3.2.1. Try to assign it to its appropriate box in Table 3.2 (p. 28).

Having read this far, you may be wondering whether the comparatively detailed reconstruction of the two fossils you have been studying (FS I and FS C) can be extended into some principles of general use in palaeontological work.

Activity 3.3 A look at *Micraster*, FS J

This activity aims to show how understanding one example from a fossil group allows you to deal with almost any other example from that group.

Because of the various modifications to the body plan that you have just been observing in *Micraster*, *Micraster* is an example of an **irregular echinoid**, in contrast to the **regular echinoids** such as *Pseudodiadema*.

- ■ Do you think *Micraster* fed in the same way as *Pseudodiadema*?
- □ Almost certainly not; it would be difficult to imagine large jaws like those in *Pseudodiadema* (Figure 3.14c) or *Paracidaris* (Figure 3.15) protruding from the mouth and being able to seize food.

In fact, *Micraster* was probably a deposit feeder. The lower lip was equipped with spines against which specialised feeding tube feet scraped in food.

The difference in positions of mouth and anus between *Pseudodiadema* and *Micraster* results from the forward migration of the mouth, and rearward migration of the anus, from their originally central positions, during successive stages in the evolution of *Micraster* from earlier forms. Some of the stages in this evolutionary change in position of the anus and mouth from species to species may be seen in other fossil echinoids. Interestingly, in the development of modern irregular echinoids this change is repeated in the early stages of each individual's growth, as the anus moves from its central aboral position. In studying the aboral surface of *Micraster* where the five ambulacra meet, note that the small plates, including the perforated plate, have not accompanied the anus on its migration, but have remained central.

- Study the pore pairs in the ambulacra of *Micraster* and decide how they differ from those of *Pseudodiadema*.

- The pore pairs of *Pseudodiadema* are circular and are separated by a tiny ridge. In *Micraster*, although the inner pore (nearest the midline of the ambulacrum) is more or less circular, the outer pore is generally more slit-like, so the tube foot was probably flattened in cross-section. Also, the pore pairs of *Micraster* did not continue onto the lower surface, although the ambulacra can be traced as rows of plates extending beyond the pore zones.

In comparing these details of the ambulacra, you should also have noticed that the broad area with tubercles between the two rows of pore pairs in *Pseudodiadema* is absent in *Micraster*.

- In *Pseudodiadema*, the former position of the spines is marked by the tubercles. Remembering that the size of the tubercles is roughly proportional to the original spine size, what can you infer about the size of the spines in *Micraster* and any variation around the skeleton?

- The tubercles of *Micraster* are far less prominent than those of *Pseudodiadema*, and indeed the spines were much thinner and shorter in *Micraster*, forming an almost hair-like coating over the animal. The largest tubercles of *Micraster* occur across the oral surface, and attached to them were relatively large flattened spines that were used in locomotion.

We could carry on this comparison between the two echinoids almost indefinitely, but now the principal points of similarity (the homologies) and the differences will be apparent. What, however, is the reason for the difference between regular echinoids such as *Pseudodiadema* and irregular echinoids like *Micraster*? The fundamental reason is connected to the shift from an epifaunal mode of life in some regular echinoids to an infaunal burrowing mode of life in the great majority of irregular echinoids. Consistent movement in one direction makes bilateral symmetry more efficient for burrowing than a radial symmetry. The positions of the mouth and anus in *Micraster* are correlated with feeding and waste disposal in a burrow. Recall also the evidence from tubercle size and pore-pair morphology for the size of the spines on the oral surface and the type of tube feet. Both the spines and tube feet are modified according to a burrowing existence. As noted earlier, in Section 3.1.3, this is a good example of the adaptation of a bodyplan to suit a new way of life.

> Echinoids are, and have been, entirely marine, like all echinoderms. Most live in shallow seas. They range in age from late Ordovician to Recent.

Now explore the echinoids in the Digital Kit.

3.3.3 Gastropods (fossil specimens A, *Athleta*, and K, *Pleurotomaria*)

You are probably familiar with gastropods. Today, they are among the commonest marine animals (e.g. sea snails, whelks, limpets and sea slugs) and are also abundant in terrestrial environments (snails and slugs). Gastropods range in age from Cambrian to Recent, and some species are used for global correlation in the Cenozoic.

The gastropod body (Figure 3.21) is partially covered over its dorsal surface by an over-arching sheet of tissue, the **mantle**, which lines and grows the shell (when present). The body has two main components: (i) the **visceral mass**, with the internal organs such as the gut, reproductive and excretory systems. This is permanently housed within the entire shell (when present), and is generally connected via a narrow waist to (ii) a broad fleshy **foot** and conjoined **head** (called the head-foot), which is equipped with a mouth, eyes and sensory tentacles. At rest, the animal's body is pulled into the shell, but when moving, the head and foot extend from the **aperture**. The animal creeps around by producing muscular waves that pass along the base of the foot. In limpets, which are adapted for clinging to rocks, a large sucker-like foot extends from a hugely expanded aperture (Figure 3.22).

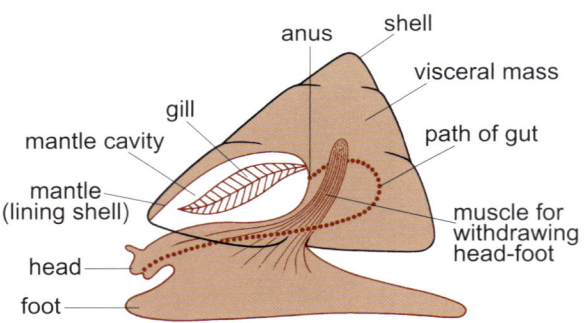

Figure 3.21 Principal features of gastropod anatomy, seen in diagrammatic side view.

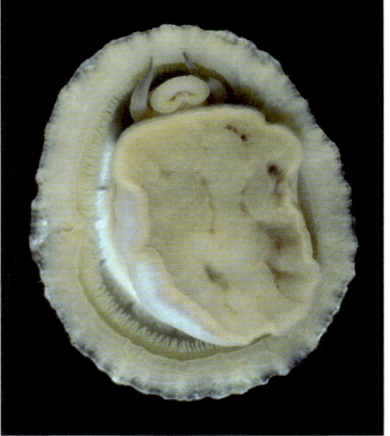

Figure 3.22 Underside of a limpet, *Patella vulgata*, showing large central sucker-like foot, a pair of large head tentacles on either side of the mouth, and many small marginal tentacles inside the edge of the shell. The mantle cavity lies outside the foot margins. Length 4 cm.

The distinction between the visceral mass and the conjoined head and foot is exaggerated by a peculiar process called **torsion** that occurs during the larval development of each individual: the visceral mass (and with it the mantle and shell) is twisted around so as to become 'back to front' with respect to the head-foot beneath. Torsion introduces a marked asymmetry to the gastropod body, making it rather difficult to visualise the orientation of the shell with respect to the foot. In Figure 3.23, plan views of the original form of the head-foot in FS K and A are shown, with the position of the overlying shells indicated by dashed lines. By placing the specimens near the diagrams in the orientations indicated, you will gain an idea of how the shells would have been borne by the two animals that produced them.

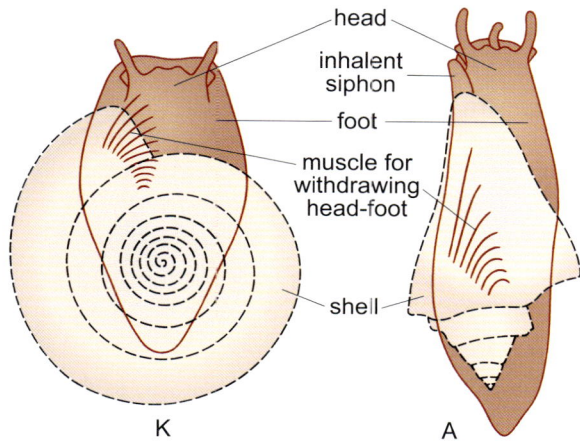

Figure 3.23 Orientation diagrams for FS K and A, seen in dorsal view.

The wide skirt-like extension of the mantle over the body produces a large space between the mantle itself and the visceral mass and upper part of the head-foot. This is the **mantle cavity** (Figure 3.21), and it connects with the outside around the free margin of the shell opening. The mantle cavity plays a vital role in the gastropod's life, for it houses the gill(s), and various sensory organs, as well as the anus. It also allows protective withdrawal of the head, followed by the foot, brought about by contraction of a large muscle (Figure 3.21) that runs from the head-foot to the central column inside the coiled shell.

Many of the differences between FS K and A are linked with the arrangement of the gills in the mantle cavity and with associated differences in their life habits. Some primitive marine gastropods have two gills, one on each side of the mantle cavity (Figure 3.24a). Water is drawn into the mantle cavity low down on both sides, flows over the gills (driven by the beating of cilia on their surfaces) and is exhaled upwards centrally, above the head. As the water leaves, it carries off faeces from the anus, which, as a result of torsion, is sited between the gills. In such gastropods, the exhalent site is usually marked by a pronounced indentation of the shell margin (Figure 3.24a). As the shell grows, former positions of this indentation are marked by a distinct track of indented growth lines around the middle part of the outer shell wall.

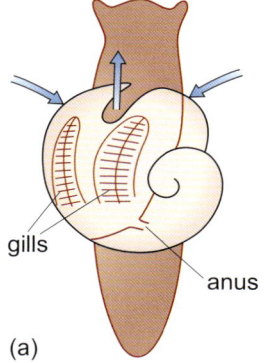

(a)

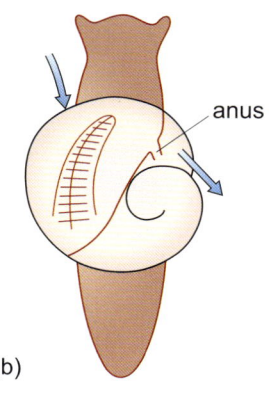

(b)

Figure 3.24
Arrangement of gills in gastropods, with water currents arrowed, shown in dorsal view: (a) two gills with exhalent indentation; (b) single gill.

■ Does either of FS K or A show such a track?

☐ There is no track in FS A, but it is present in FS K.

The presence of the track in FS K means that the shell opening originally possessed an indentation, as in Figure 3.24a. Since the indentation itself is not visible on the margin of this specimen, the shell around the opening must have been broken off. The central position of the exhalent current indicated by the track in FS K in turn implies that inhalent currents entered from either side of the animal. Therefore this gastropod possessed two gills, as in Figure 3.24a.

Question 3.6

Would the indentation of the margin in FS K have brought the exhalent current closer to, or further away from, the head than if the indentation had been absent? You can best visualise this by placing the shell over the head-foot diagram in Figure 3.23 again.

The indented margin in FS K reduced the possibility of exhalent water (which would have been depleted of oxygen and borne excretory products) returning to the inhalent sites, or to the head, in FS K.

The majority of marine gastropods have only one gill, situated on the left-hand side of the mantle cavity (Figure 3.24b). Water is drawn in anteriorly on the left-hand side of the head, passes over the single gill and thence out posteriorly to the right of the animal. The anus is situated next to the exit, on the right of the animal. Thus, neither a central indentation nor corresponding track marks the outer part of the whorl. Furthermore, in some gastropods inhalent water is channelled into the mantle cavity by a fleshy tube or **siphon** extending from the mantle cavity. This is usually housed in a scroll of the shell margin that projects in the opposite direction from the coiled apex of the shell (Figure 3.25). Such a scroll for the siphonal canal is also clearly visible in FS A. The presence of a single inhalent current associated with the siphonal canal in FS A indicates that this animal would have had only one gill, as in Figure 3.24b.

Figure 3.25 The Palaeogene gastropod, *Turricula* (3 cm), showing a long, narrow canal for the inhalent siphon.

Question 3.7

Summarise what you have been able to find out about the respiratory currents of FS K and A by placing the shells in their correct positions on Figure 3.23 again, and drawing onto the diagrams the likely inhalent and exhalent currents by means of arrows.

You may have observed that the exhalent current of FS A would have been almost in line with the inhalent current on the opposite side; there would have been even less chance for exhalent water to have returned to the inhalent site than in FS K.

Now briefly consider locomotion and position on or in the sediment.

- Do you suppose either FS A or K would have been capable of burrowing into soft sediment? Think carefully about the organisation of respiratory currents and the shell shapes (again, you will find it helpful to position the specimens on Figure 3.23 for this).

☐ It is most unlikely that FS K could have burrowed beneath the sediment surface. Submergence of its aperture would have immediately blocked the inhalent currents, and, besides, the broad shell would have presented an uncompromising impediment to burrowing. In contrast, FS A is a more likely candidate for burrowing. Its inhalent siphon would have allowed it access to overlying water even when the aperture was submerged. Note, also, the streamlined form of the shell, with the spirally coiled apex pointing backwards: it is almost torpedo-shaped, offering little hindrance to burrowing (look particularly at it in profile, when placed over Figure 3.23).

What can be said about the feeding habits of these two gastropods? Gastropods feed in a wide variety of ways, employing different modifications of the same basic feeding apparatus: the mouth contains a protrusive cartilaginous bar over which a tooth-studded ribbon of tissue (Figure 3.26a) called the **radula** is run to and fro. When the mouth and radula are applied to a food-bearing surface, food material is brushed or scraped into the mouth.

Many gastropods use the radula to graze on algal material growing on surfaces or on loose organic detritus and microscopic organisms on the sea floor. Many others hunt vagrant prey, however, which they kill by smothering them with the foot or even by driving into them specialised radula teeth furnished with poison. Certain predatory forms use the radula rather like sandpaper to drill holes through the shells of their prey (Figure 3.26b). A tubular trunk with a terminal mouth is then used to tear out the tissues of the shell's occupant. Most of the advanced gastropods that possess a well-developed siphon (as in FS A) are predators or scavengers, feeding in one or other of these specialised ways. As noted above, the siphon probably evolved in the first instance as an adaptation enabling gastropods to penetrate soft sediment. However, since the gill it supplies is typically equipped with a sensory gland that can 'sniff' the incoming water, the manoeuvrable siphon allows the gastropod to locate the source of any 'smell' with accuracy. Thus, in predatory or scavenging forms this apparatus seems to have become secondarily adapted as a means of locating and tracking their food.

Question 3.8

How would you interpret the likely feeding habits of FS K and A? Enter them in the appropriate boxes in Table 3.2 (p. 28).

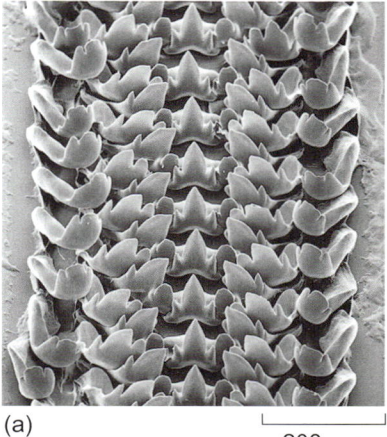

Figure 3.26 (a) Electron microscope view of a gastropod radula (from a grazing form). (b) Boring (2.5 mm in diameter) made in a bivalve shell by the radula of a predatory gastropod. Notice the much smaller circular depression near the shell margin where a boring was started and abandoned.

Book 3 Fossils and Sedimentary Rocks

Figure 3.27 Internal mould of *Aptyxiella*, a Jurassic gastropod (7 cm). The aragonitic shell has dissolved away.

When present, the shells of gastropods are usually made of aragonite rather than calcite, and in fossils the aragonite has often dissolved away, leaving a space (Figure 3.27).

> Most gastropods are marine, living in shallow seas, but many inhabit rivers, lakes and ponds; others live on dry land. They range in age from Cambrian to Recent.

Now explore the gastropods in the Digital Kit.

3.3.4 Ammonites (fossil specimen D, *Amoeboceras*)

The ammonites are an extinct group that lived in Mesozoic seas. Nevertheless, there are other representatives of the broader group to which the ammonites belong – the **cephalopods** – that are still alive today. The nearest living relatives of ammonites are unknown, but lie somewhere among the subclass Coleoidea (coleoids) that includes squid, cuttlefish and octopuses, in which the shell is reduced or lost.

The living *Nautilus*, although only distantly related to ammonites (as you saw earlier in Chapter 2), has a chambered shell which can be closely compared with that of ammonites. Figure 3.28 shows the anatomy of *Nautilus*. *Nautilus* belongs to a subclass of cephalopods, the Nautiloidea, which was much more diverse in the past. Ancient nautiloids had many different shapes – some straight, some curved and some irregularly coiled. Figure 3.29 shows some fossil nautiloids. Nautiloids range in age from late Cambrian to Recent.

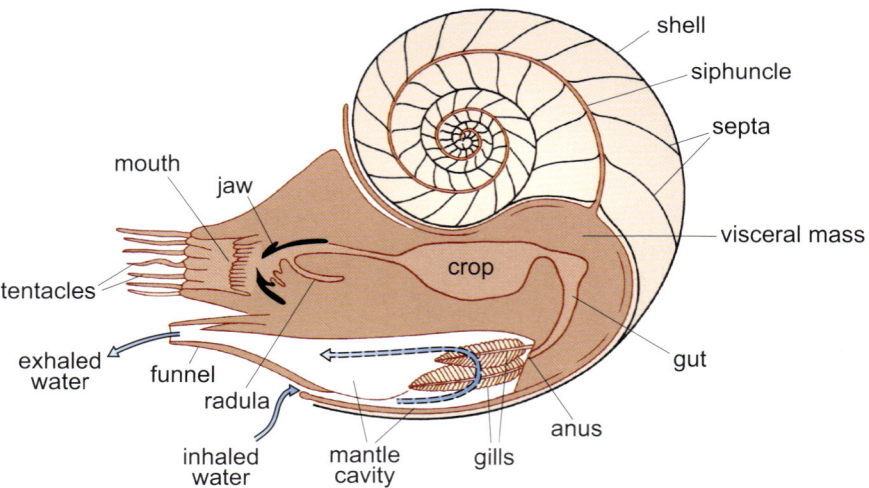

Figure 3.28 The anatomy of *Nautilus*, seen in diagrammatic section along the body.

Chapter 3 Fossil classification and palaeobiology

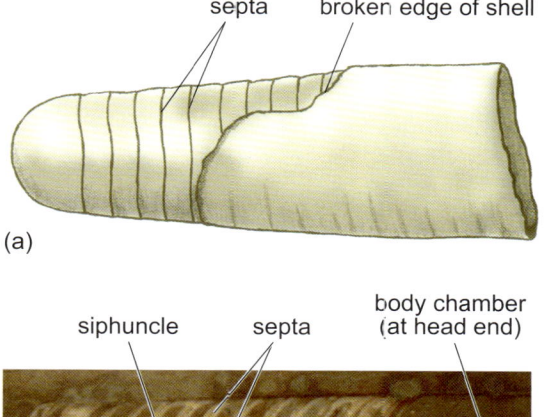

Figure 3.29 Fossil nautiloids. (a) *Michelinoceras*, Silurian (9 cm, incomplete – total length at least 20 cm). (b) Section through part of a straight nautiloid, Ordovician (12 cm). (c) *Cenoceras*, Jurassic (7.5 cm). The body chamber is missing (broken off).

Ammonites form part of a third, entirely extinct, cephalopod subclass called the Ammonoidea. In this course, we restrict the name 'ammonites' to ammonoids of Mesozoic age, but elsewhere you may sometimes see some Palaeozoic ammonoids referred to as ammonites. Goniatites were one group of Palaeozoic ammonoids; a typical example is shown in Figure 3.30.

- Compare the complexity of sutures in typical nautiloids (Figure 3.29), goniatites (Figure 3.30), and ammonites (FS D and Figure 3.31). What are the main differences?

☐ Nautiloids have straight or gently curving sutures. Goniatites have simple zigzag sutures, whereas ammonites have much more complex sutures.

Ammonite shells were made of aragonite, which in fossils has often dissolved away (as in FS D), or recrystallised to calcite. Ammonite shells vary greatly in form. Some are relatively compressed from side to side, like FS D, and others are fat. Some are coiled so that later whorls overlap earlier whorls, while others are coiled with no overlap, rather like a coiled rope. The sides of the shells are commonly ornamented with radial ridges called **ribs**. Sharp, fairly evenly spaced ribs are visible on the earlier whorls of FS D. However, they get smaller on later whorls, so that they can be seen only as vestiges on the outer shoulders of the final whorl. Remember, however, from Section 3.1.1 that FS D is an internal mould, so that we cannot be certain from this specimen that the outside of the shell was rib-free. Notice also that the ribs do not continue around the outer edge of the coil of the shell. Instead, the outer edge of the coil in this specimen is marked by a sharply projecting ridge called the **keel**.

In contrast to the gastropod body, the ammonite body occupied only a small fraction of the shell interior, the rest of the shell being divided into chambers by the septa (Section 3.1.1).

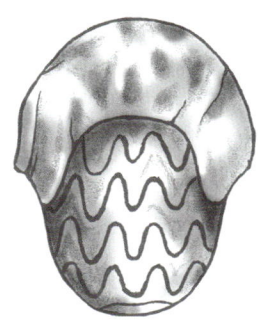

Figure 3.30 *Goniatites*, a representative of a group of Palaeozoic ammonoids called goniatites. Length 4 cm. The shell of this Carboniferous specimen is mostly broken away, revealing chambers infilled with sediment. The body chamber is missing (broken off).

47

■ How much of the shell was occupied by the body in FS D?

□ The body occupied just under half a whorl. The sutures that mark the outer edges of the septa stop just under half a whorl away from the aperture of the shell (Section 3.1.1). (You cannot see the whole of the body chamber on FS D because the specimen is incomplete near the aperture.)

In the absence of living examples, palaeontologists have reconstructed the probable anatomy and life habits of ammonites on the basis of:

- homology and analogy (Chapter 2), especially by reference to *Nautilus* (Figure 3.28), and to artificial models whose mechanical properties can be measured
- rarely preserved gut contents in ammonite fossils
- information on the distribution of fossil ammonites and the other fossils and sediments with which they are associated.

Perhaps surprisingly, no well-preserved soft parts of ammonites (e.g. tentacles) have yet been found.

In life, the chambers of the ammonite shell were not entirely sealed off, for a membrane-bound tube called the **siphuncle** ran through them, connecting with the rear part of the body. Note that unlike in *Nautilus* (and other nautiloids) in which the siphuncle is centrally placed within the whorls of the shell (Figures 3.28 and 3.29b), the siphuncle in ammonites is situated close to the outer part of each whorl (Figure 3.31b). As is the case in *Nautilus*, it is probable that the siphuncle of ammonites was capable of withdrawing water from the chambers and introducing gas. The chambered shell thus seems to have served as a buoyancy tank (as it does in *Nautilus*), so ammonites are believed to have been predominantly pelagic, like most other cephalopods.

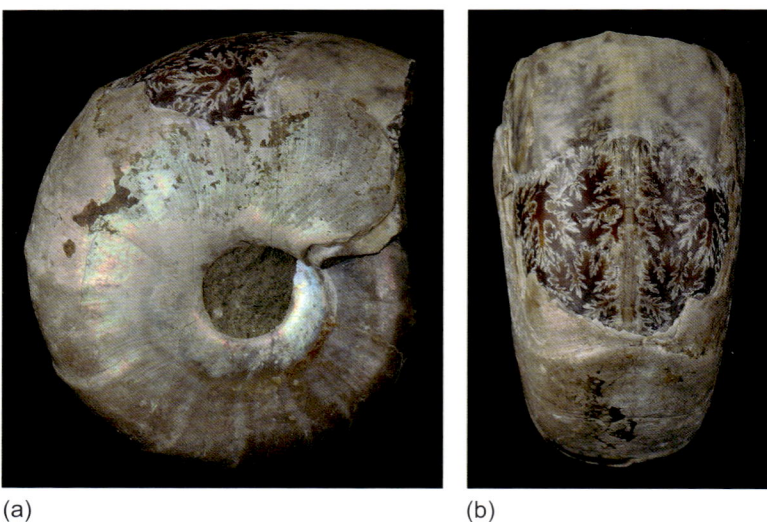

(a) (b)

Figure 3.31 (a) and (b) A Cretaceous ammonite, *Desmoceras* (3.5 cm). Its unusually well-preserved shell is peeling away in places to reveal the sutures. In (b), the siphuncle can be seen close to the outer part of the tubular shell.

We may further suppose that the ammonites, like *Nautilus*, possessed a well-developed head with eyes and a mouth surrounded by a ring of tentacles. The mouth was certainly equipped with beak-like jaws as well as a narrow radula, comparable with, but simpler than, that in most gastropods; both jaws and radulae are known from ammonite fossils. Behind the head would have lain the visceral mass, occupying the part of the body chamber on the inside of the coil. Most of the outer side of the body chamber was probably taken up by the mantle cavity, as in *Nautilus* (Figure 3.28). The gills, anus and various sensory and other organs would have been situated here, beneath the body. In living cephalopods, the mantle cavity serves as a sort of bellows into which water is drawn around a wide inhalent margin and then pumped out via a tubular **funnel** situated immediately beneath the head. This provides for a form of jet propulsion at the same time as supplying water to the gills. It is thought likely that many, if not all, ammonites were similarly capable of at least weak swimming by jet propulsion, though work on experimental models shows that even well-streamlined ammonites (such as FS D) would have fallen far behind today's squid and high-speed fish in terms of swimming efficiency.

The range of shell forms in ammonites, implying widely different swimming abilities, shows that their life habits were correspondingly diverse. Ammonites were presumably carnivorous, like all cephalopods today. Many species are interpreted as having been predators on small prey or nektonic suspension feeders (catching zooplankton with their tentacles), while others were possibly large-scale predators, scavengers or detritus collectors swimming near the sea floor. These interpretations are, however, still largely speculative. By way of comparison, the living adult *Nautilus* lives in moderately deep water (up to a few hundred metres depth), swimming close to the bottom, and scavenging on the corpses of animals such as crustaceans and fish. It is also occasionally predatory.

One aspect of ammonite palaeobiology – that of shell development – can be studied in detail. Since the shell was grown incrementally throughout the later larval and post-larval life of the animal, it reveals a 'potted history' of its development. As the shell opening grew forwards through the addition of new material, so the body migrated forwards to keep pace with it, and laid down the successive septa behind it. Forward growth of the shell opening and migration of the body evidently slowed down in mature individuals, for the final septa in these became more closely spaced.

■ Using the criterion of septal spacing, does FS D represent a mature individual?

☐ It seems reasonable to suppose that FS D comes from a mature individual because the last ten or so sutures (and therefore the septa they represent) are more closely spaced than earlier ones.

We are now in a position to look for differences between juvenile and adult features. To study the juvenile features of the shell, you need to look, of course, at the inner whorls, which were grown first.

■ What is the most striking difference in ornamentation between the visible part of the juvenile shell and the adult part of the shell?

☐ As you saw earlier, the juvenile shell (inner whorls) has fairly pronounced ribs, while the adult shell is much smoother with very subdued ribs. Note, however, that for a full comparison between the juvenile and adult portions, you would either have to find a juvenile specimen or strip away the outer whorls. You may have also noticed that the sutures on the inner whorls are less complex than those in the outer whorl, although this is not surprising as the septa are smaller.

These observations illustrate an important point when dealing with fossils of this kind: different growth stages can look very different and, since ammonite species are typically identified on the basis of shell shape, ornamentation and sutural form, there is considerable scope for spurious separation of different growth stages into different species. To make matters worse, at least some ammonites were like some other animals in showing differences between the adult shells of the two sexes (i.e. **sexual dimorphism**, Figure 3.32).

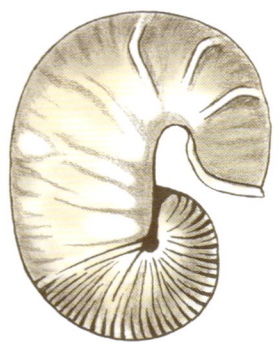

(a)

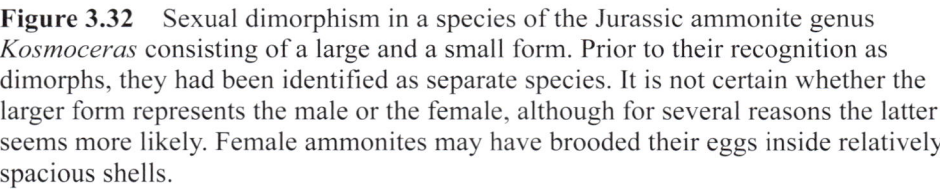

Figure 3.32 Sexual dimorphism in a species of the Jurassic ammonite genus *Kosmoceras* consisting of a large and a small form. Prior to their recognition as dimorphs, they had been identified as separate species. It is not certain whether the larger form represents the male or the female, although for several reasons the latter seems more likely. Female ammonites may have brooded their eggs inside relatively spacious shells.

■ How might a consideration of shell development help to pair up males and females from dimorphic species?

☐ The juvenile shells (inner whorls) might be expected to be closely similar before sexual maturity.

Unlike the other specimens studied so far, we cannot confidently assert what the feeding habits of FS D itself were, though its streamlined shape suggests that it was a relatively good swimmer (in ammonite terms) and thus nektonic. Now place FS D tentatively in the appropriate box(es) in Table 3.2 (p. 28). Because many ammonites were nektonic, their shells tended to become widely distributed. This, coupled with relatively high rates of evolution, make them particularly suitable candidates for stratigraphical correlation (Section 3.8). Ammonites are thus widely used for dating Jurassic and Cretaceous marine deposits.

(b)

Figure 3.33 Cretaceous ammonites with unusual shapes: (a) *Scaphites* (8 cm); (b) *Turrilites* (8 cm).

Some ammonites, especially in the Cretaceous, became uncoiled or coiled into strange shapes (Figure 3.33).

Chapter 3 Fossil classification and palaeobiology

■ What feature of the shell would you look for in an ammonite such as *Turrilites* (Figure 3.33b) to confirm that it was indeed an ammonite and not a gastropod, despite its helical form?

☐ The presence of septa and sutures (which are never present in gastropods).

> Ammonites were entirely marine, like all cephalopods past and present. They range in age from Triassic to end-Cretaceous.

Now explore the ammonites and nautiloids in the Digital Kit.

3.3.5 Brachiopods (fossil specimens F, *Epithyris* and L, *Quadratirhynchia*)

Brachiopods (Figure 3.34) are a relatively small group today, but they are very abundant and diverse in the fossil record, particularly in the Palaeozoic.

(a)

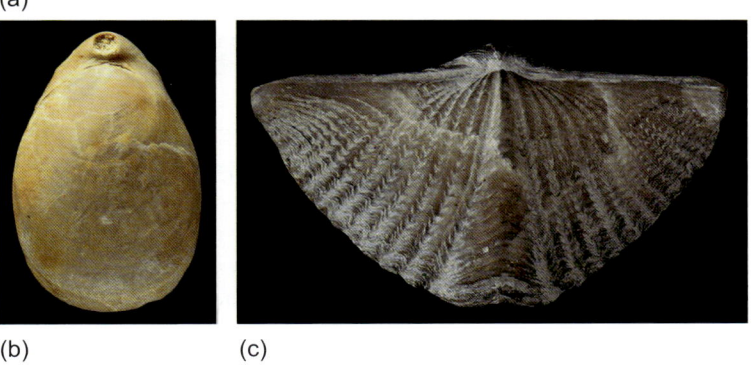
(b) (c)

Figure 3.34 (a) Living brachiopods from the Canary Islands. The complex structure of the tentacular feeding organ (the lophophore) can be seen in individuals that have opened their valves for feeding. The shells of this small (0.5 cm) species, *Pajaudina atlantica*, are cemented by their pedicle valves to the sea floor (rather than attached by a pedicle), and are coated in algae. (b) A Cretaceous example of the long-ranging brachiopod genus *Terebratula* (4 cm). (c) A Devonian brachiopod, *Mucrospirifer* (3 cm).

The brachiopod body is enclosed by the shell, which consists of two valves, one almost always larger than the other (Figure 3.35).

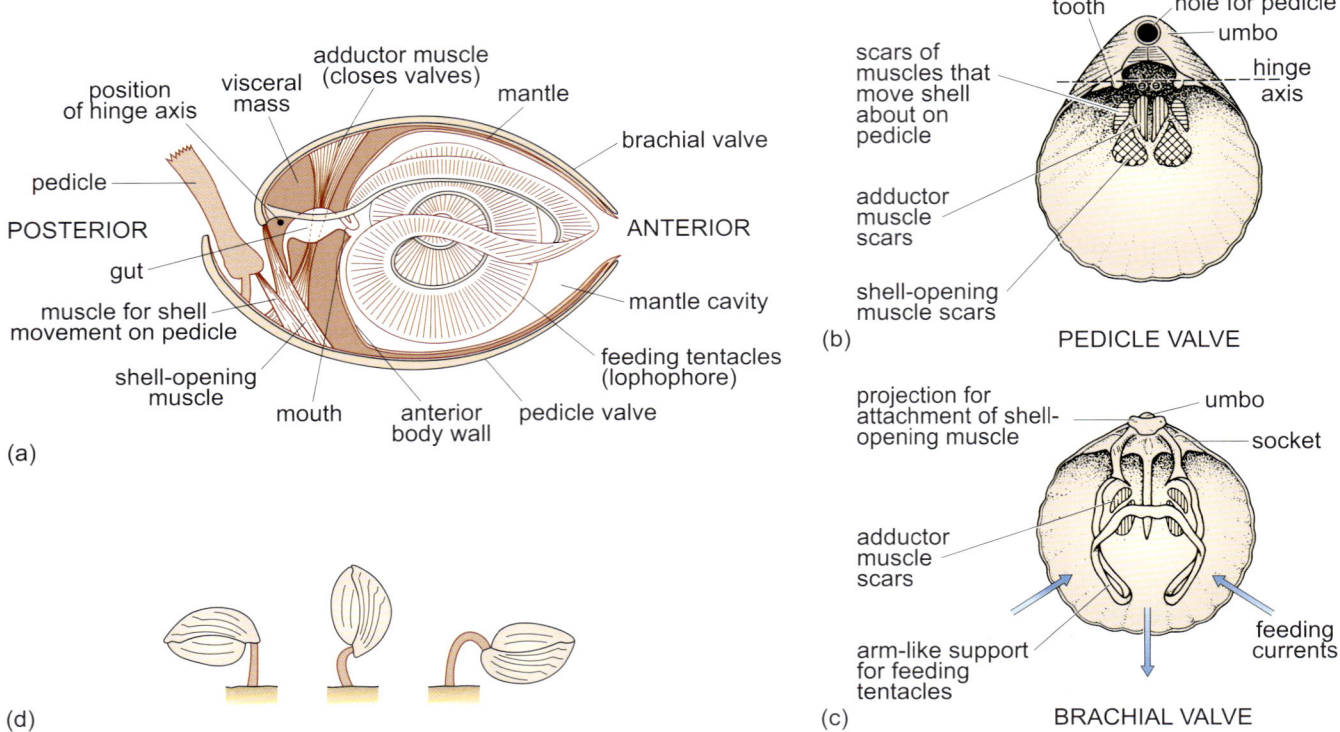

Figure 3.35 (a) Anatomy and shell of a living articulate brachiopod, in diagrammatic longitudinal section. (b) and (c) Interiors of (b) pedicle valve, and (c) brachial valve (with orientations of feeding currents indicated in (c)). (d) Variety of possible life positions.

Figure 3.36 Specimens of the inarticulate brachiopod *Lingula*, extracted from their vertical burrows on the sea floor, in which they are anchored by a long, muscular pedicle. Typical shell length 4 cm.

- How does the shell's symmetry differ from that of bivalves?

- In brachiopods, the plane of symmetry runs *through* the valves, whereas in bivalves the plane of symmetry (when present) runs *between* the two valves (Section 3.1.2 and 3.3.1).

In brachiopods, the two valves articulate *posteriorly* and, in most examples, the hinge is formed by the projection of a pair of **teeth** from the pedicle valve (Figure 3.35b) into a corresponding pair of **sockets** in the brachial valve (Figure 3.35c). Such forms are referred to as being **articulate brachiopods**. Both FS F and FS L are of this sort but, since teeth and sockets are internal features, they are not directly visible in these specimens. In fossils, the valves of some early Palaeozoic articulate brachiopods tend to have separated, but in most brachiopods the values remain attached to each other, giving whole, closed shells, in contrast to fossil bivalves where detached valves are common. Another small group of brachiopods lack hinge teeth and sockets, the valves being held in position entirely by muscles. These are the **inarticulate brachiopods** which, like articulate brachiopods, range in age from early Cambrian to Recent. In some inarticulate brachiopods (e.g. *Lingula*, Figure 3.36) the shell is phosphatic (and often dark in colour), whereas in all articulate brachiopods, the shell is composed of calcite.

Anterior to the body is a large **mantle cavity**, lined by mantle tissue, and occupied by a convoluted bar of cartilaginous material that supports an extended row of feeding tentacles called a *lophophore* (Figures 3.34a and 3.35a). In some articulate brachiopods, this bar has a shelly support (which may be folded back on itself) that is rooted in the brachial valve – the word 'brachial' refers to this arm-like support (Figure 3.35c). Note that the brachial valve is sometimes called the 'dorsal valve' and the pedicle valve is sometimes called the 'ventral valve'. However, this convention bears no consistent relationship to life position (Figure 3.35d).

Beating of cilia on the tentacles of the lophophore creates a current from which food particles are trapped and passed to the mouth. Several different feeding patterns have been described from living brachiopods and these are associated with much variation in the shape of the valves and lophophore. In FS L, it is likely that the feeding currents entered the mantle cavity on each side of the shell and left from the middle of the anterior side (as in Figure 3.35c).

- Considering the positions of the inhalent and exhalent water currents, suggest an explanation for the overall form of the valve margins in FS L. (*Hint*: cast your mind back to your consideration of the respiratory currents in gastropods.)

- ☐ The abrupt flexure of the central parts of the anterior valve margins keeps them markedly separated from those on each side. Since the central parts supposedly contained the exhalent site and those to the sides contained the inhalent areas, these large-scale flexures of the margins may be considered as an adaptation to promote separation of the inhalent and exhalent currents.

The shell is opened and closed entirely by the action of muscles (Figure 3.35a). **Adductor muscles** draw the valves together, while **diductor muscles** pull the valves open. Figures 3.35b and c show the scars of these muscles on the shell interior.

- How does this method of opening and closing the valves with two sets of muscles differ from that of bivalves, such as *Circomphalus* (FS I)?

- ☐ Bivalves are similar in that they employ adductor muscles to close the valves, but they differ in using the stress stored in an elastic ligament, rather than muscular force, to open them (Section 3.3.1).

All brachiopods are benthic. In life, most brachiopods are attached to the sea floor by means of a horny stalk (Figure 3.35a, d), known as a *pedicle*, that projects out of the shell through a hole situated on the umbo of the pedicle valve (Figure 3.35b). Muscles that insert into the pedicle allow the brachiopod to move its shell around (Figure 3.35d). Reduction or loss of the pedicle occurred during evolution in some brachiopods that lay on the bottom of the seabed, or nestled shallowly into it, and, in some cases, they became tethered there by fine branches of the pedicle projecting into the sediment. Rarely, some brachiopods are cemented onto the seabed (as in Figure 3.34a). The pedicle itself is not preserved in fossils.

FS F has a large and obvious hole for the pedicle. This indicates that it was attached to a hard surface by means of a large pedicle on which it could probably change its orientation. FS L, in contrast, has either a very small hole or no hole at all on its small umbo; it is not possible to tell which, as the umbonal tip of the pedicle valve has been chipped off. Thus, if it had a pedicle it must have been greatly reduced and could only have served to tether the shell to the sea floor. It is thus likely that FS L nestled in loose sediment.

Question 3.9

Summarise what you have learned about the life habits of FS F and L by assigning them to the appropriate boxes in Table 3.2 (p. 28).

> Brachiopods are, and have been, entirely marine. They range in age from early Cambrian to Recent.

Now explore the fossil brachiopods in the Digital Kit.

3.3.6 Crinoids (fossil specimen H, *Sagenocrinites*)

Like brachiopods, crinoids are less abundant and diverse now than in former times. They range in age from early Ordovician to Recent. In the Palaeozoic, their remains sometimes make up the bulk of whole beds of limestone, as in RS 24 and the crinoidal limestone in the VM and DK. Crinoids, like other echinoderms, cannot tolerate significant departures from normal marine salinity, and have always been marine.

Most crinoids found as fossils were attached in life to the sea floor by a stem (or stalk) with a root-like **holdfast** (Figure 3.37). The majority of living crinoids have a greatly reduced stem or no stem at all, and are capable of creeping around or even swimming. The few surviving stemmed forms generally live in deeper water.

The crinoid body is housed in an enclosed cup at the top of the stem (if present), and this cup contains most of the internal organs, including a twisted U-shaped gut, with the mouth in the middle of the cup's upper surface and the anus to one side of it (Figure 3.37). Surrounding the rest of the cup is a ring of branching arms showing a fivefold radial arrangement, although this is not particularly clear in FS H.

FS H shows the cup to have a tile-like skeleton of plates, and the stem and arms consist of stacks of plates that look like piles of coins. As with the echinoids (Section 3.3.2), these plates are made of calcite, with a sponge-like three-dimensional network of pores, and they enclose the bulk of the soft parts; but again, the plates are endoskeletal and are secreted within soft tissue, which coats the animal's outer surface.

Chapter 3 Fossil classification and palaeobiology

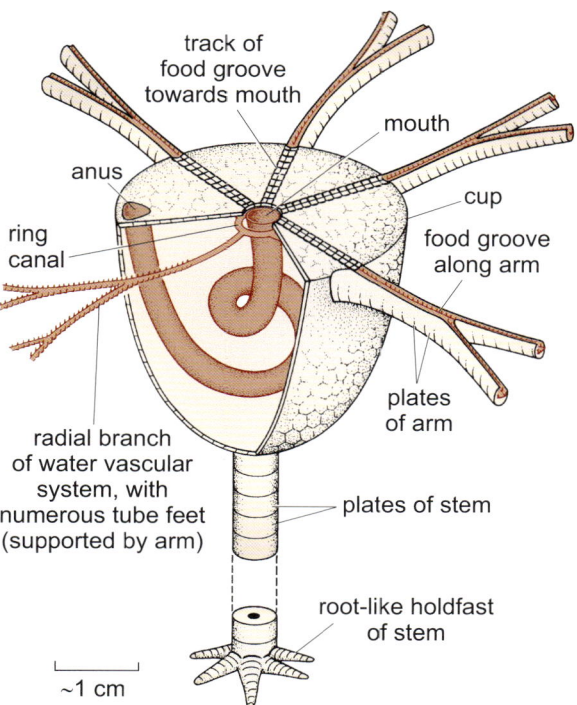

Figure 3.37 Basic anatomy of a crinoid.

The stacked plates of the arms support a muscular coating that controls arm movement, and house radial branches of the water vascular system with tube feet, which run along their upper surfaces. As in the echinoids, these branches radiate from a ring canal around the mouth, though in crinoids there is no perforated plate connecting the system with seawater. The exposure of the branches of the water vascular system on the arms also contrasts with their internal position in echinoids.

Underwater photographs of living crinoids show that feeding is usually accomplished by spreading the arms outwards and backwards into a current to form an umbrella-like filtration fan (Figure 3.38). Microscopic food particles are trapped on the tube feet arrayed along the arms. From there they are passed to the food grooves overlying the radial branches of the water vascular system (Figure 3.37), and then, with the aid of cilia, moved towards the mouth. The geometric construction of crinoids is well suited to forming an umbrella-shaped filtration fan. The ability to keep branching their arms allows the area of the umbrella-shaped device to remain filled in as it enlarges during growth. FS H is preserved with its arms bunched together, so the shape of the spread-out filtration fan is not seen.

(a)

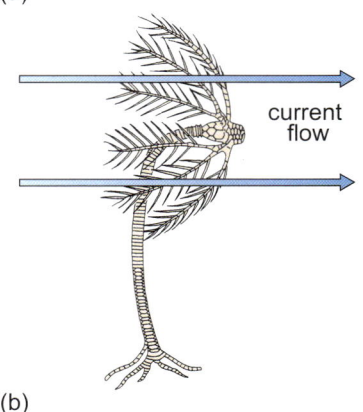

(b)

Figure 3.38 (a) A living crinoid, *Neocrinus decorus*, photographed at about 420 m depth off Grand Bahama Island. (b) The feeding posture of a crinoid in a current.

FS H is unusual in being preserved with its assembly of skeletal plates intact, as is also the case with the crinoid shown in Figure 3.39a. As with the echinoids (Section 3.3.2), the plates were connected only by organic material (collagen) in life. Many of these connections were flexible, as was clearly the case in the arms of FS H and the specimen in Figure 3.39a. Thus, on death and decay of the organic material, crinoid skeletons tend to disintegrate rather rapidly. Most of the crinoid remains in RS 24 are individual plates from stems that have become disarticulated (separated); see also Figure 3.39b and c.

■ What does the mode of preservation of FS H suggest about how this individual may have died?

□ For the plates to have been preserved intact, without falling apart, this crinoid must have been entirely, and permanently, buried, either alive or very shortly after death, by a rapid influx of sediment.

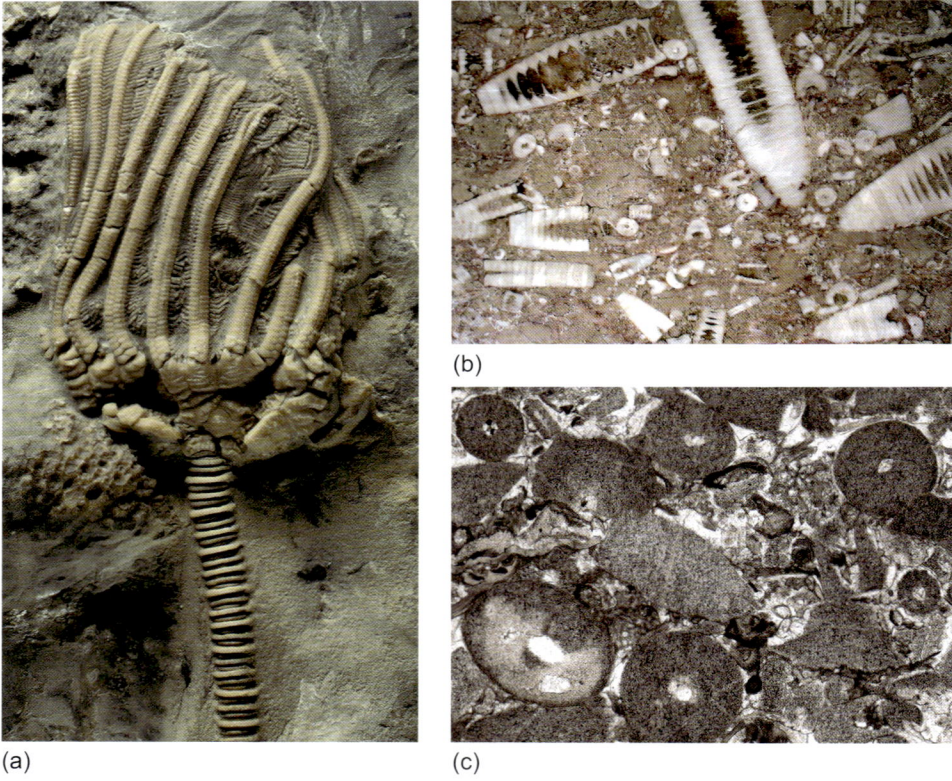

Figure 3.39 (a) *Marsupiocrinites*, a Silurian crinoid. The whole stem was longer than is seen here. The cup is 3 cm across. (b) Carboniferous Limestone with fragments of crinoid stems and isolated stem plates (mostly 1–2 cm across). (c) Thin section of limestone dominated by crinoid fragments, in plane-polarised light. Field of view 4.5 mm.

Inferences such as this, derived from the mode of preservation of fossils, can provide valuable clues to conditions in ancient environments – in this case, for example, testifying to occasional, probably storm-driven, influxes of sediment. We will explore this topic further in Chapter 4.

Question 3.10

Summarise what you have learned about the life habits of FS H by assigning it to the appropriate box in Table 3.2 (p. 28).

Now explore the crinoids in the Digital Kit.

3.3.7 Corals (fossil specimens E, *Dibunophyllum* and M, *Lithostrotion*)

A coral individual, or **polyp**, is essentially similar to a sea anemone in bodyplan, although, unlike the latter, it builds up a cup of calcium carbonate beneath itself as a sort of platform (Figure 3.40). As the outer wall of the cup grows upwards, so too do the septa (Section 3.1.1) supporting the base of the polyp. In FS E, much of the outer wall has been lost, revealing the outer edges of the septa. Successive 'false floors' are also formed beneath the polyp (Figure 3.40), but these are usually not visible from the outside. Look closely inside the cup in FS E (with the hand lens) and you might see some thin, sometimes curved, partitions called **dissepiments** between the septa, especially towards their inner ends. However, they are not clearly visible within the many cups in FS M.

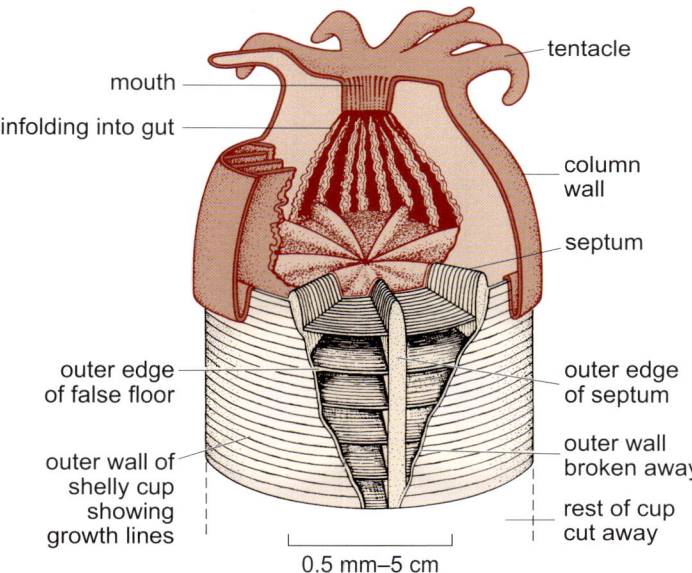

Figure 3.40 A single coral polyp showing 'false floors' beneath it. Note that polyp diameter may range from about 1 mm to 10 cm.

The polyp is of simple construction, comprising a thick-walled sac with an upward-facing opening surrounded by a ring of tentacles. The internal cavity is where it digests its food and the single opening serves as both entrance and exit (for undigested matter). The sidewalls of the cavity have numerous infoldings, separated by the septa, which greatly increase the surface area of the cavity, facilitating absorption of the products of digestion. The tentacles are armed with stinging cells, which paralyse the prey (mainly microscopic planktonic animals) as it is passed to the opening of the internal cavity.

In the groups we have considered so far, reproduction is almost always sexual. In corals, however, new individuals can arise by sexual reproduction or by asexual reproduction (budding). In some species, separation of the buds is incomplete, resulting in a more or less densely connected cluster or **colony** of polyps. The mode of budding (which is characteristic for each species) affects the way the colony grows, although the overall colony form is also subject to environmental influences. In FS M, which represents part of a colonial coral, new buds developed from the polyp wall, so that new corallites formed between existing corallites. The pattern of budding in FS M produced a massive globular colony, whose upper surface became increasingly domed as more and more polyps were packed into it. Such a colony is therefore made up of many genetically identical individuals that share a skeleton. Figure 3.41 shows a selection of colonial corals.

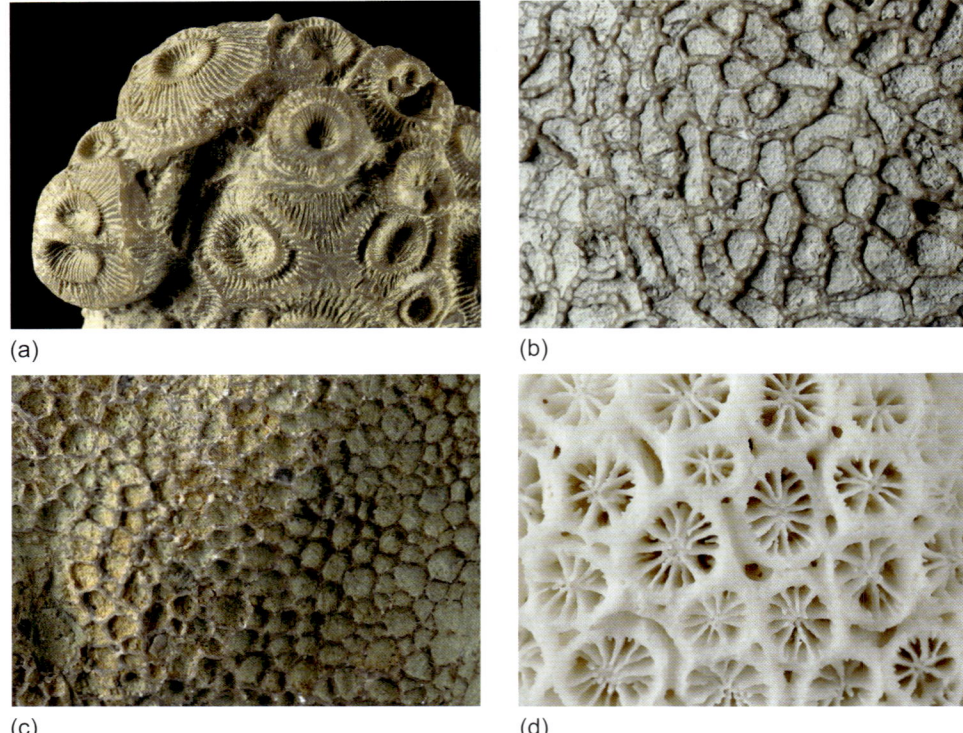

Figure 3.41 Colonial corals: (a) *Acervularia*, a Silurian rugose coral (field of view 5 cm); (b) *Catenipora*, a Silurian tabulate coral (field of view 3 cm); (c) *Favosites*, a Silurian tabulate coral (field of view 2.5 cm); (d) *Favia*, a Recent scleractinian coral (field of view 1.5 cm).

There are three major groups of corals. Two groups, **rugose** corals and **tabulate** corals, are confined to the Palaeozoic, each ranging from Ordovician to Permian. The third group, **scleractinian** corals, arose in the mid Triassic and is still flourishing today. Throughout their history, corals (of all groups) have been entirely marine, and certain forms have played a major role in the growth of reefs at various stages in geological history.

FS E and M are both rugose corals. Rugose corals may be solitary individuals (like FS E) or form colonies (like FS M). Tabulate corals, however, are always colonial (Figure 3.41b and c), and their corallites are much smaller than the corallites of rugose or scleractinian corals. In tabulate corals, the septa are usually reduced or absent.

Scleractinian corals may be either solitary or colonial. The scleractinian corals evolved from soft-bodied, sea anemone-like ancestors with a long pre-Triassic history. Thus scleractinian corals are not descended from either of the major

Palaeozoic *coral* groups, as corals, by definition, possess a skeleton. Although scleractinian corals are much younger than rugose or tabulate corals, their fossils are often much less well preserved because their skeletons are made of aragonite, rather than calcite. The soft-bodied relatives of corals, such as the group that includes today's sea anemones, are very rare in the fossil record.

The colonial habit in some corals, with its potential for generating robust interlocking skeletal frameworks on the sea floor, has been a major factor in their recent history of reef-building. Another factor, seen in living corals, is that many species have great numbers of single-celled algae (**zooxanthellae**) dwelling within their soft tissues. The zooxanthellae provide the host coral with various products of photosynthesis, which form nutrients (including carbon) for the coral, favouring its fast and efficient growth. In return, the host coral provides the zooxanthellae with protection, nutrients (especially waste products of its own digestive processes) and a supply of carbon dioxide required for photosynthesis.

This symbiotic association ensures a tight recycling of nutrients within the corals, allowing them to thrive especially in waters with low nutrient levels, where the growth of fleshy algae (which elsewhere can swamp the slower-growing corals) is kept in check by grazers. Today, such corals are prominent constituents of tropical shallow-water reefs in areas that might otherwise be described as 'nutrient deserts'. However, there is much debate as to which fossil corals did, or did not, have such single-celled algae in their tissues in life. Several lines of evidence suggest it is likely that neither rugose nor tabulate corals had symbiotic algae. It would therefore be merely speculative to attribute this ecological trait to unrelated Palaeozoic corals such as FS E and M.

Question 3.11

Enter the information on the life habits of FS E and M in the appropriate box or boxes of Table 3.2 (p. 28).

Now explore the corals in the Digital Kit.

3.3.8 Trilobites (fossil specimen G, *Dalmanites*)

Trilobites are extinct, and are entirely restricted to Palaeozoic strata. They range in age from early Cambrian to late Permian, and are particularly abundant in the Cambrian, Ordovician and Silurian. Trilobites are only remotely related to groups alive today. Thus one cannot so readily reconstruct the form and function of their soft-part anatomy on the basis of homology as it was possible to do, to some extent, with the ammonites by using *Nautilus*.

The basic features of the dorsal exoskeleton of the trilobite were described in Section 3.1.3. Like most invertebrates, trilobites grew throughout their life, albeit more slowly during adulthood. The exoskeleton, however, once mineralised, could not increase in size, and so periodic shedding of the old one, and secretion of a new, larger skeleton was necessary to permit growth, just as in crabs and lobsters today. **Moulting** involved breakage of the skeleton, along lines of built-in weakness, into distinct components, like a mediaeval knight getting out of a metallic suit of armour.

- Whereabouts can you see any such lines of weakness on the head-shield of FS G (a Silurian trilobite)?

  ☐ A fine line is visible on each side of the head-shield, running outwards from immediately behind each eye.

In fact, the line on each side continues in towards the central portion of the head-shield and runs around the back of each eye, though this is not clearly visible in this specimen. The outer parts of the head-shield (which, in most trilobites with eyes, included the eye surfaces) would thus have broken off along these lines of weakness. Moulting means that during growth each individual left behind a succession of potentially fossilisable skeletal elements. Most trilobite fossils are bits and pieces of exoskeleton cast off during moulting; complete carcasses resulting from the death of an individual are much less common.

- Is FS G more likely to represent a moulted skeleton or an individual that died?

  ☐ Since the exoskeleton is entire, it almost certainly represents an individual that died.

A distinctive feature on FS G is the pair of eyes on the head-shield. Trilobites are the oldest fossils that show evidence of a visual system. Inspect the eyes with the hand lens. You should see that the outward-facing surface of each eye consists of many rows of low, rounded domes about 0.5 mm in diameter. Each dome represents a lens. They are thus compound eyes like those of flies (insects) and crayfish (crustaceans). However, the detailed structure of the compound eye differs among these groups. Compound eyes are particularly good at detecting motion of nearby objects because the same changing pattern of light and shade is sensed many times over by the bank of lenses. The lenses of trilobite eyes were made of calcite like the rest of the exoskeleton, so they had good preservation potential.

Question 3.12

Inspect the orientation of the bank of lenses on each eye of FS G. What was the trilobite's field of view? (*Note*: The right eye in this specimen is incomplete, but its form would have been the mirror image of the left eye, which is only slightly chipped.)

Trilobites were adapted to a wide range of habitats, though *all of them were marine*. Most lived on or near the sea floor, sometimes in very deep water, but a few were free-swimming in the open ocean. Some ocean-going species had remarkably large eyes with a field of view that allowed them simultaneously to see below, as well as above and sideways, with an almost spherical field of vision.

Figure 3.42 shows two Cambrian trilobite specimens with appendages, including legs and antennae. Such preservation is

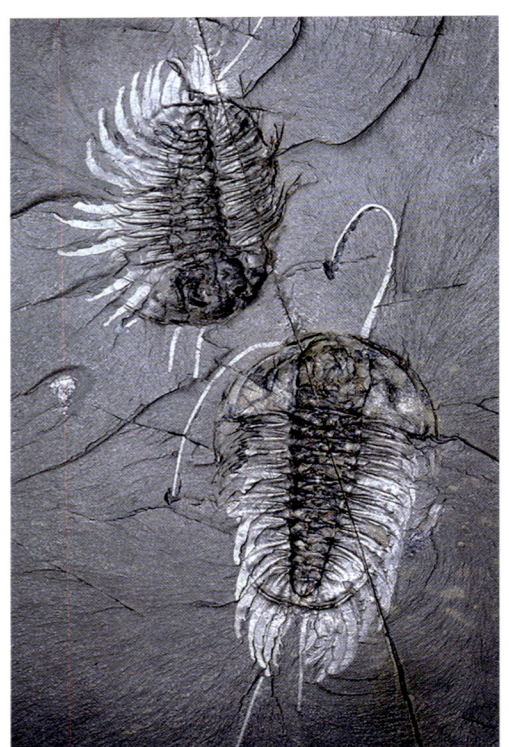

Figure 3.42 Specimens of the Cambrian trilobite *Olenoides serratus*, with appendages preserved, from the Burgess Shale of British Columbia, Canada. The larger trilobite is 15 cm long, including appendages.

extremely rare, being known from fewer than 20 trilobite species from over 20 000 species that have been described. Given such relatively little evidence from across the complete spectrum of trilobites, it is hard to claim that we know what the underside of a 'typical' trilobite looked like. Nevertheless, Figure 3.43a, a reconstruction of *Olenoides*, shows some of the main features likely to be found on the underside of most trilobites. At the front was a pair of flexible antennae. Each limb (Figure 3.43b) had two branches: a lower 'walking branch' and an upper branch with feather-like filaments that functioned as a gill for respiration. This 'gill branch' was, at least in some species, probably also used in swimming, and in some species may have worked like a sieve to strain off food particles. This would have been an efficient arrangement, as movement of the walking branch also moved the gills, causing water to pass across them for gas exchange and possibly for food gathering too. At the rear of *Olenoides* were also two antennae-like sensory appendages called cerci, though these are known only from this trilobite species.

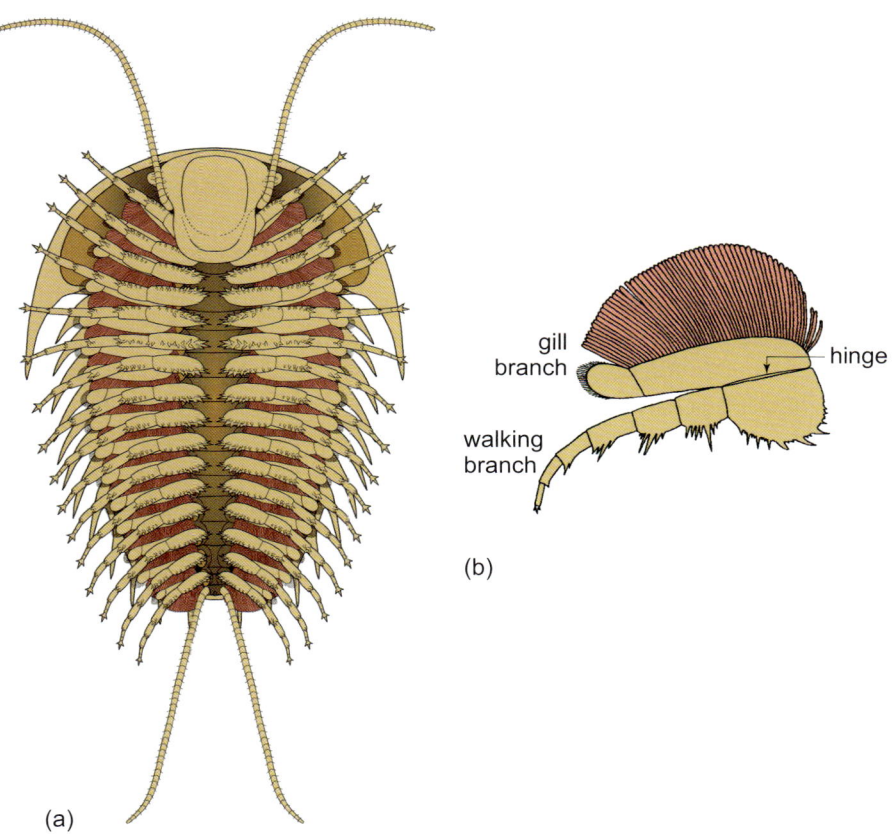

Figure 3.43 Reconstructions of the Cambrian trilobite *Olenoides serratus*, based on exceptionally well-preserved specimens like those shown in Figure 3.42. (a) Ventral surface (underside) showing appendages. (b) A single limb. See text for discussion.

Figure 3.44 A rolled-up specimen of the Silurian trilobite, *Calymene* (2.5 cm).

It is likely that trilobites such as FS G spent much time scuttling over the sea floor. Such locomotion was probably typical of most trilobites, though some were evidently free-swimming, while others may have made shallow burrows. The turret-like projection of the eyes in FS G might also mean that the animal was capable of shallowly submerging itself in sediment, with only the eyes above the surface. Some trilobites are found fossilised in a rolled-up position, like miniature startled armadillos (Figure 3.44). FS G itself could roll up: notice how the articulation of the segments lying between the cephalon and the pygidium would have allowed the animal to do this – the outer parts of each segment sliding under those of the one in front. This was probably a defence posture.

So how did trilobites feed? Unlike many crustaceans, trilobites apparently did not have specialised mouthparts in the head region. Look again at Figure 3.43a. Running down the centre of the animal's underside was a sort of channel, the sides of which were formed by the basal segments of the limbs. In the limbs of some exceptionally well-preserved trilobite species (like *Olenoides*), these basal segments were furnished with bristles, which projected into the channel (Figure 3.43a and b). This arrangement is analogous to that found in certain living primitive crustaceans, which feed on sea-floor detritus brushed into the central channel by scuffling movements of the appendages. So the 'mouthparts' effectively occupied the length of the body behind the head. Food would have been passed forwards to the mouth by the paddling action of the bristles as the basal segments of the limbs moved backwards and forwards. The stomach lay in the head region, below the bulbous central part of the head called the glabella, and the gut passed backwards down the axis to the rear of the pygidium.

In some trilobites, such as *Olenoides*, the bristles of the walking branch were sharp spines and probably acted as teeth and jaws, grabbing and processing prey (e.g. worms) and occasionally dead organisms. A few trilobite groups, including the one to which FS G belongs, were probably predator-scavengers of this type, like the majority of marine crustaceans today. Many, perhaps most, trilobites, however, probably did not actively seek prey but were deposit-feeders, ingesting particles of organic detritus along with mud, though perhaps also scavenging larger, juicy morsels as opportunity permitted. Other likely feeding modes adopted by some trilobites include suspension-feeding and grazing on algae. Trilobites living in low-oxygen environments may have even obtained nutrients directly from symbiotic sulfur-eating bacteria that, it is suggested, lived on their gills.

Some of the speculations about trilobite lifestyle and function may never be confirmed, but it is clear that trilobites spanned a wide variety of marine habitats, and probably occupied most, if not all, of the ecological niches that crustaceans do today. Figure 3.45 shows a small selection from the huge range of trilobite morphology, all variations on the trilobed theme. Note that not all trilobites had eyes; a number of groups became blind, probably in response to living in deep parts of the sea with no light, or living more or less burrowed in mud. Figure 3.45c and d show examples of trilobites lacking eyes. Trinucleid trilobites such as the one shown in Figure 3.45c had a strange pitted fringe around the head, the function of which is still a mystery; no organism today has a structure like it.

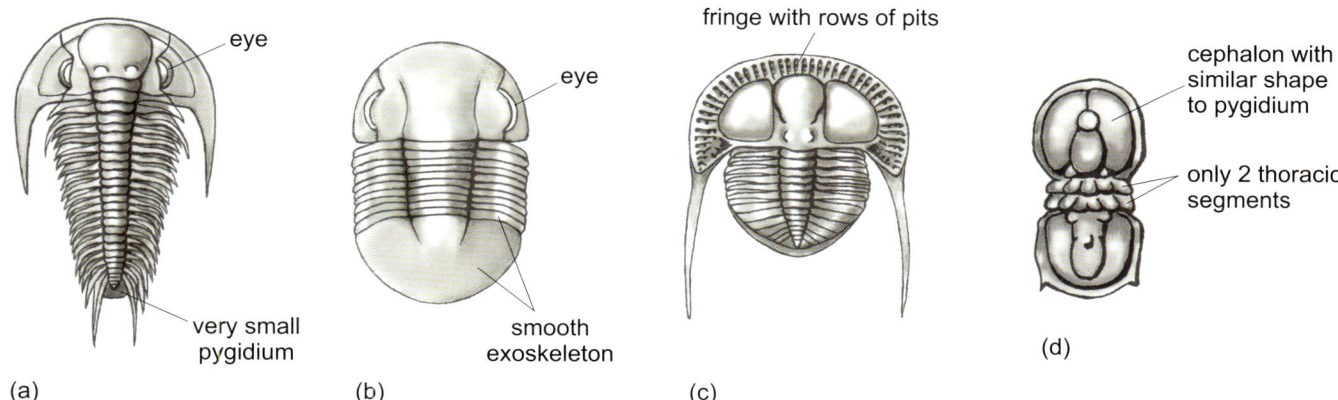

Figure 3.45 Selection of trilobites, with lengths as follows: (a) *Paradoxides*, Cambrian (20 cm); (b) *Illaenus*, Ordovician (4 cm); (c) *Trinucleus*, Ordovician (3 cm); (d) *Agnostus*, Cambrian (6 mm).

Some trilobite species are useful for global correlation, especially in parts of the Cambrian and Ordovician.

Question 3.13

Summarise the data on the life habits of FS G by assigning it to the appropriate box(es) in Table 3.2 (p. 28).

Now explore the trilobites in the Digital Kit.

3.4 Relationships and classification

This chapter began with a simple but practical key system for classifying the specimens in the Home Kit. Now that you have learned something about the biology of the original organisms, we shall take a brief look at a hierarchical biological classification.

It is easy to group the familiar large multicellular organisms into animals, plants and fungi. These are three of the major groupings, or **kingdoms**, into which life can be classified. Other major groupings can be detected among single-celled organisms, though we need not be concerned with them here. Within the kingdoms, organisms naturally fall into broad groups, each of which is characterised by a distinct bodyplan differing fundamentally from that of other such groups. Such a group is termed a **phylum** (plural: phyla). For example, humans belong to the Phylum Chordata. There are about 35 animal phyla, though experts differ on the precise number. Within each phylum there are subgroups called **classes**. Each class has its own characteristic modification of the basic phylum bodyplan. As with phyla, classes are thus recognised by their morphological distinctiveness. Note, however, that they need not contain large numbers of species (though many do).

Within classes, a nested series of smaller subgroups may be recognised until finally the level of single **species**, the natural basic unit of classification, is reached. From the class downwards, these are: **order**, **family** and **genus** (plural: **genera**)

(Table 3.3), and various intermediate divisions are denoted by such prefixes as 'sub-', as in 'subclass' or 'suborder'. The successive groupings, generally known as **taxa** (singular: taxon), thus form a nested series, or **taxonomic hierarchy**, which ideally should reflect evolutionary relationships. The taxonomic hierarchy has many uses. It provides an efficient means of identifying a specimen. For example, knowing that FS 1 is a bivalve means that the search can be narrowed to the Class Bivalvia (to give it its formal name, which, by convention, is always latinised). One can then try to decide to which order the bivalve belongs, and so on, until the appropriate species description is found. The available information on the species can then be used to determine, for example, its likely geological age (for stratigraphic correlation), or to infer its life habits and hence what the original environment was like.

Table 3.3 The taxonomic hierarchy.

Taxon	Example	Contents
Kingdom	Animalia	all animals
Phylum	Chordata	all vertebrates plus some minor groups
Class	Mammalia	all mammals
Order	Carnivora	all dogs, cats, badgers, seals etc.
Family	Felidae	all the cat family (e.g. lions, tigers, lynx)
Genus	*Felis*	all wild and domestic cats
Species	*catus*	domestic cat

Note that, formally, the names of all taxa should begin with a capital letter, except for the species name, which always begins with a lower-case letter. Moreover, the species name is always given together with the genus name, in italics (or underlined if written), so the domestic cat is therefore referred to as *Felis catus*. Our own species is, of course, *Homo sapiens*.

3.4.1 Recognising bodyplans and phyla

The classes Echinoidea (Section 3.3.2) and Crinoidea (Section 3.3.6) are both placed in the Phylum **Echinodermata**.

Question 3.14

What are the three principal features in common between these two classes?

Although echinoids and crinoids look quite different, they have the same bodyplan and are constructed from the same basic components. The fundamental similarities (homologies) place them in the same phylum whilst the differences (relating to radically different general life habits) dictate that they belong to distinct classes.

Apart from the Echinodermata, only one other phylum, the **Mollusca** (informally called 'molluscs'), is represented in the Home Kit body fossils by more than one class. The following features are shared by practically all molluscs: a dorsally attached calcareous shell that grows incrementally by a lining of mantle tissue; a mantle cavity, containing gills, beneath the shell and next to the body; a dorsally situated visceral mass within the body, beneath which is a muscular organ of locomotion. All but one of the molluscan classes (the bivalves) possess a radula inside the mouth.

Question 3.15

Considering this list of characters, which of the groups represented by specimens in the Home Kit belong to the Phylum Mollusca?

Some of the features mentioned above may be greatly modified, or even lost, in particular forms. The octopus, for example, lacks a shell, but it is still recognisable as a cephalopod mollusc because it retains other typical features of that class (e.g. beak-like jaws, radula and tentacles).

You may have been tempted to include brachiopods in your answer to Question 3.15, because they too have an incrementally grown calcareous shell, mantle tissue and a mantle cavity. Brachiopods, however, are quite separate with a different orientation of the shell, possession of a distinctive tentaculate feeding organ (the lophophore) and absence of both a foot and radula. They belong to a distinct phylum (the **Brachiopoda**).

Corals (Section 3.3.7), together with sea anemones, are placed in the Class **Anthozoa** (informally called 'anthozoans'). Together with various other creatures such as jellyfish, seawhips and sea fans, they belong to the Phylum **Cnidaria** (pronounced 'nigh-dare-rhea' and informally called 'cnidarians'). The cnidarian bodyplan is simpler than those of the other phyla considered here, and tends to contain an abundance of gelatinous material, especially in the jellyfish. The body usually has radial symmetry, expressed in the cylindrical body and ring of tentacles around the opening of the internal cavity. The tentacles bear characteristic stinging cells that are a unique cnidarian feature. The gut (internal cavity) is one-ended, combining the functions of mouth and anus.

The final group to consider is the **Trilobita**. These comprise a class within the Phylum **Arthropoda** (informally called 'arthropods'), an enormous assemblage that today consists of over three-quarters of all known living animal species (mostly insects, but including crustaceans such as crabs and shrimps, and many other groups). The characteristics of arthropods include a segmented body with paired limbs. The body and limbs are covered with a stiff, jointed exoskeleton that is periodically moulted. Compound eyes are often present.

The taxonomic information given in this section is summarised in Table 3.4.

Table 3.4 Taxonomic position of specimens A to M.

Phylum	Class	Specimen FS
Echinodermata	Echinoidea	C, J
	Crinoidea	H
Mollusca	Bivalvia	B, I
	Gastropoda	A, K
	Cephalopoda (subclass Ammonoidea)	D
Brachiopoda	Articulata	F, L
Cnidaria	Anthozoa (corals)	E, M
Arthropoda	Trilobita	G

Activity 3.4 Revision of the groups represented by body fossils in the Home Kit

This activity develops your understanding of the groups covered so far in Chapter 3.

3.5 Body fossils of other organisms

So far we have concentrated on commonly encountered kinds of marine invertebrate body fossils represented by replicas in the Home Kit. However, macrofossils from other groups of organisms, including other invertebrates as well as vertebrates and plants, can be abundant and/or of great interpretative value in certain deposits. Microfossils are also widely distributed, often abundant, and come from many different groups of organisms.

3.5.1 Invertebrates

Belemnites

Belemnites are an extinct group of cephalopods that in many ways were probably rather like squid. They had a unique, bullet-shaped, internal shell called a **guard** that, being made of calcite, was easily fossilised. At the wider (head) end of the guard (Figure 3.46) was a chambered structure made of aragonite called the **phragmocone**. In fossils, this has often fallen out or dissolved away leaving a cone-shaped hole. Sometimes a very large number of belemnites occur together in the same bed of rock, possibly representing post-mating death events similar to those that occur in modern squid. Some small, isolated patches of belemnite guards are probably the regurgitated, indigestible remains of belemnites eaten by marine reptiles. Belemnites range in age from early Jurassic to the end of the Cretaceous.

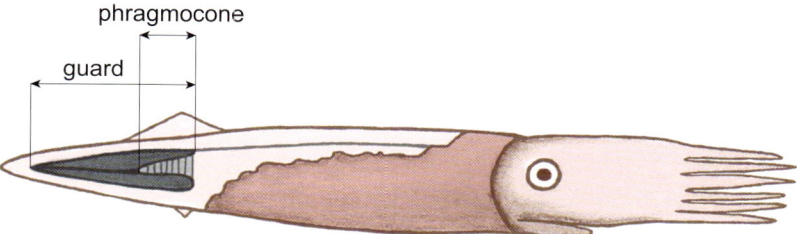

Figure 3.46 Reconstruction of a living belemnite. The soft tissue is shown as if partially removed to reveal the internal skeleton (the bit found fossilised) at the rear.

Graptolites

Graptolites are an extinct group of marine colonial organisms distantly related to chordates. Many look like small saw blades a few centimetres long, with 'teeth' on one or both sides of the 'saw'. The 'teeth' were actually tiny cups (**thecae**, singular: theca) that housed the individuals which made up the colony and possessed suspension-feeding tentacles. Only the resistant skeletons (originally made of collagen-like proteins) occur as fossils. Graptolites with only a few branches (Figure 3.47) – a sub-group known as **graptoloids** – ranged from Ordovician to early Devonian. They are often found in shales laid down in quiet, oxygen-poor conditions on the sea floor. Many species were short-lived, making them very useful zone fossils for correlation during that part of the Palaeozoic (like the ammonites in the Mesozoic). Graptoloids were pelagic zooplankton, either drifting or possibly swimming feebly, and are found widely distributed in marine sediments. Primitive graptolites, dating from the mid Cambrian but persisting until the late Carboniferous, had bush-like colonies with numerous branches; many of these lived attached to the sea floor.

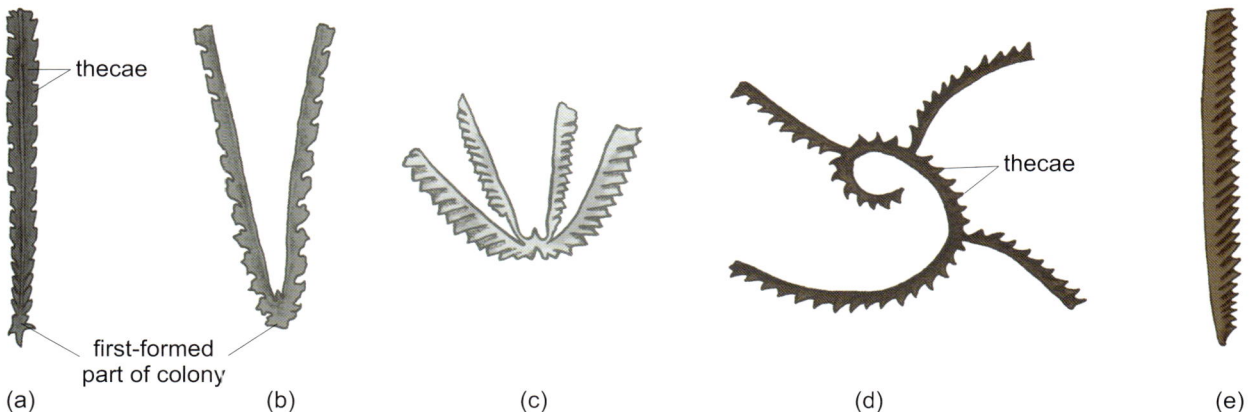

Figure 3.47 Planktonic graptolites (all graptoloids). (a) *Climacograptus*, Ordovician (typical length 2 cm); (b) *Dicellograptus*, Ordovician (1.5 cm); (c) *Tetragraptus*, Ordovician (1.5 cm); (d) *Cyrtograptus*, Silurian (3 cm); (e) *Monograptus*, Silurian (2 cm). Single-branch varieties with thecae on both sides, such as (a), are typical of the middle Ordovician to early Silurian. Single-branch varieties with thecae on one side only such as (e) are limited to, although widespread in, Silurian to early Devonian strata.

Bryozoans

Bryozoans are a phylum of colonial animals. Bryozoans can often be seen on modern beaches, encrusting seaweed, rocks and shells. They are common fossils, but being rather small and often delicate, are relatively unfamiliar. They range in age from early Ordovician to Recent. Most live in shallow seas, some live in freshwater. All are aquatic. Bryozoan colonies range from a few millimetres to 1 m across, but the individuals (called zooids) that make up the colonies are tiny, usually less than 1 mm long. Each zooid builds a tube or box (zooecium) of calcium carbonate, with an aperture (opening). Colonies vary greatly in form, for example encrusting sheets ('sea mats'), net-like fronds (Figure 3.48a) or branching twigs (Figure 3.48b). Some bryozoans look superficially like small corals. The zooids' tentacles filter plankton from the water. Bryozoans often occur in limestones.

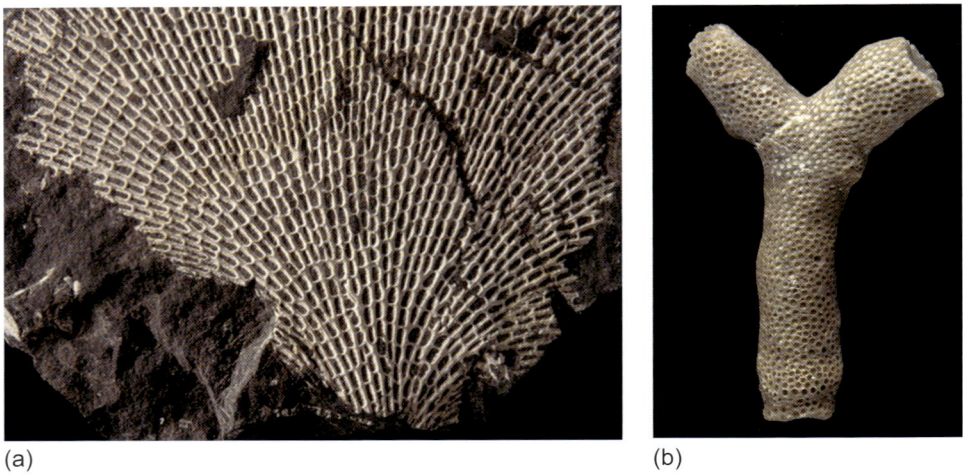

(a) (b)

Figure 3.48 (a) The Carboniferous bryozoan *Fenestella* – a fan-shaped, net-like colony (4 cm wide). The apertures cannot be seen here. (b) *Hallopora*, Silurian (2 cm). Large, round apertures are visible in this cylindrical, branching colony.

Sponges

Sponges (Figure 3.49) are the simplest multicellular animals. They lack definite tissues and organs, for example they have no nervous system. They are all aquatic, benthic, sessile and mainly marine; some live in freshwater. Sponges have a skeleton of slender, pointed elements called spicules. The spicules are made of calcium carbonate, silica, or, as in some modern bath sponges, horny organic material. Sponges are suspension-feeders. Water passes in through the sponge's many surface pores, often to the central cavity of a sack-like body, and out through a large hole at the top. Sponges vary greatly in shape. Some have a stalk; others are encrusting and irregular. Sponges range in age from early Cambrian to Recent.

Chapter 3 Fossil classification and palaeobiology

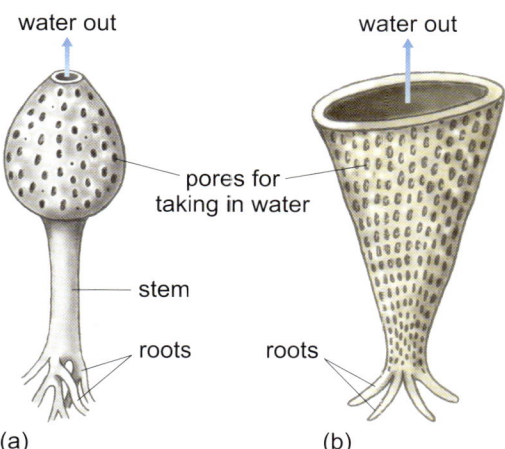

Figure 3.49 Two Cretaceous sponges: (a) *Siphonia* (10 cm); (b) *Rhizopterion* (12 cm).

Worms

'Worm' is an informal name for various invertebrate groups belonging to different phyla. Most types of worm are entirely soft-bodied, and trace fossils may be the only evidence of their existence. Some worms, especially a group of polychaetes (bristleworms) called serpulids (in the Phylum Annelida), secrete a tubular shell for living in, usually made of calcium carbonate (calcite or aragonite) (Figure 3.50). Some tube-secreting worms are free-living, whilst others cement themselves to hard surfaces such as shells, shell fragments and pebbles. Many worm tubes are rather irregular in shape. Worms date from the early Cambrian or possibly earlier.

Now explore the belemnites, graptolites, bryozoans, sponges and worms represented in the Digital Kit.

Figure 3.50 *Serpula*. This worm tube, which varies in shape, is often found attached to shells or other hard substrates on which the worm grew. The drawing shows a Jurassic specimen of this long-ranging polychaete genus. Typical length 2–5 cm.

3.5.2 Vertebrates

Fossil vertebrates (animals with a bony or cartilaginous skeleton and a skull) are much rarer compared with invertebrates, mainly due to smaller populations, so are treated only briefly here. Land-dwelling vertebrates, and those of freshwater rivers and lakes, are rarer as fossils than are marine vertebrates. The five major groups (classes) of vertebrates today, in order of evolutionary appearance, are fish, amphibians, reptiles, mammals and birds. Apart from trace fossils (e.g. footprints), vertebrates are mostly represented by bones and teeth. Bone tends to have a rather irregular, spongy texture, lacking the very fine, closely parallel lines or 'grain' that fossil wood often possesses.

Fish first appeared in the early Cambrian, but remained rare until the Devonian. The earliest fish lacked jaws; some had massive bony armour. Sharks' teeth are relatively common fossils, as are fish scales, which are usually brown, shiny, often

69

1–2 mm in size, and often rhombic in shape. The scales of many ancient fish were thicker than those of familiar fish today (Figure 3.51a).

Figure 3.51 (a) Detail of part of a fossil fish, *Palaeoniscus*, Permian (field of view 2 cm). (b) Ichthyosaur vertebra, Jurassic (6.5 cm). (c) Upper molar tooth of a woolly rhinoceros, *Coelodonta*, Pleistocene (6 cm across).

Amphibians first appeared in the late Devonian, evolving from a group of fish. Some Palaeozoic forms were several metres long, unlike modern frogs, toads and newts. Fossils are very rare.

Reptiles first appeared in the mid Carboniferous, evolving from a group of amphibians. Fossils are relatively common, especially in Mesozoic rocks. Mesozoic land- and air-dwelling groups such as dinosaurs and pterosaurs are much rarer than Mesozoic marine reptiles, such as ichthyosaurs and plesiosaurs, isolated vertebrae of which often get washed out of clays (Figure 3.51b).

Mammals first appeared in the late Triassic, evolving from a group of reptiles. Initially, in the Mesozoic, they were small (mostly shrew-sized) and rare. Mammal diversity expanded greatly in the Palaeogene, increasing towards the present day. Primates generally, and especially early hominid fossils, are exceptionally rare worldwide. The most common mammal fossils in Britain are from the Quaternary, such as ice age mammoths and woolly rhinos (Figure 3.51c), as well as deer, horse and ox.

Birds first appeared in the late Jurassic, evolving from a group of small carnivorous dinosaurs. During the Cenozoic they expanded greatly in diversity. Fossils are very rare.

Fish, reptiles and mammals are represented in the Digital Kit; you should explore these next.

3.5.3 Microfossils

Microfossils are derived from many types of organism, both large and small. They are generally less than about 3–4 mm in size, and many are much less than 1 mm. The distinction between microfossils and macrofossils is somewhat arbitrary, though microfossils are often considered to be discrete remains whose study requires the use of a microscope throughout. Bulk collecting (i.e. taking whole chunks of rock) and processing (e.g. dissolving the matrix in acid) are usually required to concentrate microfossils. The somewhat specialised techniques required preclude detailed discussion here, and only some of the most important groups are mentioned.

Microfossils from larger organisms include plant pollen and spores (see Section 3.5.4), minute teeth and scales from animals, and the siliceous or calcareous spicules that form the skeletons of sponges. However, many microfossils, by contrast, are the remains of *whole* microscopic organisms, most of which are single-celled (e.g. foraminifera). Microfossils come from a wide variety of settings, including the land (e.g. pollen), the plankton (e.g. many single-celled algae (phytoplankton) and free-floating foraminifera) and the benthos (e.g. other foraminifera and tiny crustaceans called ostracods). **Nannoplankton** are plankton of minute size, generally between 2 μm and 20 μm, and include coccoliths (see below).

Microorganisms, especially marine plankton (both phytoplankton and zooplankton), are the source of most petroleum – oil and natural gas. The transformation of planktonic remains into the chemical residues that form petroleum requires an anoxic sedimentary environment on the seabed, and a subsequent history of burial in a subsiding sedimentary basin where the local geothermal gradient is sufficiently high for hydrocarbon generation. Ancient life is also vitally important for the petroleum industry in another way, as microfossils are used extensively for correlation of strata during exploration for rock reservoirs into which oil and gas have migrated.

- ■ Suggest two advantages that microfossils have over macrofossils in petroleum exploration.

- ☐ Being so small, there is a good chance of finding them whole in small rock chips obtained from boreholes and whole rock cores, where the diameter is always limited. Microfossils also tend to be much more abundant than macrofossils. As an extreme case, imagine how difficult it would be to correlate borehole chippings and cores on the basis of dinosaur or ichthyosaur remains.

The small size and abundance of microfossils favour many uses for them in addition to biostratigraphy (Section 3.8). As large, statistically significant, numbers of specimens can be readily obtained from closely spaced stratigraphic intervals, microfossils such as foraminifera have provided high-resolution evidence concerning evolution in individual lineages and the formation of new species. Our understanding of the early evolution of life, from its origin to the late Precambrian, comes almost entirely from microfossils; apart from stromatolites (see p. 75), Precambrian fossils are all microscopic until the Ediacaran fauna appeared about 600 Ma ago. Microfossils can be used to determine the provenance of archaeological artefacts, especially when only minute samples are available for analysis. They also play an increasing role in forensic science (Box 3.1).

Box 3.1 Cretaceous microfossils that helped convict a killer

Microfossils can play an important role in forensic science. The conviction of Ian Huntley, a school caretaker, for the murder in 2002 of two schoolgirls from Soham in Cambridgeshire, partly depended on foraminiferal and nannoplankton evidence. Huntley had denied driving along the track where the girls' bodies were found in a remote location. Under Huntley's car, sitting on the suspension arm, were found bits of Chalk containing an assemblage of microfossils, including the benthic foraminifera, *Lingulogavelinella jarzevae* (Figure 3.52a), *Pseudotextulariella cretosa* and *Marssonella ozawai*, plus an association of several coccolith species including *Corollithion kennedyi* (Figure 3.52b).

The stratigraphic ranges of these species are different, and where they overlap they are characteristic of a very narrow, poorly exposed interval near the base of the Lower Chalk (to be precise, the UKB3 foraminiferal biozone and the UC1a – UC3a nannoplankton biozones). A farmer had recently repaired the track's surface with Chalk obtained from a temporary excavation. With the help of expert witnesses, the prosecution showed that, at the time, the only place in the country where there were piles of 'soil' containing Chalk of that precise age and mineral composition that could be easily driven over, was the very track where the bodies were found.

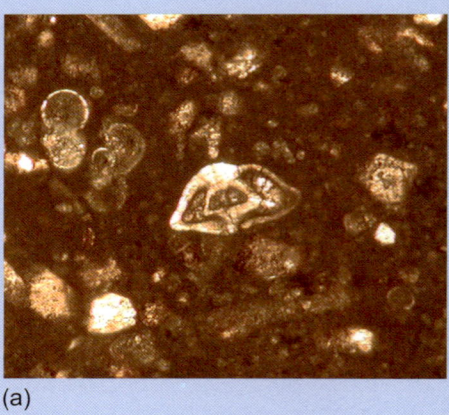
(a)

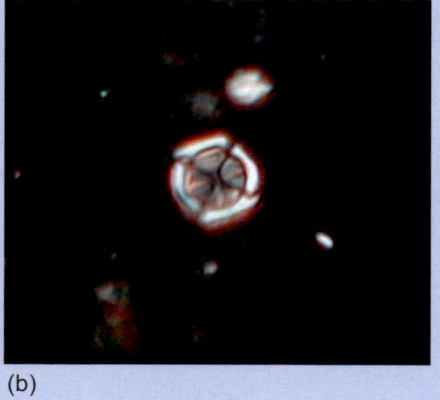
(b)

Figure 3.52 (a) The benthic foraminiferan *Lingulogavelinella jarzevae* in thin section (~200 μm in diameter). (b) The coccolith *Corollithion kennedyi* (5 μm in diameter). See text for explanation.

Foraminifera

Foraminifera (often abbreviated informally to forams) are arguably the most important microfossils. They are aquatic, mainly marine, mostly benthic, single-celled organisms that have a shell (Figure 3.53). Most forams are tiny, less than 1 mm across, but some are much larger, reaching several centimetres in diameter, with a highly complex, chambered construction. In species that attain a large size, the cell contains many nuclei. Forams may be very abundant, forming the bulk of some limestones. They have a wide environmental range, from inland lakes to the deep sea, and from polar to tropical areas. Most foram shells are calcareous; a few are composed of silica or organic material, whilst others construct their shell by selecting grains such as sand or shell fragments from the sea floor and cementing them together. Although there are relatively few species of planktonic forams (zooplankton), they are not only widespread, but small (mostly <0.1 mm) and abundant, and so are extremely useful in stratigraphic correlation, especially in the petroleum industry. Benthic forams, which are usually larger than planktonic forams, tend to have narrower geographic distributions, but are very useful in regional correlation. Forams provide some of the best evidence of climate change in the Mesozoic and Cenozoic because certain isotopes and trace elements within their shells record changes in temperature, salinity and ocean chemistry. They have also provided high-resolution evidence about evolutionary patterns. Benthic forams range in age from early Cambrian to Recent. Planktonic forams range from early Jurassic to Recent.

Figure 3.53 Fossil foraminifera. These are planktonic species, of Neogene age, from the tropical west Pacific, and are shown at different magnifications; their size ranges between 200 and 600 μm.

Coccoliths

Coccoliths are individual calcite plates formed by coccolithophores, which are aquatic, single-celled algae. The vast majority of species are marine and live as phytoplankton in the photic zone of the open ocean, where they are a major source of food and a significant producer of oxygen. The soft photosynthetic tissue of the coccolithophore is generally surrounded by a few dozen of the transparent, protective coccolith plates (Figures 3.52b, 8.2 and 8.5). Coccoliths are so small (typically 2–5 μm in diameter) that a scanning electron microscope is needed to study them. What they lack in size they make up for in volume. Coccoliths form the bulk of the Chalk, and today contribute to calcareous oozes on parts of the ocean floor. They range in age from Triassic to Recent and are extremely useful in Mesozoic and Cenozoic biostratigraphy. Coccolithophore diversity and abundance peaked in the late Cretaceous, but they almost became extinct at the Cretaceous–Tertiary boundary, after which they eventually recovered, though never quite to their former glory.

Ostracods

Ostracods (Figure 3.54) are small, bivalved crustaceans, and are the most abundant fossil arthropods. Many look like little oval-shaped seeds or miniature beans, typically about 0.5–3 mm in length. The animal's soft parts (e.g. appendages) can be withdrawn within the calcitic shell, which may be smooth or bear complex patterns of ridges, bumps and pimples. Ostracods are abundant and widespread in aquatic environments. Most are marine, and of these most are benthic, though some are pelagic, swimming in the open ocean. Some are freshwater; yet others live in brackish waters. Some living species are tolerant of a wide range of salinities. Some even live in damp soil and leaf litter. Ostracods are used to some extent in biostratigraphy, but are of most use in reconstructing ancient environments (e.g. the location and nature of shorelines and lakes) and in various aspects of climate change. Ostracods range in age from early Ordovician (or possibly Cambrian) to Recent.

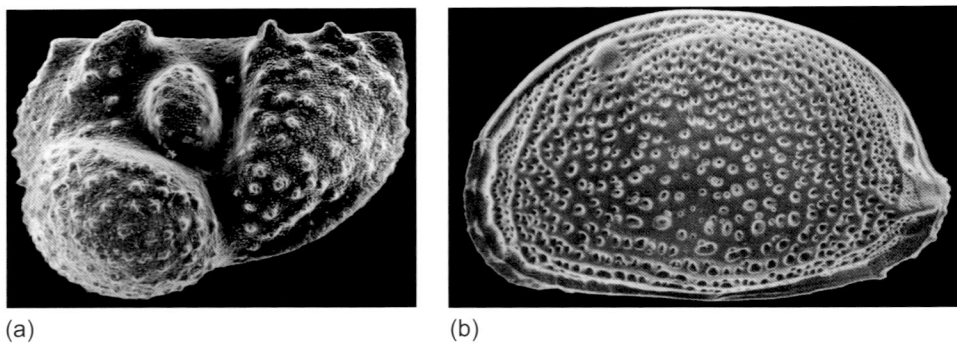

(a) (b)

Figure 3.54 Fossil ostracods, both female. (a) *Beyrichia clausa*, Silurian. Length 2 mm. (b) *Mutilus speyeri*, late Neogene/early Quaternary. Length 1 mm.

Radiolarians

Radiolarians (or radiolaria) (Figures 3.55a,b and 8.7) produce intricately shaped skeletons made of silica, typically 50–300 μm in size. They are single-celled, marine zooplankton, and their remains accumulate on the deep ocean floor as radiolarian ooze. Being siliceous, they are particularly useful in dating sedimentary rocks formed at such great depths that calcareous foram shells and coccoliths dissolve away before reaching the sea floor, and so are not preserved. They are also useful in studies of palaeoclimate and ocean temperatures. Radiolarians range in age from early Cambrian to Recent.

 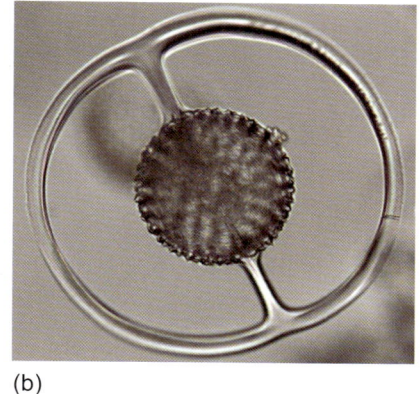 

(a) (b) (c)

Figure 3.55 (a) and (b) Fossil radiolarians from the early Palaeogene. (a) Length of largest individual 0.35 mm. (b) This specimen of the genus *Saturnalis* (appropriately named after the planet Saturn, given its shape), is 0.3 mm across. (c) Conodonts: four elements of Silurian to Carboniferous age on a pinhead for scale.

Conodonts

Conodonts (Figure 3.55c) are tiny tooth-like fossils from a group of small, extinct, primitive, eel-like, jawless vertebrates. Fossil evidence of the soft parts is very rare. Typically 0.25–2 mm in size, fifteen or more individual 'teeth' originally fitted together into a complex feeding apparatus, the individual elements of which varied in shape and were usually dispersed after death. Not only useful in biostratigraphy, conodont elements are also valuable in petroleum exploration to indicate the degree of thermal alteration to which sedimentary rocks have been subjected: under higher temperatures the calcium phosphate of which conodonts are made undergoes quantifiable, permanent, color changes. Conodonts range in age from late Cambrian to Triassic, and were entirely marine.

Stromatolites

Stromatolites are finely layered, often crinkled or lumpy-looking accumulations of carbonate muds or sands, trapped by surface-dwelling mats of microbes, principally cyanobacteria (Figures 8.6 and 9.18d). The cyanobacteria live in shallow water and require strong sunlight to photosynthesise. The trapping of sediment grains as the mats of cyanobacteria grow upwards results in structures that are easily visible to the naked eye, sometimes reaching a size of several metres. Although macroscopic in size, stromatolites are included here within microfossils, though the microbes themselves are rarely preserved. Stromatolites were widespread and common in the Proterozoic, during which diverse and intricate forms were produced. In the Phanerozoic, by contrast, they became largely confined to restricted environments, such as intertidal to supratidal flats in hot and arid areas, excessively salty bays and lagoons, and hidden cavities in reefs. This restriction seems to have been largely due to the destruction of the mats by grazing animals in most normal marine environments.

3.5.4 Plants

Plant fossils are far more abundant and useful than you might at first think. They chart not only the history of plant evolution, but also the history of climate change and the carbon cycle. They are invaluable in relative dating, especially in

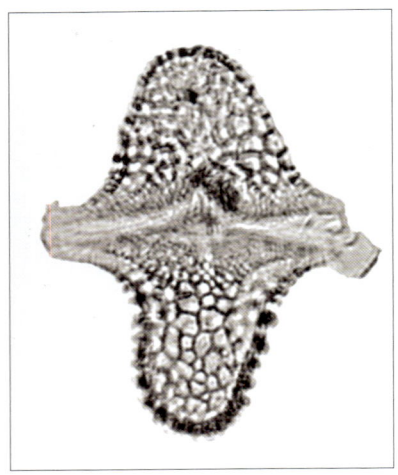

(a)

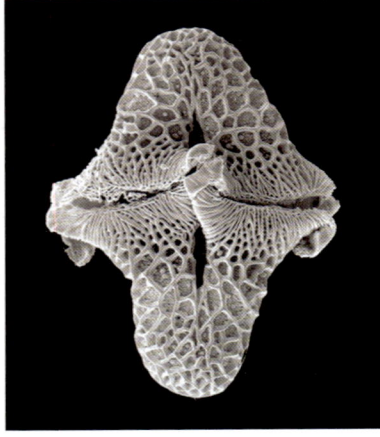

(b)

Figure 3.56 Images of a late Cretaceous pollen grain, *Aquilapollenites chlonovae*. (a) a light microscope view; (b) the same grain under a scanning electron microscope. Length 60 μm.

the petroleum industry, where their preservation state can be used for measuring the burial history of rocks and thus the potential of those rocks to yield oil and gas. The study of plant fossils is called **palaeobotany**.

If you pick up a fist-sized chunk of mudstone or shale formed on or anywhere near land at any time in the last 400 million years, you will probably be holding thousands of plant fossils. Such fine-grained sediments usually contain vast numbers of pollen grains and spores. Pollen grains come from plants that reproduce by means of seeds (like apple trees and pine trees), whereas spores are produced by seedless plants such as ferns. Pollen and spores, collectively called **sporomorphs**, have an outer coat that is extremely resistant to decay and even to chemical treatment. Strong acids (e.g. hydrofluoric acid) can be used to dissolve sedimentary rocks, removing minerals such as quartz and other silicates, leaving only the sporomorphs. Figure 3.56 shows the same fossil pollen grain, about 70 million years old, under an ordinary light microscope and under a scanning electron microscope. Sporomorphs can be exquisitely preserved in great detail, allowing palynologists (people who study sporomorphs) to distinguish between different sporomorph species.

- Give one drawback for palynologists caused by the strong resistance to decay of pollen and spores.

☐ Individual grains can survive not only transport by wind and water far away from the parent plant, and persist through burial and the passage of time, but can also survive erosion and redeposition in another place and time.

There are also advantages. The small size of sporomorphs, usually between 10 μm and 100 μm, means that large numbers can be extracted from small rock samples. This allows samples to be taken at closely spaced intervals through a sedimentary succession, giving a high temporal resolution.

This ability to be reworked from older into younger sediments is easy to spot if, say, a Devonian sporomorph is reworked into a Cretaceous deposit, but less easy to detect if reworking occurs over shorter time intervals of a few tens of thousands of years or even a few million years. This reworking tends to 'smear' the temporal resolution, but palynologists have developed ways to eliminate many of the problems this might otherwise introduce.

Being small, sporomorphs can be transported easily. As hay fever sufferers know, the air can sometimes be full of pollen grains. Plants such as grasses and conifer trees produce huge amounts of pollen and depend on the wind to carry the pollen from the male parts of one plant to the female parts of another. Some plants, usually those with bright, showy flowers, produce far less pollen and rely on animals such as insects, birds and bats to transfer the pollen in a more targeted fashion. Pollen from wind pollinated plants can be blown long distances (sometimes >1000 km). Grains produced by plants pollinated by animals are far rarer, but because transport is more limited (typically <5 km) they can indicate which plants were growing close to a lake or pond. By looking at the changes in sporomorph types and abundance, palynologists can build up a picture of changing vegetation patterns with time.

■ Suggest two main reasons why matching a fossil pollen grain to its parent plant is not always easy.

☐ The further we go back in time the more difficult it gets, because we start dealing with extinct species, genera or even families of plants. Also, the fossil pollen is usually detached from the parent plant, which is, in addition, much less likely to get preserved.

Fossil pollen grains can often look quite unlike anything living today. Unless the sporomorphs are found inside the sporomorph-producing parts of a plant that, in turn, are attached to stems that are attached to leaves or other parts of the plant, it is often impossible to say what plants produced the grains – except at the most basic level (e.g. a fern, conifer or a flowering plant). Many sporomorphs are known only as dispersed grains, and cannot be linked to a whole plant; these are often named without any implication as to their biological identity.

As plants shed their organs (e.g. leaves, pollen or spores, seeds and flowers) on a seasonal basis, plant fossils are rarely found as whole plants. Each plant's isolated organs have different characteristics in terms of fluid dynamics (such as size, weight and shape), so wind and water will transport them different distances. Only rarely is a fossilised leaf found attached to a branch that is attached to a flower or woody trunk. Nevertheless, a number of complete plants have been successfully reconstructed from their component parts. Until that happens, the individual parts have to be described separately and given a genus name and a species name, just as would a whole plant or animal. So, for example, the pollen grain in Figure 3.56 has been named as *Aquilapollenites chlonovae*, even though this name does not relate to a whole plant but just to the pollen grain.

Until a whole plant can be reconstructed, there will be several, perhaps many, parts with different genus and species names. If a whole plant can be put together, the *whole* plant is named after the part that was either found first or is most characteristic. Its individual organs, however, retain the names given to them when they were still not recognised as part of the whole plant. For example, *Lepidodendron* (Figure 3.57) was a large (up to 30 m) clubmoss tree

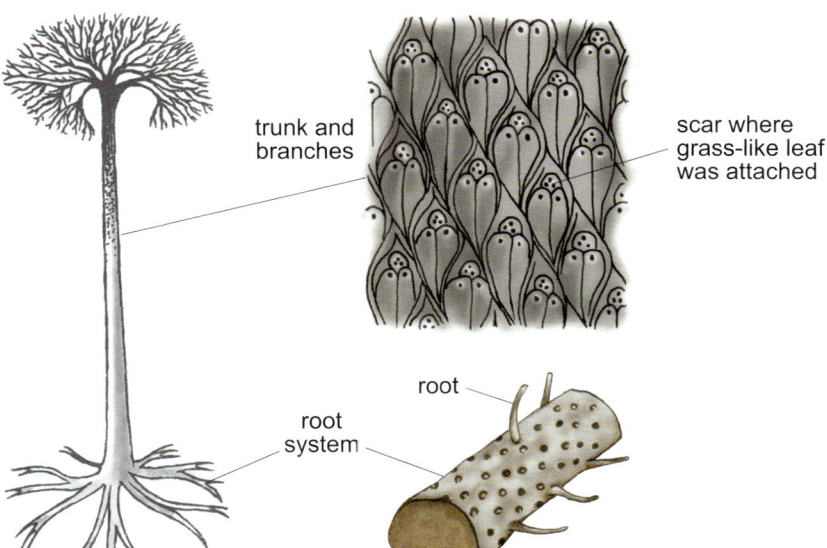

Figure 3.57 Reconstruction of *Lepidodendron*, a tall (up to 30 m) clubmoss tree from the Carboniferous. Different parts of the tree are given different names: the root system, for example, is called *Stigmaria*. See text for discussion.

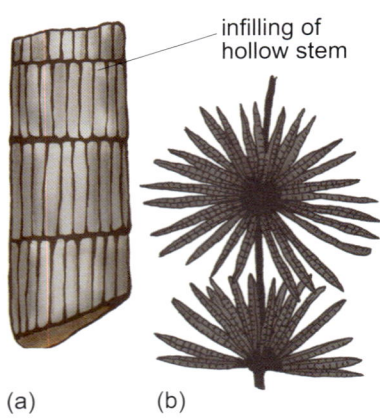

Figure 3.58 The Carboniferous horsetail *Calamites*. The whole plant and the stem (shown in (a), here 5 cm across, but often much larger) are given the name *Calamites*, whereas the rosettes of leaves (6 cm across) shown in (b) are separately called *Annularia*.

Figure 3.59 Cross-section of a branch of a Cretaceous tree that grew near the North Pole 70 million years ago. Note the variation in ring widths as growth proceeded from top to bottom. Vertical field of view about 2 cm.

that lived in Carboniferous swamp forests. The whole *Lepidodendron* tree gets its name from the name originally given to the parts of the trunk and branches with characteristic diamond-shaped markings where grass-like leaves were attached. The root system (Figure 3.57) is, however, separately called *Stigmaria*, whilst other parts, such as the leaves themselves, the cones and the spores, are given yet other names. Similarly, *Calamites* (Figure 3.58a) is the name given to a whole horsetail plant that lived along rivers in Carboniferous swamp forests, and to isolated fossils of its stem – whereas its rosettes of leaves (Figure 3.58b) are separately called *Annularia*. Although seemingly confusing, this is a useful system because if we changed all the names of the parts it would obscure an important feature of plant evolution. As the leaves have a different function to the flowers and the flowers have a different function to the seeds, and so on, different parts are subject to different selection pressures and so evolve at different rates. The leaves have features that allow them to photosynthesise efficiently, while the flowers have to attract pollinators or shed and capture pollen transferred by wind. So, in the case of an insect-pollinated flower its evolution might be linked with that of its pollinator, while the leaf might evolve features more related to climate.

This largely separate evolution of structures is called **mosaic evolution**, and is especially common in plants. It means that a particular dispersed pollen species might have quite a different stratigraphic range to a leaf species, even though at one point in time they were both borne on the same plant. If the names of all the different plant parts were changed to reflect their host plant at any one given time (even if that were possible), then the ability to track mosaic evolution would be lost.

Leaves, like pollen and spores, are produced in large numbers throughout the lifetime of a tree and are quite common fossils. Fossil leaves are particularly useful as indicators of ancient climates because in order to photosynthesise most efficiently they have to have an architecture (i.e. a form and structure) that is specifically adapted to the immediate environment. For example, leaves tend to be small in deserts and large in rainforests, independent of the species involved.

In flowering plants in particular, evolutionary adaptation, honed by selection over many generations, has resulted in vegetation, and especially leaves, being highly characteristic of the particular climates in which they grow. Leaf form becomes an optimal, or near optimal, engineering solution to a variety of often-conflicting environmental and structural constraints. This engineering solution reflects the laws of physics relating to fluid flow, gas diffusion and mechanical strength, so to a large extent the relationship between leaf form and environment is stable over time. Leaf architectural information is retained even in impressions when no organic matter from the original leaf survives (Section 4.2.4) and, providing the leaf architecture can be interpreted correctly in terms of environment, fossil leaves can tell us a lot about ancient climates long before present-day species evolved.

Fossil wood (Figure 3.59) can occur as large logs many metres long or small branches. Where it is well preserved, growth rings can often be seen if the tree grew in a seasonal climate. Seasonality might have been due to variations in temperature, rainfall or even light, but in all cases variations in ring widths, and even variations in cell size within a ring, reveal that growth conditions were not constant. In many cases it takes about three days to produce a single wood cell so, potentially, tree ring analysis can provide information on almost day-to-day changes in ancient weather. By matching up ring sequences between different trees, not only can we get a clear idea of regional climate, but when one ring sequence partially overlaps another and a third partially overlaps the second, it is possible to build up records of climate change ranging over thousands of years. Analysis of isotopes of carbon, oxygen and hydrogen in fossil wood can also provide climate information.

Geological history of plants

Although evidence from spores suggests that some plants were growing on land in the Ordovician, it is not until the Silurian that there is evidence for what the whole plants looked like. These plants, initially just a few centimetres high, colonised the edges of land via freshwater rivers and lakes (rather than directly from the sea). The first trees with woody trunks appeared in the late Devonian. The first really abundant plant fossils occur in coal-bearing, late Carboniferous rocks. Among the groups flourishing then in equatorial forests and low-lying swamps were giant clubmosses, horsetails and seed ferns (Figure 3.60a).

Conifers were the dominant forest trees throughout most of the Mesozoic. Conifers, seed ferns, true ferns (Figure 3.60b), cycads (Figure 3.60c) and ginkgos (Figure 3.60d) are relatively common, for example, in Jurassic sandstones and shales.

Flowering plants (angiosperms) first appeared in the early Cretaceous. Initially rare, they expanded through the late Cretaceous and Cenozoic, and today make up the vast majority of plants both in numbers and diversity. Most trees, shrubs, grasses, hedgerow and garden plants, and food crops are angiosperms.

Now explore the fossil plants in the Digital Kit.

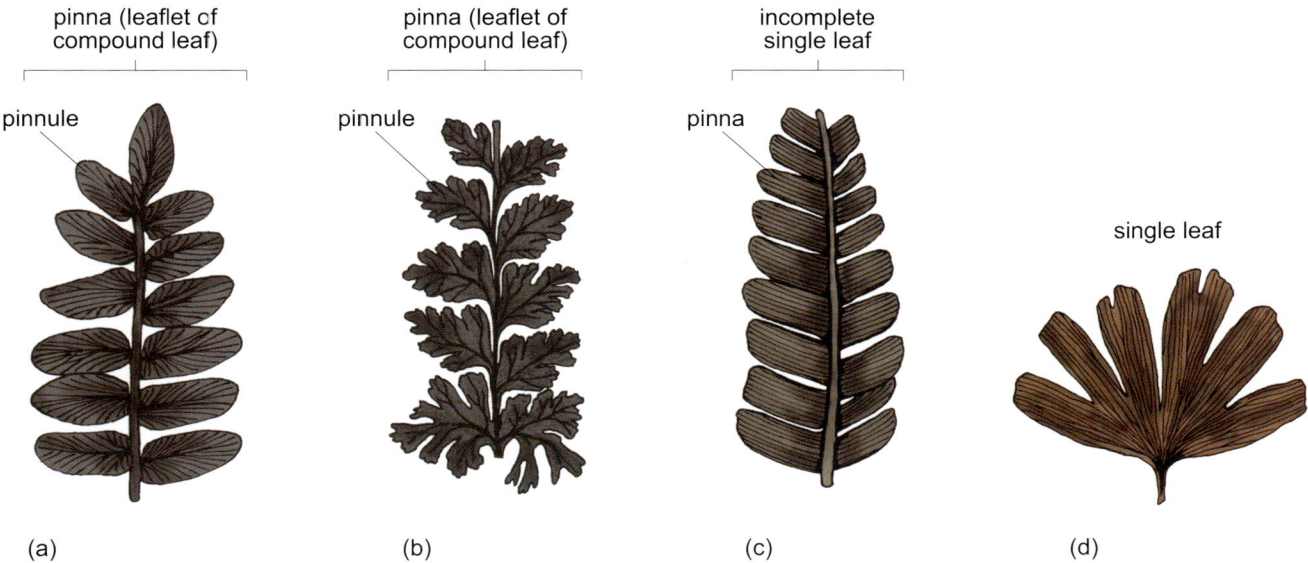

Figure 3.60 (a) A Carboniferous seed fern, *Neuropteris* (5 cm). (b) A Jurassic fern, *Coniopteris* (4 cm). (c) A Jurassic cycad, *Nilssonia* (5 cm). (d) A Jurassic ginkgo, *Ginkgo* (7 cm). A pinna is a leaflet of a compound leaf, and may itself be subdivided into pinnules either side of a central stem.

3.6 Trace fossils

You will need FS N–P and the hand lens from the Home Kit for your study of this section.

Palaeontologists have been studying body fossils systematically for well over 200 years, but only in the last 50 years or so have trace fossils been analysed in any great detail. Indeed, for a long time some branching trace fossils (Figure 3.61) were identified as fossil seaweeds. The study of trace fossils is called ichnology (from the Greek *ichnos*, meaning a trace). Today, sophisticated methods, including computer modelling, enable trace fossils to provide a wide range of information, ranging from estimating speeds of dinosaur movement during herding behaviour – based on well-preserved, parallel trackways – to predator/prey relationships based on matching bite marks on bones and shells, to microstructures on the teeth of predators.

Figure 3.61 The trace fossil (branching burrow system) *Chondrites* (pronounced 'Kon-dry-tees'), which superficially resembles a plant. Jurassic.

■ What do trace fossils record that body fossils do not?

□ Evidence of the activity of organisms, which tells us something about their behaviour.

Trace fossils may be classified according to the mode of behaviour represented – for example, moving, feeding, preying, dwelling, escaping, resting and breeding. Behaviour, and the ways that traces are produced, are closely related to the local conditions experienced by the organism. Most trace fossils are direct, *in situ* evidence of the environment at the time and place that the organism was living. This is because, unlike many body fossils, which may end up a long way from where the organism originally lived, most trace fossils are simply rearrangements of the sediment particles and cannot be transported without being destroyed.

(The few exceptions include borings in shells and pebbles that may survive transport and reworking.) Among the more common types of trace fossil are burrows in soft sediment, borings into hard surfaces, tracks and trails across the sea floor, footprints of land-dwelling vertebrates, and droppings (coprolites and (smaller) faecal pellets).

The study of trace fossils is challenging for several reasons. It is often impossible to assign trace fossils to a specific maker, and the detective work is difficult. Only on very rare occasions are culprits found, red-handed as it were, at the end of their tracks (Figure 3.62). Even if a burrow contains a fossil (e.g. a bivalve shell), it may be that it was washed in and is simply there by chance. Many trace fossils reflect the activities of entirely soft-bodied organisms, e.g. burrowing worms. Trace fossils are often the only evidence left of such organisms, as their body fossils are extremely rare.

Figure 3.62 A Jurassic horseshoe crab, *Mesolimulus*, that died at the end of its track. It is 15 cm long.

Trace fossils are given generic and specific names for ease of reference, but it is essential to realise that they are *not* equivalent to genera and species of living organisms or their body fossils. Each type of trace fossil is given its own name, irrespective of what organism made it. There are two key points here:

1. One animal may produce several different kinds of traces in different situations or at different times of its life.
2. One type of trace fossil may be made by several completely different types of organisms that share the same type of behaviour.

Although trace fossils are common from the start of the Cambrian onwards, they are, not surprisingly, of limited use in biostratigraphy.

Most trace fossils in sedimentary rocks deposited in water were made by infaunal animals, especially deposit feeders. This leads to problems when trying to find modern equivalents of ancient traces. Trails made today on sediment surfaces are easy to observe, especially in shallow water and intertidal areas, but these are liable to erosion by currents and are relatively rarely fossilised. Infaunal traces have a far higher potential for preservation, though modern examples tend to be inaccessible. Various techniques have been developed to overcome this problem. One method is to inject quick-setting resin into burrow systems, but this has the disadvantage of only filling spaces: any filled areas, perhaps stuffed with faecal pellets, will remain undetected. Another method is to X-ray a block of modern sediment. For many trace fossils, however, modern equivalents are not known. The contrast in appearance and physical structure between the trace and the surrounding sediment, often slight in newly formed traces, is usually accentuated during fossilisation. For this and other reasons, it is often easier and more informative to study fossil traces rather than modern ones.

As many traces are originally formed as hollow burrows, borings, footprints or tracks excavated into sediment, the trace fossils that result are commonly hardened infillings of the sediment that later filled the hollows. Thus, when trails were cut into a muddy sea floor or lake bottom and then buried by sand, it is the infill standing proud in relief on the base of the resulting sandstone bed that is commonly found. The same is often true of fossil footprints from terrestrial settings. The rock tends to split along the natural zone of weakness between the mudstone and the overlying sandstone, and the mudstone erodes away much more easily.

You are now about to look at some distinct, easily identifiable examples of trace fossils. It is important to realise, however, that many trace fossils lack such diagnostic and identifiable features, and can only be classed, for example, as a 'burrow' or a 'trail'. In some cases the sediment may be completely reworked by burrowers, which destroy primary sedimentary structures such as cross-stratification and lamination. This churning of the sediment, known as **bioturbation**, leaves few clear individual traces. Nevertheless, even this effect is informative for interpreting ancient environments. Strongly bioturbated sediments indicate both well-oxygenated conditions capable of supporting a rich infauna, and either relatively slow rates of deposition that allow thorough mixing of the sediments or a prolonged absence of deposition. Where there is a thriving infauna, but also more or less frequent current activity – as in many shallow marine settings, for example – the changing balance of preservation of sedimentary lamination *versus* bioturbation can provide a detailed record of subtle variations in, say, depth of deposition, sediment influx or storm activity.

3.6.1 Mining buried food in quiet environments

Some of the most interesting traces come from sediments deposited in fairly deep water, below the **photic zone**, which is the layer of water penetrated by sunlight. At these depths, there is no photosynthesis by algae, and food, in the form of dead plankton and other detritus raining down from shallower levels, is usually in comparatively short supply. Many of the traces, therefore, show regular patterns formed during the systematic search by deposit feeders through sediment for the food that has accumulated there, relatively undisturbed by currents. It is possible to replicate such regular traces by programming a computer with a series of simple commands, such as 'turn after a given distance', 'avoid crossing an earlier trace', and so forth (Figure 3.63). The original trace-makers were more sophisticated than a simple computer programme, and if a patch of sea floor proved unproductive they stopped the systematic search and moved elsewhere.

Figure 3.64a shows a complex horizontal burrow system known as *Nereites* (pronounced 'Nerry-eye-tees'). This is a Palaeozoic trace, probably formed on or near the sediment surface as the animal meandered through the sediment. In this example, you are looking at the actual trail of disturbed sediment itself, as seen on the top of a bed. It consists of a median zone with lobate edges, and each lobe contains fine curved lines. The lobate areas are interpreted as patches of sediment searched for food particles and then back-filled, while the central groove appears to have been filled with a faecal string of sediment that has passed through the body (Figure 3.64b). *Nereites* evidently represents a trail produced by a deposit feeder, and it probably lived in deep water. But what was the original animal like? As with many trace fossils, the nature of the original animal remains mysterious. In the case of *Nereites*, it seems most likely that some form of sediment-eating worm was responsible, but no firm evidence is available.

Figure 3.63 A complex, meandering Cretaceous trace fossil (above, 15 cm across) and analogous computer-generated simulation (below).

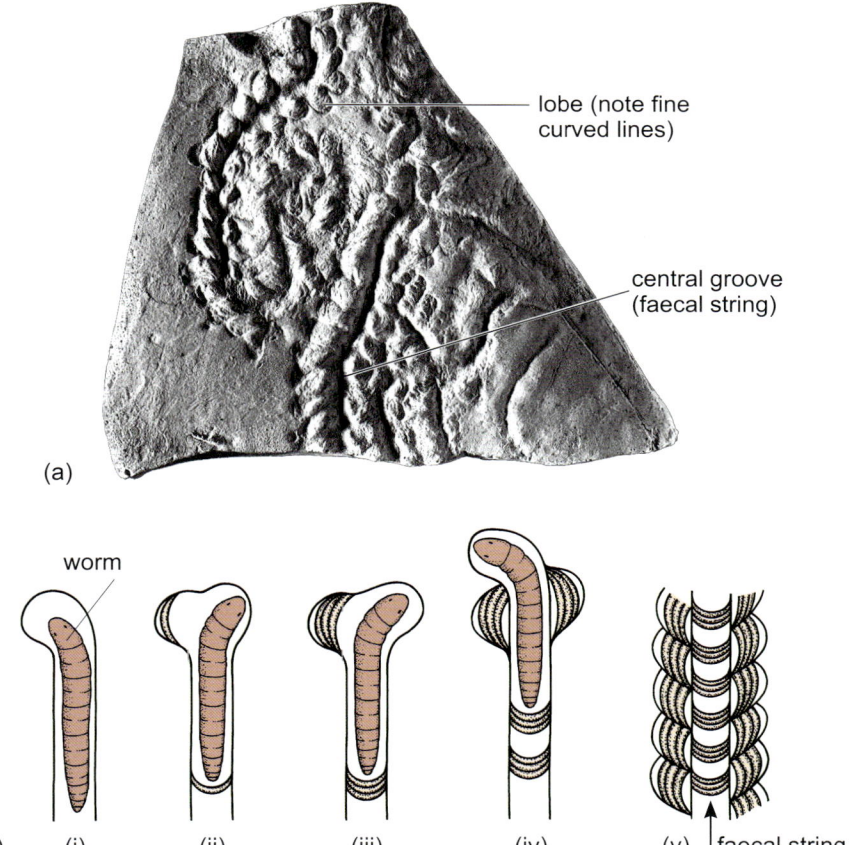

Figure 3.64 (a) The trace fossil *Nereites* (specimen is about 7 cm long). (b) Possible mode of formation of *Nereites*, seen from above. A soft-bodied animal (thought to be a worm) excavates areas on each side of its anterior end, sorts sediment for food and replaces the processed sediment in the lateral excavations. As the animal crawls forward and repeats this process, a faecal string fills the burrow behind, as shown in stages (i)–(v).

For an example of extreme regularity in a trace fossil, turn to FS N. It is the trace fossil *Paleodictyon*, which forms a characteristic honeycomb-like network (the finer network visible on the slab – the larger and more irregular network represents the activities of another animal). Unlike *Nereites*, in which the trace was formed by being packed with sediment, *Paleodictyon* was originally a system of hollow tubes beneath the muddy sediment surface (Figure 3.65). However, it is now preserved in positive relief on the base of an overlying bed of sandstone, with the tubes filled with sediment from above. How did this filling occur? One interpretation is that *Paleodictyon* was formed in mud just beneath the sea floor, with short vertical tubes connecting the hexagonal network with the overlying seawater. A current laden with sediment (a turbidity current, Section 7.6.3) flowed across the sea floor and scoured away the roofing mud and vertical tubes, laying open the trace. Coarser-grained sediment dropped from the current then filled the tubes to give an internal mould of the original trace on the underside of the resulting bed.

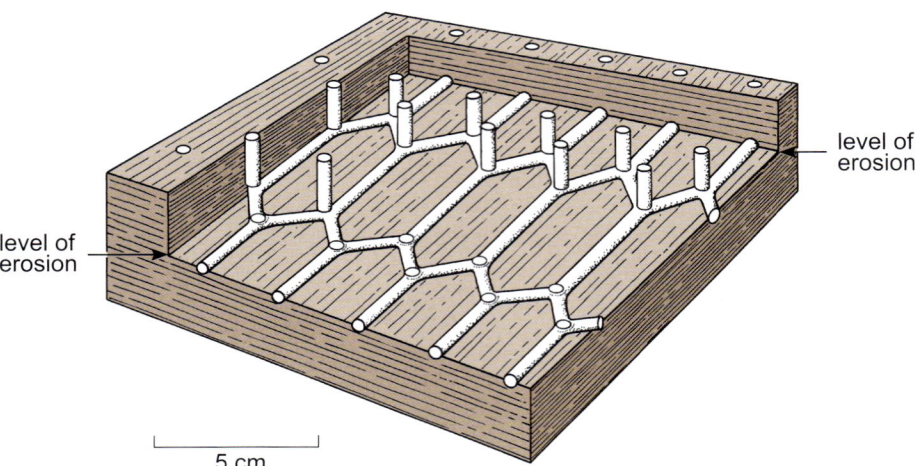

Figure 3.65 Interpretation of the trace fossil *Paleodictyon*. Erosion by a sediment-laden current to the arrowed level destroyed some of the vertical tubes leading to the sediment surface and exposed the hexagonal network, which was then filled with sediment.

Question 3.16

Inspect FS N with this model in mind. Notice how one end of the *Paleodictyon* network stands proud on a slight bump on the surface of the slab, ending along a well-defined boundary, while the other end is more subdued, seemingly fading into the slab. How might you explain these differences in relation to the model described above? (*Hint*: remember that you are looking at relief on the *underside* of a bed; you might find it helpful to hold the slab with the trace fossils facing down, and to look up at them from beneath, in order to imagine the processes of scour and fill that took place.)

As with *Nereites*, the maker of *Paleodictyon* is not known. Fossil FS N is from the Cretaceous, but *Paleodictyon* is found from the Cambrian onwards. It may be that the net-like pattern originally arose from the methodical search for food by a deposit feeder, but that the structure was later exploited as a sort of microbial 'garden' or 'farm'. Food material that drifts down into the deep sea has already been stripped of much of its available organic content and what remains is difficult for deposit feeders to digest. Bacteria and other microbes living in the *Paleodictyon* network, however, may have been able to accelerate the breakdown of such material to the benefit of a deposit feeder that ate the microbes and organic residues. This is not a far-fetched idea, as some modern marine burrowing crustaceans are also known to be 'composters'. Recently, observations have been made of a strikingly similar hexagonal pattern at abyssal depths on the ocean floor. What made these patterns remains a mystery; the creature has not been seen or caught. It is even possible that rather than being a trace, *Paleodictyon* may represent an unusual type of sponge with a highly compressed body shape; in which case, *Paleodictyon* may actually be a body fossil.

Both the *Nereites* and *Paleodictyon* specimens were formed in deep-water deposits of predominantly fine sediments with influxes of coarser-grained material in turbidity flows. Their morphology reflects the need for efficient exploitation of scarce food supplies.

3.6.2 Feeding in current-swept environments

The remaining trace fossils to be studied were probably formed in shallower water than *Nereites* and *Paleodictyon* and, furthermore, can be reasonably attributed to known organisms. Look first at FS O. This specimen again shows the underside of a sandstone bed, so that the features projecting from its surface are the fills of depressions in the underlying mud. In this case, however, you are not looking at the sand fills of pre-existing traces exposed by scour, but instead those formed at the interface between mud and sand layers by animals that had burrowed down through the sand to this level.

The most obvious feature in FS O is the large five-rayed trace.

■ Casting your mind back to the discussion of bodyplans in Section 3.4, of which phylum is this trace-maker likely to have been a member?

□ The fivefold symmetry of the trace strongly hints at an echinoderm.

There can be little doubt that this trace was formed by the digging movements of a starfish, which is a member of another echinoderm class, the **Asteroidea**. This trace is called *Asteriacites*. In some cases, successions of these traces are found one above the other; they represent the escape of the starfish from an influx of sediment. The coarse striations arising from the central zone along each arm were formed by the digging action of the tube feet, which ran along the underside of each arm.

The other traces on this slab are the numerous oval, pod-like projections, where the sand has filled excavations made by infaunal animals. Bivalves are the most likely candidates. In burrowing, infaunal bivalves form a pod-shaped chamber in the sediment. When the bivalve leaves this, it is filled with sediment from above to produce the trace fossil.

These pod-like trace fossils are called *Lockeia.* Every bivalve trace cannot have been simultaneously occupied, as some traces cut across others. Some may have been formed at the same time as the starfish trace, but others must have been made earlier or later than the starfish trace, as they are either cut by, or cut across, it. Whether the bivalves were suspension feeders (most likely), or surface detritus collectors, their food was taken directly from above the sediment surface, unlike that mined by the makers of *Nereites* and *Paleodictyon*. Food particles were probably replenished at the surface by current activity, which either transported material in, or resuspended detritus from, the sediment itself.

The association of traces formed by starfish and bivalves is interesting because many starfish prey on bivalves either by ingesting them whole, or by wrapping their arms around them and pulling the valves apart using their tube feet as suckers to grip them. Fossil assemblages comprising distinct clusters of bivalves and starfish, with some of the starfish still wrapped over the bivalves, have also been found.

FS P is a Cambrian example of a trace fossil known as *Cruziana*. It represents a crawling trace, quite likely made during a search for food (so *Cruziana* is often also classified as a feeding trace too). The original trace was a furrow in the muddy seabed, but the trace fossil is again preserved as coarser-grained sediment filling the hollow. Much of the trace consists of pairs of scratch or rake-like marks that come together to form a chevron pattern.

- ■ Given this information, which kind of organism represented in the Home Kit is most likely to have produced this trace?

- ☐ A trilobite (FS G), which possessed numerous pairs of jointed legs (Figure 3.43) well suited for scratching and raking as the animal ploughed through the topmost layer of sediment.

The filling of the original furrow by coarse-grained sediment, which was presumably carried in a relatively fast current, raises the question of why the trace itself was not eroded by the current. Fine-grained sediment is cohesive, and needs quite high-speed currents to erode it (Section 7.1.2). Thus it may still be resistant to erosion, even in a current that may carry coarse-grained sand. Furthermore, experiments have shown that if the mud loses water and becomes more consolidated, its resistance to erosion further increases. Thus, the mud over which the animal moved may already have been somewhat consolidated, perhaps through compaction beneath a subsequently eroded layer of sand.

In forming *Cruziana*, the walking legs of the trilobite swept backwards and inwards so that the chevrons point in the opposite direction to that of movement. The angle formed by the two sets of striations is not constant; it may vary from about 50° to 140°. The variation in the angle is probably related to the speed of locomotion, smaller angles being produced at somewhat higher speeds. Much of this specimen of *Cruziana* is occupied by the scratch marks of the walking limbs. There are, however, two sets of fine 'wispy' structures in a few areas along the edges of the chevron pattern, just within the sharp ridges that run down its sides.

- What part of the trilobite body could have made these markings?
□ Possibly the gill branches (Section 3.3.8 and Figure 3.43) as they dragged over the sediment.

In other specimens of *Cruziana*, the marks made by the gills form a well-defined zone on each side of the scratch marks. The other notable feature in FS P is the ridge running along each side.

- Remembering that each ridge marks an original furrow in the underlying mud, which parts of the trilobite body could have produced the grooves (taking FS G as a model)?
□ The most likely parts are the spines that trail from the posterior corners of the head-shield.

Note, however, that FS G is a Silurian species of trilobite and the trace is Cambrian. The original trilobite that produced the *Cruziana* in FS P would certainly have been from a different species.

While studying *Cruziana*, you may have noticed that at one end of the specimen there is a ribbed oval area with a coarse unfurrowed texture. This is also a trace fossil. The same animal made both this trace and *Cruziana*, but because the traces differ, the oval trace is given a separate name. It is called *Rusophycus* and was made while the trilobite rested on the sea floor. Imagine that the walking legs, instead of striking backwards and inwards to produce the raking marks of *Cruziana*, scooped together to excavate an oval hollow into which the trilobite settled. FS P is highly unusual in that *Rusophycus* and *Cruziana* are directly associated, whereas in nearly all other cases they occur separately.

Question 3.17

Which of the two traces was made first in FS P?

The *Cruziana* in FS P was clearly made by a trilobite that had its body close to the sediment surface, with the limbs excavating a pronounced trough and the head spines being dragged through the sediment, leaving a pair of grooves. Evidence from other fossils suggests that some trilobites walked with only their legs on the sea floor and the body held well clear, so that the resulting trace fossil is a paired array of scratch marks where only the ends of the legs had come into contact with the sediment.

According to the discussion so far, *Cruziana* and *Rusophycus* were apparently made by trilobites. However, the body fossil record shows that trilobites declined in numbers and diversity during the late Palaeozoic and the class finally became extinct during the Permian. Well-preserved Triassic *Cruziana* have, however, been described – for example from marine deposits in Canada and even freshwater deposits in Greenland. It is thus certain that not all *Cruziana* were made by trilobites, although the correlation between trilobite abundance and *Cruziana* in the early Palaeozoic suggests that most *Cruziana* specimens of that age were formed by trilobites. Other arthropods, such as crustaceans, may have been responsible for some *Cruziana*. This example emphasises again that almost identical traces can be made by very different animals.

Some trace fossils reflect the migration of burrow systems through the sediment. Figure 3.66a shows a U-shaped burrow known as *Rhizocorallium*, which could attain a length of about 1 metre. The burrow lay almost parallel to the sea floor and was connected to the overlying seawater via two, more steeply inclined, tubes (Figure 3.66b).

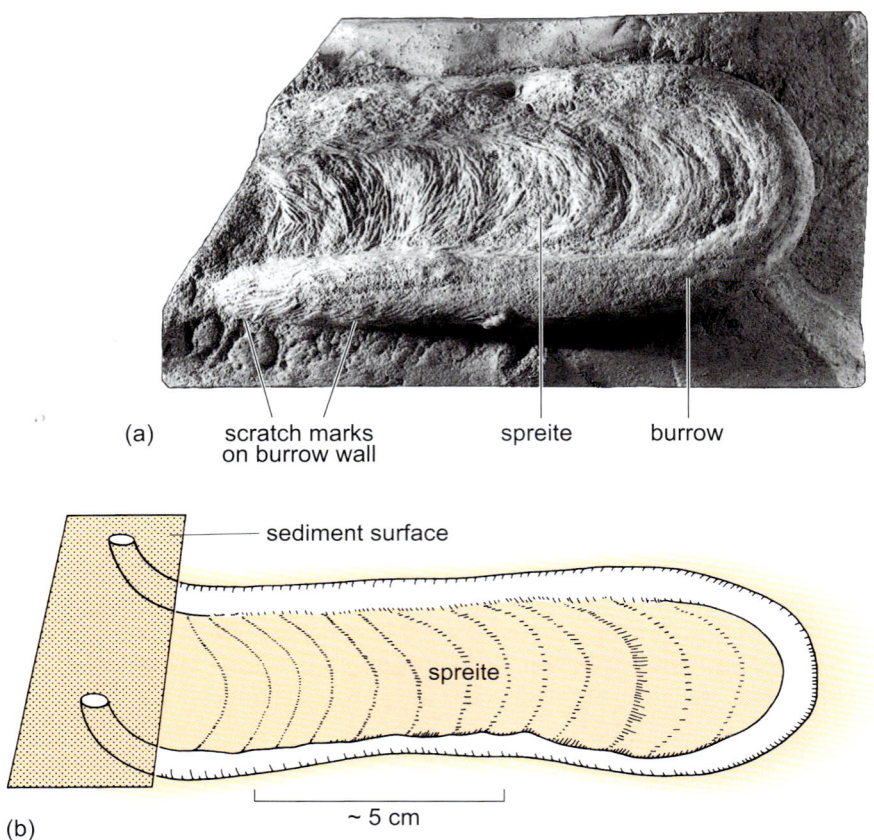

Figure 3.66 (a) The trace fossil *Rhizocorallium*. (b) Diagram showing the relation of the burrow to the seabed, seen obliquely from above.

The U-tube was originally hollow, but has subsequently been filled. What sort of animal constructed this trace? A clue comes again from the scratch marks along the length of the tube walls (Figure 3.66a), where the burrow-maker had evidently scrabbled material away. As with *Cruziana*, such scratches are likely to have been made by another stiff-legged arthropod, and in the case of *Rhizocorallium* possibly, or even probably, by a shrimp-like creature (crustacean). Alternatively, or even additionally, *Rhizocorallium* may have not been made by an arthropod, but perhaps by an annelid worm. The jury is still out.

Between the sides of the tube is a web-like area known as a **spreite** (pronounced 'spryter'; plural: spreiten, from the German for 'spread out', e.g. like the web between the toes of a duck), filled with a series of curved markings. These markings formed as the burrow was extended, with the animal scratching away the outer surface of the U and plastering layers of sediment against its inner surface. The animal that built *Rhizocorallium* probably occupied the burrow for some time. It has been suggested that the animal was a deposit feeder while it excavated the sediment to form the burrow and, having exhausted this food supply, was thereafter a suspension feeder, drawing in water through one of the openings. *Rhizocorallium* is thus both a dwelling trace and a feeding trace.

Another spreite-bearing trace fossil is *Diplocraterion*. This trace fossil is similar to *Rhizocorallium*, but here the U-tube is vertical. It is interpreted as being the dwelling trace of a suspension-feeding organism, perhaps an annelid worm or possibly a crustacean. The spreiten of *Diplocraterion* are believed to reflect the animals' responses to sediment deposition or erosion rather than their feeding behaviour. Downward movement, in response to erosion of the overlying sediment, resulted in a spreite between the two branches of the U-tube, whereas upward movement following sediment deposition resulted in a spreite beneath the tube (Figure 3.67). Such structures must have been associated with fairly extensive current-driven fluxes of sediment as may occur, for example, in tidal deposits. They also tend to show some alignment with presumed currents. By studying sequences of the two kinds of *Diplocraterion*, it is possible to reconstruct the history of sediment influxes and erosion.

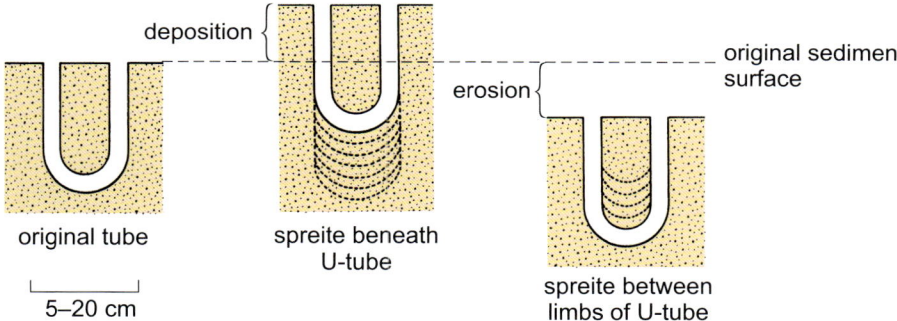

Figure 3.67 Spreiten in the trace fossil *Diplocraterion* formed in response to deposition and erosion.

3.6.3 Traces and water depth

You have just studied some traces made in relatively deep water (*Nereites* and *Paleodictyon*) and in shallow water (*Cruziana*, *Asteriacites*, *Lockeia*, *Rhizocorallium* and *Diplocraterion*). The relationship with water depth arises not because of any direct response by the trace-makers to depth *per se*, but as a consequence of their responses to environmental conditions that tend to vary with depth. The nature of the food supply is a major factor, though this in turn is affected by, for example, nutrient supply, light levels and current activity. Because of variations according to local circumstances, it is not possible to put absolute values on the water depths inferred. By looking at Figure 3.68, however, you can assign some of the trace fossils you have studied to broadly defined zones of water depth. Each suite is named after a characteristic trace fossil, but note that the actual genus, e.g. *Cruziana*, does not have to be present for the depth zone to be recognised, nor is the genus necessarily confined to that depth zone. The depth zone name is just a convenient label for a given commonly observed association. The differences between the suites of trace fossils are largely connected with feeding habits, as discussed above. Another influence, especially in shallow water, can be a need to burrow, for protection from predators or environmental stress. In shallow water, where plankton and detritus resuspended by frequent current activity are abundant, suspension feeders commonly predominate, often producing vertically oriented burrows and borings. In the quieter and muddier sediments of deep water, by contrast, particulate edible material is mostly incorporated in the bottom sediments, where thriving bacteria reprocess it, often enhancing its food value to consumers. Here, deposit feeders tend to dominate. In the deepest water, many traces reflect a methodical search for food, usually with predominantly horizontal burrow systems.

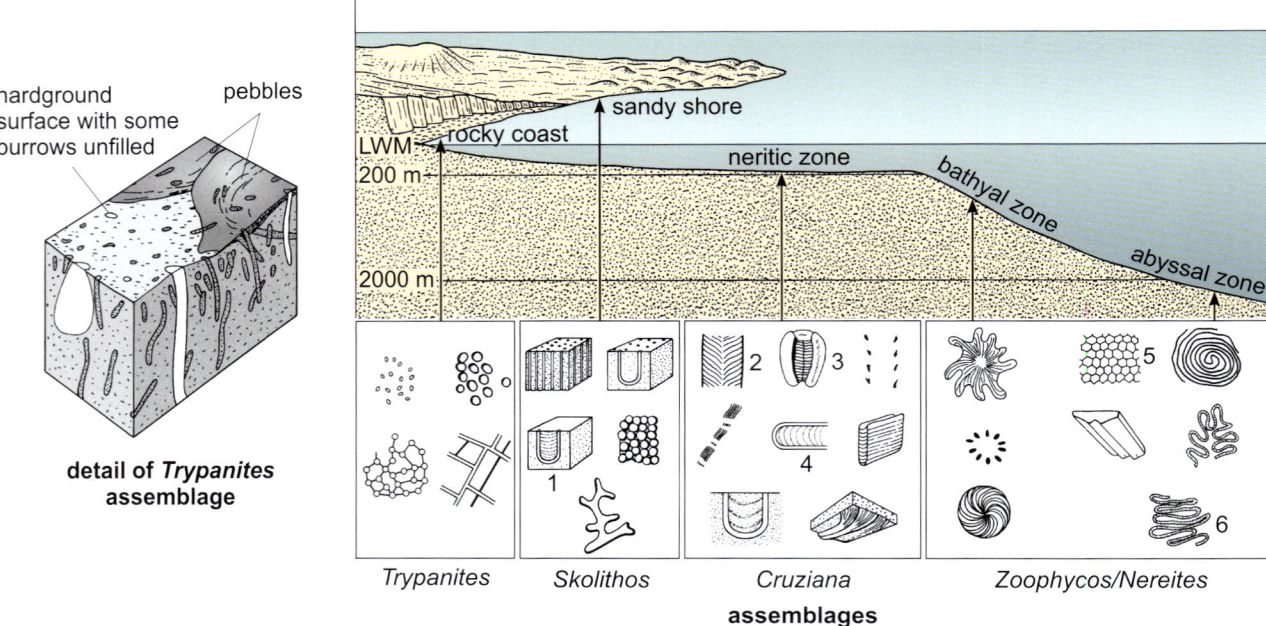

Figure 3.68 Assemblages of trace fossils and their interpreted relationship to environment, especially relative water depth. Each assemblage is named after a characteristic trace fossil (which need not be present, nor confined to that zone; see text for discussion). Trace fossils mentioned in this section are: 1 *Diplocraterion*; 2 *Cruziana*; 3 *Rusophycus*; 4 *Rhizocorallium*; 5 *Paleodictyon*; and 6 *Nereites*. Inset on left shows detail of *Trypanites* assemblage borings into pebbles and a hardground (lithified sea floor). *Trypanites* itself is a narrow, cylindrical, unbranched boring with an approximately circular cross-section.

The foregoing discussion of trace fossils and water depths is generalised. *Not all deep-water traces are complex, nor are all shallow-water traces simple.* Lagoonal muds deposited in shallow but quiet conditions may be dominated by deposit feeders capable of making relatively complex traces. *Nereites*, for example, has been recorded in such deposits associated with coals (formed from peats). In contrast, localised areas in the deep sea may be subject to relatively rapid deposition and reworking of sediment, and 'shallow-water'-type suites of trace fossils have been found in ancient deposits from such environments. Therefore, when studying trace fossils it is essential to study entire assemblages rather than particular specimens, and to test your interpretations for consilience with other depth-related (e.g. sedimentological) criteria.

A few trace fossils are represented in the Digital Kit; look at these now.

Activity 3.5   Revision of some fossil groups in Sections 3.5 and 3.6

This activity mainly develops your understanding of body fossils of various groups that are not represented in the Home Kit.

3.7   A brief perspective on the history of life

Throughout this chapter it has been clear that some of the different groups you have been studying existed at different times in Earth's history – for example, trilobites in the Palaeozoic and ammonites in the Mesozoic. This final section looks very briefly at some aspects of the distribution of organisms through time; a detailed survey of the history of life is beyond the scope of the course.

The oldest known sedimentary rocks, 3850 Ma old, contain carbon in a form interpreted by some as evidence of biological activity. This interpretation assumes that biological processes caused the relative enrichment in these rocks of the lighter, stable isotope of carbon, as they do today. If genuine, these 'chemical fossils' are thought to have originated from marine, bacteria-like organisms. Much still remains to be understood about the origin of life, which may have occurred some time before the oldest known sedimentary rocks, perhaps as long ago as 4000 Ma.

The first structures interpreted as visible fossils by most (but not all) palaeobiologists date from about 3500 Ma ago, and are stromatolites. The photosynthetic cyanobacteria that created them are believed to have produced most of the oxygen that eventually formed a permanent accumulation in the atmosphere by about 2200 Ma ago. About 1500 Ma ago (but possibly much earlier), complex cells with a nucleus first appear in the fossil record. Multicellular algae had evolved by about 1200 Ma ago. It was at about this time that sexual reproduction originated, enabling the production and inheritance of more genetic variation, and promoting diversification of species. Late in Precambrian times, from around 600 Ma ago, a variety of large, soft-bodied, marine organisms appear widely in the fossil record. Collectively called the Ediacaran fauna, these mysterious worm-like, jellyfish-like and sponge-like creatures, along with others bearing no apparent relationship to modern phyla, were the first large multicellular animals. By the end of the Precambrian, most of the Ediacaran fauna had disappeared, and all life was apparently still confined to the sea – though perhaps some bacteria may have begun to invade the porous surfaces of bare rock.

During a sudden burst of evolution called the Cambrian explosion, starting about 542 Ma ago, many animal groups acquired the ability to secrete hard parts, and some became capable of making complex trace fossils. From their fossil record, almost all animal phyla seem to appear at this time, but many may have diverged from each other much earlier; previously lacking shells or teeth, and possibly being a lot smaller, their preservation potential before the Cambrian was minimal.

During the Ordovician, there was a massive increase in diversity *within* phyla, giving many new classes and orders of shallow-marine organisms. The evolution of life onto land only got fully underway in Silurian times – not directly from the sea but via freshwater. Initially led by leafless, spore-bearing plants only a few centimetres tall, and by detritus-eating arthropods only a few millimetres long, the emergence of life onto land opened up a vast new range of niches. Jawless fish appeared in freshwater rivers and lakes, to be followed by jawed fish, which were much more efficient predators. In land plants, competition for light soon favoured the evolution of various structural improvements. By mid-Devonian times, some plants were a metre high; by the end of the Devonian, the first trees with woody trunks had evolved. The first vertebrates to stumble onto land appeared in the late Devonian.

By the end of the Palaeozoic, which encompassed several mass extinctions, including the most severe of all at the end of the Permian (Box 3.2), many of the major and subsequently important groups of animals and plants had evolved. Significant exceptions that were to appear in the Mesozoic include the modern amphibians (frogs, toads and newts), dinosaurs, crocodiles, lizards, snakes, turtles, tortoises, mammals, birds and flowering plants. (A brief outline of the history of vertebrates and plants has already been given in Sections 3.5.2 and 3.5.4, respectively.)

The mass extinction at the end of the Mesozoic opened up opportunities for various groups that had previously been only minor components of the biota. Groups that greatly increased in numbers and diversity during the Cenozoic included flowering plants, herbivorous and pollinating insects, lizards, snakes, mammals and birds. As far as marine invertebrates are concerned, gastropods underwent a spectacular surge in diversity during the Cenozoic, and other groups that diversified considerably include bivalves, reef-building scleractinian corals and echinoids, whilst the brachiopods never recovered their Mesozoic abundance.

All the available evidence suggests that by the beginning of the Quaternary, nearly 2 Ma ago, there was a greater diversity of life, and a higher total number of species, than ever before in Earth's history. Despite the climatic upheavals of glacial and interglacial cycles in the Pleistocene, few groups suffered much extinction. We do not actually know how many species there are today in total, not even to the nearest million (estimates range from about 5 million to 30 million). Shockingly, however, an estimated 5–50 species are now becoming extinct *every day*, mainly through the activities of our own uniquely destructive species. If current trends continue, this present time in Earth history will appear as another mass extinction in the geological record – the sixth extinction to add to the previous 'Big Five' mass extinctions in the Phanerozoic (Box 3.2).

Box 3.2 Mass extinctions in the Phanerozoic

At any time in the history of life, some species will be originating and others will be in the process of becoming extinct. Extinction, like death, is a normal aspect of the history of life, and the vast majority of species that have ever existed have become extinct. Extinction is the complete, global end of the line for a species; it leaves no descendant individuals anywhere. The average duration of a marine invertebrate species in the fossil record (from origin to extinction) is about 5 Ma, though there is much variation about this mean. The fossil record shows that there has always been a normal, 'background' rate of both speciation (the formation of new species) and extinction of species. The majority of past species extinctions have been part of this 'background' rate of extinction. However, at various times, many groups disappear from the fossil record more or less together, never to be found again in younger rocks. Many of the boundaries between one geological period and another were erected by early geologists for this reason. Geologically rapid, major reductions in the diversity of life on a global scale are called **mass extinctions**.

Two of the most severe mass extinctions mark the end, not just of periods, but also of eras – the Palaeozoic Era and the Mesozoic Era (i.e. the end of the Permian Period and the end of the Cretaceous Period). Although the mass extinction that ended the Cretaceous Period, 65 Ma ago, is the most famous of them all (because dinosaurs were its most notable victims), the mass extinction that ended the Permian Period was even more severe. The loss of marine animal species in the late Permian has been estimated to be as high as 95%, compared with around 70% loss of marine animal species for some others of the Big Five, including the late Cretaceous.

There have been five especially severe mass extinctions in the Phanerozoic – known as the **Big Five**. These, and some of their casualties, are as follows:

- *Late Ordovician* Many types of trilobites, brachiopods, graptolites, echinoderms and corals.
- *Late Devonian* Many marine families, especially those of tropical reef-dwelling organisms such as corals, brachiopods, bivalves and sponges.
- *Late Permian* 57% of marine families, especially those from low latitudes. Both the rugose and tabulate corals became extinct, and reefs were eliminated. Trilobites disappeared forever. Crinoids, brachiopods, bivalves and gastropods suffered huge losses. The goniatites (ammonoids) became extinct. Many groups of amphibians and reptiles perished; 70% of vertebrate genera living on land vanished. Land plants, by contrast, were relatively little affected.
- *Late Triassic* Major losses among cephalopods, gastropods, brachiopods, bivalves, sponges and marine reptiles. On land many insect families became extinct, as did most mammal-like reptiles and large amphibians.
- *Late Cretaceous* Whole groups that became extinct near (and not necessarily at) the end of the Cretaceous included ammonites, large marine reptiles (e.g. plesiosaurs) and, on land, dinosaurs and pterosaurs (flying reptiles). Groups suffering major losses included microscopic marine plankton

(e.g. coccolithophores), brachiopods, bivalves and sea urchins. Vertebrate groups affected relatively *little* included fishes, amphibians, crocodiles, snakes, turtles and mammals. Flowering plants, including hardwood trees, suffered also, but mostly in the Northern Hemisphere.

It seems likely that different mass extinctions had different primary causes, a discussion of which is beyond the scope of this course. The positive and negative feedback mechanisms in the Earth's ocean–atmosphere system, and in its ecosystems, are, of course, immensely complex. Establishing the full chain of cause and effect during extinctions, and precisely which biological attributes – or lack of them – led to the demise of a particular species, is a difficult if not impossible task, even for most extinctions taking place today. It is also important to realise that some of the groups lost in mass extinctions were already far from flourishing. For example, the decline of trilobites was well underway by the start of the Carboniferous, long before their eventual disappearance at the end of the Permian.

In the case of the mass extinction at the end of the Mesozoic, a massive meteorite certainly struck the Earth at the end of the Cretaceous, 65 Ma ago, forming a large impact crater (the Chicxulub Crater) off what is now the Mexican coast. The many environmental disturbances, including a lengthy darkening of the skies, appear to have led particularly to the mass extinction of plankton (Figure 3.69). A massive reduction in phytoplankton caused by enfeebled photosynthesis (due to gloomy conditions), in addition to a wide range of physical and chemical effects resulting from the impact, are believed to have promoted a cascade of extinctions affecting zooplankton and then larger organisms higher up the food chain. These short-lived (in geological terms) global events following the impact occurred within a context of much longer-term environmental change (including huge eruptions of basalt in India), and the meteorite impact seems to have had little or nothing to do with some of the other extinctions occurring around this time. Some groups, such as ichthyosaurs, had already become extinct long before the impact. For others, the impact appears to have been the final blow. Ammonites, for example, had been slowly declining for millions of years, though what was left of the group became abruptly extinct at the boundary. The fossil record of some groups such as dinosaurs is too poor to be sure of their precise extinction pattern.

Figure 3.69 The clay layer at the Cretaceous–Palaeogene boundary at Gubbio in Italy, where the chemical evidence for a meteorite impact at this level was first discovered. A rich variety of fossil plankton occurs below the boundary, but few forms survived into the layer above it. A pencil for scale rests on the thin dark clay layer in these tilted beds.

3.8 The use of fossils in biostratigraphy

Biostratigraphy is the use of fossils for relative age determination and stratigraphic correlation. Sedimentary rocks of identical age can look very different because of local variation: think of the sediments forming today in environments ranging from glacial lakes to deserts, and coral reefs to abyssal depths. Similarly, two sedimentary rocks of very different age but formed in similar environments can look more or less identical. Fossils, if present, can be used to sort out both such cases. Fossils are often the easiest, and usually the most accurate, means of establishing the relative ages of unmetamorphosed strata from the start of the Cambrian onwards. No organisms suitable for correlation span all environments and periods, so different schemes using different groups of organisms at different times have been developed.

The stratigraphic column is divided into **biozones** that are characterised by one or more particular fossil species called **zone fossils**. Ideally, to be useful, zone fossils should have a short time range (i.e. belong to a rapidly evolving group), a distinctive appearance with many easily recognisable characters, wide geographic distribution, wide environmental tolerance, and high abundance. Fossils that possess most, though rarely, if ever, all these attributes include the graptolites, ammonites and some groups of microfossils including pollen, planktonic foraminifera and coccoliths.

The utility of other groups in correlation depends on circumstances. Some groups with a long time range may be useful only occasionally for global correlation, such as gastropods in the Cenozoic and crinoids in the late Cretaceous. Some groups may be confined to certain sedimentary settings, but occur worldwide, and so are useful for global correlation. An example of this is a group of bivalves called rudists, which occur widely in Cretaceous limestones. Yet other groups may be restricted to particular climatic belts or oceans – within which their high abundance and wide distribution make them ideal, but outside which they occur little if at all. These groups may form valuable zone fossils for *regional* correlation, such as corals and brachiopods in parts of the Palaeozoic, and echinoids in the late Cretaceous. In practice, the geologist uses whatever groups prove to be effective in the circumstances.

Painstaking fieldwork and documentation have established the relative time range of many thousands of species in strata, allowing **range charts** to be drawn up showing times of first and last occurrences, i.e. from origination to extinction, against a standard stratigraphic column of periods and their subdivisions.

The level of precision afforded by biostratigraphy can be impressive. For example, the Silurian Period, which spans 28 Ma, is divided into 31 graptolite zones, averaging about 900 000 years in duration. At the regional level, where subzones are recognised, the precision is better still. The Jurassic Period, which spans 54 Ma, is divided into about 80 ammonite zones, averaging about 680 000 years. These ammonite zones are further subdivided into subzones with an average duration of only about 300 000 years. This degree of discrimination is often better than can be achieved with radiometric dating, although the latter ultimately provides the framework of absolute ages.

There are several types of biozone, the main ones of which are illustrated in Figure 3.70). The simplest form of information for correlation is the vertical range of a single species in a local succession, which defines a *local range zone*. However, this is often only a partial record of the overall stratigraphic range of the species, either because it was only intermittently present in the area or because of local preservation failure. By combining all known local ranges of the species, a *total range zone* can be obtained (Figure 3.70a). This zone describes the boundaries within which any strata bearing the species should be correlated. There is always the possibility that such a zone is based on a **diachronous** deposit, i.e. one in which the age of a rock layer varies systematically as it is traced laterally, such that its base does not represent a uniform time plane. This

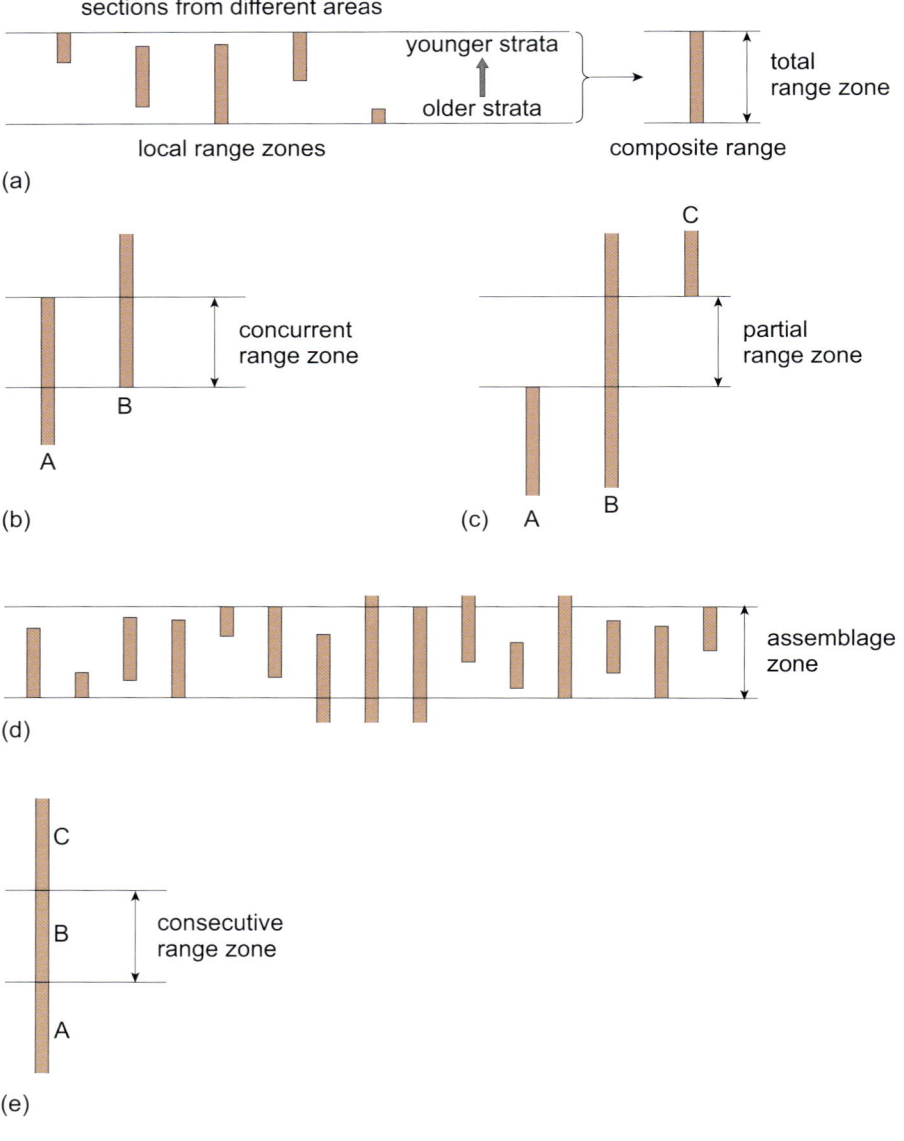

Figure 3.70 Various kinds of biozones used in biostratigraphy: (a) total range zone; (b) concurrent range zone (overlapping range zone); (c) partial range zone; (d) assemblage zone; (e) consecutive range zone. See text for explanation.

can happen, for example, where a shoreline moves laterally as sea level rises. A sandstone bed deposited along the shoreline could end up as a continuous layer across a wide area, but be younger at one end than at the other. Such correlation problems can be reduced by correlating with pelagic species (either nektonic or planktonic ones) rather than benthic ones. Benthic species tend to track their favoured environment, such as a sandy shoreline, over time, migrating with that particular setting as it changes geographic position.

Zonation schemes can usually be improved by combining the ranges of several species. There are various ways of doing this, and different methods prove suitable in different circumstances.

- A *concurrent range zone* (*overlapping range zone*) is very simple in principle (Figure 3.70b). The zone is defined by the overlap of two species that have different total ranges. The presence of both species can very significantly narrow down the age of the strata.

- A *partial range zone* is defined by that part of the total range of a species that lies above the last appearance of one species and below the first appearance of another species (Figure 3.70c).

- The *assemblage zone* is very widely used (Figure 3.70d). It is defined on the basis of several species with overlapping ranges, some of which mark the base and some of which mark the top of the zone. One advantage of an assemblage zone is that strata can still be assigned to them even when some of the species are missing. A common procedure is to place the base of a zone in a particular section at the *first* appearance of a distinctive species, though each zone will contain other appearances and disappearances that help to identify the zone. Although assemblage zones rely on the ranges of a number of fossils, they are formally named after a common and distinctive species, called an **index fossil**. However, this index fossil does not have to be present in every locality for the zone to be recognised, nor need it range through the entire zone.

- Yet another type of zone, the *consecutive range zone*, uses the fact that some individual lineages of organisms evolve into several consecutive species, each lasting a certain period of time. When one species finishes, the next species, its direct descendant, begins and is given a different name (Figure 3.70e). Note that, in this case, the species that 'finishes' is not truly extinct, but transformed – a phenomenon called *pseudoextinction*.

There may be complications, however. For various reasons, first and last appearances of species may turn out not to be synchronous across all areas, though the notion of synchrony depends on the level of time resolution considered. Whether or not an event – such as the global spread of a species outwards from its place of origin – takes, for example, 100 years or 1000 years may be irrelevant if time intervals in strata cannot otherwise be resolved to less than 100 000 years, which is often the case in rocks older than the Cenozoic.

- Another problem is that observed stratigraphic ranges are always less than the true ranges of species. Think of a reason why the extinction level of a species seen in strata will usually be earlier than the true extinction of that species.

□ Sampling is always incomplete and preservation never perfect; we will presumably never find the last fossil individual of a species. Also, as a species approaches extinction, the number of individuals in the population declines, making fossils even more difficult to find.

Another problem is that fossils may sometimes be derived ('reworked') into younger strata than the true last occurrence of the species concerned. Careful study of sedimentary successions to identify, for instance, episodes of erosion and burrowing, can reveal the possibility of reworking (see also Chapter 4). Again, the principle of using range data from as many fossil species as possible tends to reduce the chances of erroneous interpretation.

Once the relative age of a rock is known, the absolute age can be obtained by referring to an up-to-date timescale based on radiometric dating. Several sophisticated correlation techniques for sedimentary rocks are available in addition to biostratigraphy and radiometric dating, ranging from magnetostratigraphy (using reversals of polarity in the Earth's magnetic field); cyclostratigraphy (using the rhythmic expression within sediments of astronomically forced climate cycles caused by variations in the Earth's orbit); and chemostratigraphy (using chemical and, particularly, isotopic signals that have changed over geological time). The applicability of these techniques, however, like biostratigraphy, is not universal and varies in the stratigraphic record from place to place and from time to time. All stratigraphic correlation schemes have levels of uncertainty attached to them. Generally speaking, the more lines of evidence that can be used simultaneously, the better both accuracy and precision are likely to be. Dating and correlation schemes are continuously revised as fresh evidence comes to light and new techniques are developed. For example, new fossil discoveries during the life of this course may well change some of the stratigraphic ranges of groups given here.

3.9 Summary of Chapter 3

Shelly marine invertebrates yield the commonest fossils. Eight frequently encountered classes, in five phyla, are represented by fossil specimens (FS) A–M in the Home Kit:

1   Molluscs (including bivalves, gastropods and cephalopods) are characterised by a dorsal calcareous shell (providing protection) and a ventral muscular foot (assisting locomotion), with a mantle cavity between the body and the shell, housing gills.

In bivalves (FS B and I), the shell is divided into two enclosing lateral valves that are opened by a dorsal ligament and closed by adductor muscles (with shear between the valves prevented by hinge teeth and sockets). Most species have externally symmetrical valves (with the plane of symmetry running between them), and burrow with the foot, circulating water via siphons from above the sediment surface, for suspension feeding on expanded gills. There are also epifaunal and boring forms, as well as deposit feeders. Bivalves are entirely aquatic. Most are marine, living on shallow sea floors, though some inhabit freshwater.

Gastropods (FS A and K) have a broad creeping foot, which can be retracted into the shell. Some forms retain a pair of gills, and water enters on either side and is exhaled between them, via an indentation of the outer shell margin. But most have one gill, and water enters, and exits, at opposite ends of the aperture; the inhalent site may then be extended as a siphon, housed in a shelly scroll (the siphonal canal). Most gastropods are grazers, although there are also detritus collectors and scavengers or predators. Most gastropods are marine, living in shallow seas, but many inhabit rivers, lakes and ponds; others live on dry land.

In cephalopods (FS D), the foot is wrapped forwards under the body to form (in part) a funnel for water squirted from the mantle cavity, enabling swimming by jet propulsion. The shell, where retained externally (e.g. in *Nautilus* and the extinct ammonites), is chambered, serving as a gas-filled buoyancy device. Grasping tentacles around the mouth allow the capture of prey (large and/or microscopic), and scavenging. *Nautilus* belongs to the subclass Nautiloidea, whereas the Mesozoic ammonites form part of an extinct subclass called the Ammonoidea, among which are also the Palaeozoic goniatites. All cephalopods are, and have been, entirely marine.

2 Echinoderms (including echinoids and crinoids) have an enclosing endoskeleton consisting of numerous calcite plates, each of which is a single crystal with a microscopic spongy texture. A water vascular system has radial arms that support exposed flexible tube feet, used in feeding, respiration, locomotion and, in some species, burrowing. Most echinoderms show a fivefold arrangement (if not symmetry). All are, and have been, entirely marine.

Echinoids (FS C and J) have a globular test, consisting of interlocking plates, covered in spines that are articulated on tubercles on its outer surface. Tube feet project from rows of pore pairs on five radial sectors (ambulacra) of the test. Regular echinoids show fivefold symmetry, with the mouth centred on the lower surface and the anus on top. They are epifaunal grazers to predators. Irregular echinoids show bilateral symmetry, with relocation of the mouth and anus, and are detritus collectors; the great majority are infaunal. Most echinoids live in shallow seas.

In crinoids (FS H), the mouth is central on the upper surface of the cup, with the anus alongside, surrounded by a ring of branching arms supporting the tube feet. The latter trap suspended food particles, and pass them into the mouth. During feeding, the arms are usually spread outwards and backwards into a current to form an umbrella-like filtration fan. Fossil crinoids were commonly attached to the sea floor by a stem, though most living forms are stemless.

3   Brachiopods (FS F and L) have a two-valved shell, but differ from bivalves in having a plane of bilateral symmetry bisecting each valve. Muscles control both opening and closure. Articulate brachiopods, which have a calcite shell, are hinged posteriorly by teeth (in the pedicle valve) and sockets (in the brachial valve), features that are lacking in inarticulate brachiopods. Most are (or were) attached to the sea floor by a stalk (the pedicle) emerging from a hole in the umbo of the pedicle valve, although some were unattached. Suspension feeding is by means of a tentaculate feeding organ housed within the mantle cavity. All are, and have been, entirely marine.

4   Cnidarians most notably include corals (FS E and M). Each individual coral polyp builds a cup-like calcareous skeleton under it, divided internally by radial walls (septa), which lie between infoldings of the polyp's internal cavity. The polyp (like an anemone) has a ring of stinging tentacles, which trap microscopic prey and pass them to the single orifice of the internal cavity. Many living examples have symbiotic single-celled algae in their tissues, which allow the hosts to thrive in waters with low nutrient levels. Corals may be solitary, or, as a result of budding, colonial. The rugose and tabulate corals are confined to the Palaeozoic, whilst the scleractinian corals arose in the Triassic and are still flourishing today. Corals are, and have been, entirely marine.

5   Arthropods are the largest (most species-rich) animal phylum, mainly through the inclusion of insects. The most important fossil group is the extinct, entirely marine trilobites (FS G). The bilaterally symmetrical exoskeleton of trilobites consists of numerous articulated segments, and was periodically moulted. Its calcified dorsal part comprises a head-shield (cephalon) that bore compound eyes in most species, a central trunk region (thorax) and a tailpiece (pygidium), all divided longitudinally into three lobes. The unmineralised underside of the animal included numerous paired jointed limbs, flanking a central channel along which food particles (whole prey, bits of prey or organic detritus) were passed forward to the mouth.

6   Other important kinds of body fossils include the belemnites (extinct squid-like cephalopods with a bullet-shaped internal guard); graptolites (extinct marine, mainly planktonic, colonial suspension feeders), bryozoans (aquatic, mostly marine, colonial animals); sponges; worms; vertebrates (fish, amphibians, reptiles, mammals and birds); and land plants, fossils of which include sporomorphs (pollen and spores, classified as microfossils), branches, leaves, seeds and flowers. Isolated, different parts of the same fossil plant have often initially been given different generic names, and these names are retained even when they are known to belong to the same organism. As well as sporomorphs, other important microfossils (also body fossils) include foraminifera (aquatic, mainly marine, mostly benthic, single-celled organisms with a shell); coccoliths (minute calcite plates from phytoplankton called coccolithophores); ostracods (tiny bivalved crustaceans); radiolarians (zooplankton with siliceous skeletons); conodonts (tiny phosphatic teeth from primitive vertebrates); and stromatolites (macroscopic, finely layered carbonate structures produced by sediment-trapping microbes, mainly cyanobacteria).

7   Trace fossils are usually named to indicate particular types of behaviour or activity, not the organism that made them. They reflect environmental conditions, and inferences from marine examples can thus be linked, to some extent, with water depth. Meandering (usually more or less horizontal) trails and burrows, reflecting systematic deposit feeding and/or 'microbial gardening', tend to characterise deeper, more offshore deposits (e.g. *Nereites* and *Paleodictyon* (FS N)). Simpler traces (often vertically extended), reflecting a variety of behaviours including responses to rapid sediment gain or loss, tend to characterise shallower waters (e.g. *Asteriacites* and *Lockeia* (FS O), *Cruziana* and *Rusophycus* (FS P), *Rhizocorallium* and *Diplocraterion*). Trace fossils are not foolproof evidence of water depth, however, as exceptions occur. They must be interpreted in conjunction with other sedimentological data.

8   Although the oldest evidence for life extends back to nearly 4 billion years ago, it is not until the Cambrian explosion, starting about 542 Ma ago, that organisms acquired the ability to secrete readily fossilised hard parts. Many phyla first appear in the fossil record at this time. Life was confined to the sea until the mid-Palaeozoic (except perhaps for some land-living bacteria). The main invasion of the land got underway in the Silurian, initially led by plants and invertebrates, and followed only much later by vertebrates (the first amphibians) near the end of the Devonian. During the Phanerozoic there have been five big mass extinctions, when the diversity of life was rapidly and severely reduced on a global scale. The two most severe of these were at the end of the Palaeozoic and at the end of the Mesozoic. Nevertheless, by the beginning of the Quaternary, there was a greater diversity of life, and a higher total number of species, than ever before in Earth's history. However, many species are now daily becoming extinct, due mainly to human activities, and this present time in Earth history could eventually appear as another mass extinction in the geological record.

9   Given the limited age range of most fossil groups and individual species, fossils are extremely useful in relative age determination. The stratigraphic column is divided into biozones that are characterised by one or more particular fossil species called zone fossils. These species should have a short time range, distinctive appearance, wide geographic distribution, wide environmental tolerance, and high abundance. Among the most useful groups for biozonation are graptolites, ammonites and some types of microfossils including pollen, planktonic foraminifera, and coccoliths. Other groups may be useful for global zonation only at certain times in their total age range. Some groups that are confined to particular lithologies, climatic belts or oceans, may nevertheless be valuable in correlation, albeit in some cases for regional rather than global correlation. There are various different types of zonation scheme, which usually combine the time ranges of two or more species.

The feeding/habitat classification of the specimens from the Home Kit (the completed version of Table 3.2) is shown in Table 3.5.

Table 3.5 Combined feeding and habitat classification system for marine animals represented by Home Kit fossils A–M (completed Table 3.2).

Habitat		predator	scavenger	deposit feeder collector	deposit feeder swallower	suspension feeder	grazer
pelagic	nekton	?D				?D	
pelagic	plankton						
epifaunal	vagrant	(?C) G	(?G)	(?K)			C K
epifaunal	sessile	E M				(E) (M) F L H	
infaunal	vagrant	A	(?A)	J		intermittently vagrant	
infaunal	sessile					I B	

3.10 Objectives for Chapter 3

Now you have completed this chapter, you should be able to:

3.1 Make simple but accurate drawings to record your observations of given fossil specimens.

3.2 Compare and contrast body fossil specimens from the major groups of shelly marine invertebrates, and classify them to the level of class, at least, according to the homologies of their bodyplans.

3.3 Make observations on body fossils of commonly occurring marine invertebrates, identify and label certain features of them, and interpret aspects of the palaeobiology of the original organisms (such as their habitats and mode of feeding).

3.4 State, in general terms, some of the varied uses of fossil plants, including sporomorphs, and outline some of the difficulties surrounding the naming of plant fossils and the reconstruction of whole plants.

3.5 Outline the main characteristics of some major groups of microfossils.

3.6 Interpret, in broad terms, the likely modes of formation of given trace fossil specimens, and the conditions in which they probably formed.

3.7 Place in chronological order some key events in the history of life, particularly during the Phanerozoic, including the first appearance of some major groups and the timing of mass extinctions.

3.8 Name some of the groups most useful in biostratigraphy, state the attributes required of good zone fossils, and recognise the main types of biozone.

Chapter 4 Fossilisation

In order to interpret fossils correctly, and to make maximum use of the information they contain, the processes involved in their formation must be understood. This is true whether using fossils to reconstruct ancient environments, investigate the biology of extinct species, establish patterns of evolution, or undertake any other type of study involving past life. You will occasionally need to refer to some of the Home Kit fossil replicas and the Virtual Microscope while reading this chapter, and Activity 4.1 involves the Digital Kit.

How do fossils form? This is a deceptively simple question. In essence, fossilisation involves eventual entombment, usually in sediment, but occasionally in other materials such as natural tars and resins, ice, volcanic ashes or even lava flows (Chapter 2). However, fossilisation often entails several complex processes. Many kinds of obstacles, acting like a set of filters, successively reduce the chances that any individual organism will get into the long-term fossil record. Among these obstacles are morphological, ecological, sedimentary and metamorphic filters. Eventually human influences may affect *apparent* preservation, such as biases during fossil collection that make some fossils less likely to be found and retrieved than others.

When an organism dies, it is normally destroyed by scavengers and various decay processes, and its component parts are recycled. The chemical constituents of the dead are usually made available for the living, and for the vast majority of organisms that have ever lived no evidence remains of their lives and habits. Under certain circumstances, however, this recycling is at least partially prevented, and parts of the organism end up as a body fossil. The soft parts soon decay or are eaten (unless highly exceptional conditions permit their preservation, Section 4.2.5). Any remaining hard parts usually change in some way; for example, they may be recrystallised, dissolved, replaced by a new mineral, or deformed. They may also be completely destroyed, either by processes acting immediately after death, or after burial, for example during metamorphism (Section 4.3).

The study of the various processes that affect the formation of fossils, from the death of organisms onwards (or, in some cases, from the shedding of parts during life), is called **taphonomy** (*taphos* is Greek for 'tomb' or 'grave'). An understanding of taphonomy not only identifies biases that must be taken into account when interpreting the fossil record, but enables many different kinds of information to be obtained from fossils.

The processes that occur between the death of an organism and its discovery as a fossil can be considered in two stages: (i) those that occur between death and final burial; and (ii) those that occur between final burial and discovery. Many processes occurring after final burial are due to **diagenesis**, which embraces all the physical and chemical processes occurring in a sediment after its deposition, including those that turn a sediment into sedimentary rock, but excluding any metamorphism should it occur later. The emphasis on *final* burial is important, as the remains of organisms may be buried and exhumed several times before eventual long-term burial in a sedimentary succession.

Most of this chapter is about the formation of body fossils in the marine settings that dominate the fossil record. Bear in mind, though, that trace fossils also undergo various preservation processes. These include burial and diagenetic changes such as lithification of the sediment due to compaction and cementation. Quite often, such changes have the effect of making the trace fossil easier to distinguish visually, and separate physically, from the surrounding matrix.

4.1 From death to final burial

4.1.1 Death

Although the great majority of body fossils record the *deaths* of organisms, there are exceptions.

- ■ Give some examples of these exceptions.
- □ Arthropod moults, shed from growing individuals, were mentioned when discussing trilobites in Section 3.3.8. Other examples include teeth, scales and feathers, as well as leaves, flowers and branches from plants.

Biological causes of death include predation, disease, starvation or simply old age. For some species, predation wields by far the largest scythe. Many common bivalves, for example, are favourites on the menu of various predators, some of whom leave distinctive traces on the shells of their victims, allowing the palaeontological sleuth to detect 'whodunit'.

- ■ Give an example of predation that was illustrated in Chapter 3.
- □ The distinctive circular drill hole of a predatory gastropod on a bivalve, shown in Figure 3.26b.

Physical causes of death include severe climatic conditions (e.g. drought, flood and frost), sudden burial under a thick layer of sediment, or changes in salinity, oxygen level or some other vital environmental variable. Such events may affect different species unequally, though occasionally an entire fauna may be virtually annihilated in a **mass mortality**, i.e. the simultaneous killing of huge numbers of individuals by some common cause. **Catastrophic burial** by sediment (or sometimes volcanic ash or lava) is one important mechanism of mass mortality. Other natural mechanisms include poisoning by volcanic gases, upwelling of oxygen-depleted seawater containing toxic hydrogen sulfide (H_2S), and the formation of 'red tides', where blooms of single-celled algae (dinoflagellates) produce potent toxins.

A large influx of sediment can kill sessile animals, such as brachiopods and benthic crinoids, and preserve them *in situ* where they lived. If the covering is not too thick, some mobile animals, especially burrowers, may be able to escape.

- ■ What record of the escaped animals might nevertheless remain in the sediment?
- □ Their escape behaviour could leave behind trace fossils.

Greater thicknesses of deposited sediment (say, half a metre or more), perhaps the result of a major storm or turbidity current, may prevent animals escaping altogether. Some may be preserved in the position in which they lived, while others may make futile movements before expiring. The catastrophic burial of a fauna is revealing for three reasons:

1. It is geologically instantaneous, preserving an assemblage of organisms that lived together at the same time, whereas in sediments that accumulate gradually, or are repeatedly disturbed by strong currents, many generations may be represented within even a thin bed. Catastrophic burial is thus of great value in reconstructing ancient communities and ecological relationships.

2. Some, especially sessile, individuals may be preserved exactly where they lived, if not in their living orientation. Examples include crinoids and other echinoderms attached to the sea floor (Figure 4.1), and bivalves still in their burrows.

3. Rapid burial (unless rough transport is involved) will help to keep the remains of organisms intact, especially animals with many skeletal elements, before they rot and fall apart.

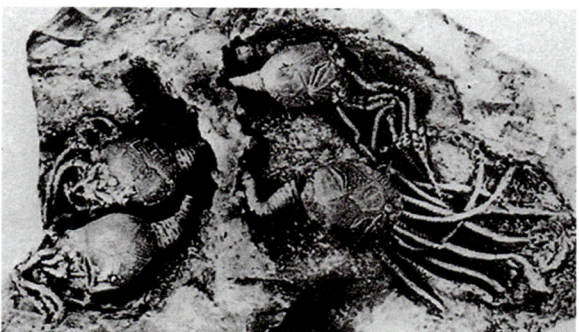

Figure 4.1 Complete skeletons of primitive stemmed echinoderms (called cystoids, an entirely Palaeozoic group distantly related to crinoids), attached to an ancient erosion surface, that have been preserved by catastrophic burial. Field of view 24 cm across.

Question 4.1

Which of the common groups of marine invertebrate fossils in the Home Kit would have been prone to skeletal break-up, unless catastrophically buried?

Infaunal organisms might, of course, die in their burrows and hence stand a chance of intact preservation without the need for catastrophic burial. This probably accounts for the relatively good fossil record of infaunal irregular echinoids compared with the epifaunal regular echinoids (Section 3.3.2).

Rapid erosion of sediment can also cause death. Again, sessile animals are particularly vulnerable – they can, for example, be ripped from their moorings. Even burrowers may not be able to escape. Downward-digging bivalves, for instance, may still be exposed if they meet an impenetrable layer. Drastic erosion is often followed by rapid burial: animals may be washed out of their burrows, transported with, and then reburied in, the originally surrounding sediment. These scenarios involving erosion and/or sudden burial typically accompany storm activity on shallow marine shelves.

4.1.2 Decay and disintegration

Unless an organism is completely consumed by predators, death is followed by bacterial decay and often disintegration by scavengers. Bacteria that cause decay can be found in all marine environments, where they are most numerous in the topmost sediment layers (cm to tens of cm down). Decay may be either **aerobic**, in which oxygen is consumed during the breakdown of organic molecules (analogous to burning in air), or **anaerobic**, in which breakdown occurs without oxygen consumption. Different bacteria specialise in each type of decay. Indeed, aerobic bacteria cannot survive in the absence of oxygen, whilst oxygen is toxic to anaerobic bacteria. In anaerobic decay, some bacteria generate hydrogen sulfide (H_2S) and methane (CH_4).

The processes of aerobic and anaerobic decay have different consequences. Aerobic decay is rapid, yielding readily lost products (mainly CO_2 and H_2O), and in high temperatures, such as on land in the tropics, the soft parts may entirely disappear in a few days. Aerobic decay is slower at lower temperatures, and in conditions of high salinity or desiccation. Anaerobic decay is usually not as rapid as aerobic decay, but more importantly it sometimes yields finely crystalline products, such as pyrite, that can permeate the decaying remains and replace soft tissues (Section 4.2.5). Conditions of anaerobic decay may arise in several ways. Some bodies of water, such as the Black Sea, are permanently anoxic for some distance above the sea floor, because of restricted circulation. Organisms that sink into an anoxic zone, or are transported into it, are often well preserved. Occasionally, as sometimes occurred during the Jurassic and Cretaceous, extensive parts of fully open oceans have become anoxic for short periods – during 'oceanic anoxic events'.

Even in oxygenated conditions, burial by sediment may lead to anaerobic decay, as the oxygen can be rapidly used up by aerobic bacteria. If you turn over a spadeful of mud on a sheltered stretch of coastline, the sediment is often brownish to yellowish at the surface, but dark grey to black just a few centimetres down and smells unpleasant. The blackness is due to finely dispersed pyrite (FeS_2) and related chemical compounds produced by the reaction of ferrous (Fe^{2+}) ions and hydrogen sulfide (H_2S – the source of the smell). The hydrogen sulfide itself is a product of anaerobic respiration involving the reduction of sulfate ions in seawater (SO_4^{2-}, Equation 4.1). The blackness thus reveals the presence of anoxic water in the sediment. A similar process may follow a mass mortality, or the death of a large organism, when oxygen is soon depleted as large amounts of organic matter begin aerobic decay. Thus, in many cases, anaerobic decay may follow aerobic decay:

$$SO_4^{2-} + \underset{\text{organic matter}}{2CH_2O} + 2H^+ \rightarrow H_2S + 2CO_2 + 2H_2O \qquad (4.1)$$

Another important aspect of decay, especially in oxygenated conditions, is the production of gas bubbles that, if trapped within the body, can cause it to float. Decay continues during flotation and parts of the carcass may drop off, scattering bits over a wide area of the sea floor. This is one reason why, for example, single vertebrae or other bones of large marine reptiles (such as ichthyosaurs and plesiosaurs) quite often occur on their own in Mesozoic strata. The destruction of dead organisms is also hastened by scavengers, such as crustaceans, which may gather around a corpse, tear it apart and scatter the remains – an important dispersal process in quiet-water environments lacking strong currents.

Distinct stages can be observed in the disintegration of a modern corpse, some of which have recognisable equivalents in the fossil record that indicate how far decay went before burial. For example, the decay sequence in a modern echinoid has been observed to include the shedding of spines by day 7 after death, and the dropping out of the small plates around the anus by day 12. Such taphonomic clues provide insights about rates of sediment flux on the sea floor.

Skeletal materials usually contain minor amounts of organic material, which will eventually rot. In some cases, unless the precipitation of additional minerals bonds the crystalline components together, the hard parts will themselves crumble.

Plant material is also subject to decay. The main structural materials in plants include cellulose in cell walls, which surrounds all plant cells, and lignin (in woody tissues). Despite their apparent toughness, these are readily attacked by a variety of organisms, especially fungi and some bacteria. The striking exception is a substance called sporopollenin, which forms the coat of spores and pollen (sporomorphs) and is remarkably resistant to decay (Section 3.5.4). Further aspects of plant decay are discussed in Section 4.2.4.

4.1.3 Fossilisation potential

The chance that an organism has of getting into the fossil record is called its **fossilisation potential** (or **preservation potential**). This varies a great deal according to factors such as whether its body has any durable parts, where it lives and dies, whether it becomes buried in sediment and whether that sediment later becomes part of the rock record. The fossil record is dominated by marine organisms, because the chances of permanent burial are greater in the sea than on land (Section 3.1). Clearly, in almost all circumstances, individual soft-bodied animals are much less likely to be fossilised than those with durable hard parts. Nevertheless, even the fossilisation potential of hard parts is far from uniform. In marine organisms, two important factors are their position relative to the sea floor and whether the skeleton is a single piece or made of many elements (Question 4.1).

- ■ The likelihood of any particular *species*, rather than just an individual member of it, being preserved in the fossil record, is affected by two crucial factors. What are they?
- ☐ The number of individuals in the species and its geographic distribution.

Species with a small average size tend to be much more abundant than very large species, and, other things being equal, are more likely to be preserved somewhere in the fossil record (although they are less conspicuous to collectors). As usual, though, there are complicating factors. Shells and bones of smaller species are usually less durable than those of larger species, so smaller *individuals*, though generally more numerous, may be less likely to be preserved. Wide geographic distribution generally increases the chances of a species being found in the fossil record, but again other factors may negate this.

- ■ Imagine a thick-shelled species that lives only on the deep ocean floor, beyond the continental shelf, but is widely dispersed, and abundant in several oceans. How likely is it to become part of the *long-term* fossil record?
- ☐ Despite a thick shell, high abundance and wide distribution, the likelihood of its survival in the long-term fossil record is limited because the vast majority of sediment that accumulates on the deep ocean floor gets destroyed in subduction zones within about 200 million years.

Question 4.2

Look at the information you recorded in Table 3.2 (or the completed version in Table 3.5), and consider the habitats, habits and skeletal construction of the organisms represented by FS B, G, H, K and J. Comment on the relative potential for the *intact* preservation of their hard parts under 'normal' circumstances (i.e. without catastrophic burial).

4.1.4 Biological attack

Skeletal material is often broken down by predators or scavengers that chip away or smash shells. This usually results in angular shell fragments (Figure 4.2a). An easy trap to fall into, when in the field, is to attribute all scattering and breakage of shells to current activity; remember to consider the alternative possibilities of predation and scavenging. Predation, however, does not necessarily result in shell destruction. Some predators swallow whole animals, digest their soft parts and then eject the empty shells, whereas others deftly extract the soft parts without damaging the shell.

Figure 4.2 Taphonomic factors contributing to the breakdown of shells: (a) predation and scavenging – in this case predation of mussels by seabirds, which have regurgitated these remains (field of view about 10 cm); (b) boring organisms; (c) shattering and rounding by currents; (d) dissolution by acidic groundwaters (these shells were in a coastal soil). Largest shell in (b) 4 cm; (c) 3.3 cm; (d) 3.5 cm.

Another form of biological attack is the invasion of the shell by boring organisms (Figure 4.2b). Although this mainly affects the shells of dead organisms lying on the sea floor, it can also seriously damage the shells of *living* epifaunal organisms. The most common shell borers include sponges, algae and fungi. Some of these borers may derive nourishment from the organic matrix in shell material, but others appear to use their borings simply as protective refuges. Extensive boring weakens the host shell and speeds up its destruction, not only by physical abrasion and/or the secretion of chemicals, but grazers may rasp away the shell to find the boring organisms themselves.

4.1.5 Physical influences

The main physical influences on skeletal material before final burial result from transport and breakage. Skeletons that separate into two or more pieces – such as the shell of a bivalve or the plates of a crinoid – often show variable hydrodynamic properties, and sorting of similar elements may occur during transport. Subtle effects may operate: for example, bivalves with holes, such as borings drilled by gastropod predators (Figure 3.26b), can respond to currents differently from unbored shells. In subaerial environments, freeze–thaw cycles can quickly break up skeletons.

Transported shells may come to rest with a preferred orientation. This is best seen in very elongate shells, such as belemnites (Figure 4.3a). Usually, unless banked up against an object, elongate shells end up with their long axes more or less parallel with the main current direction. Sometimes, currents are too weak to transport the skeletal remains themselves, but will winnow away the surrounding finer sediment, leaving a condensed **lag deposit** of shells, called a **shell bed** (Figure 4.3b).

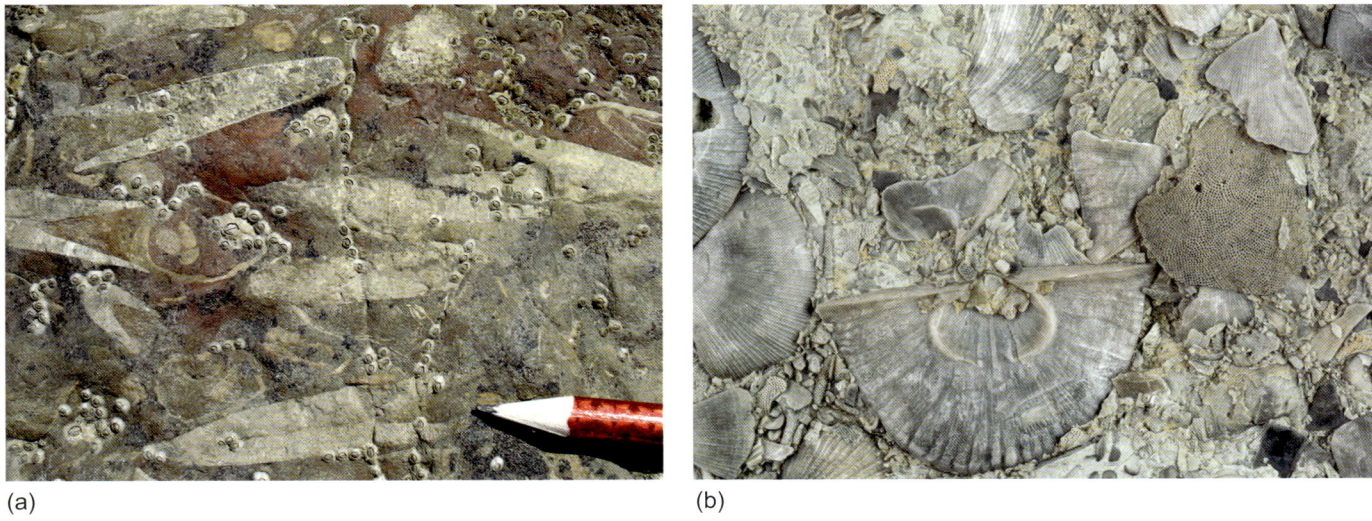

Figure 4.3 (a) Current-orientated Jurassic belemnites in an ironstone bed that has been eroded by the sea and encrusted with modern barnacles. (b) An Ordovician shell bed formed mainly from brachiopods, bryozoans and crinoids. Field of view 5.5 cm across.

Abrasion commonly occurs during transport, giving rounded shell fragments (Figure 4.2c). In subaerial environments, the abrasive action of wind, carrying sediment in suspension, can be intense. In marine settings, the complete destruction of shell material by abrasion is only likely in high-energy beach environments. On rocky shores, shells may not only be abraded but fractured by impact against stones. Often, a shell will fracture preferentially along natural planes of weakness. In quiet, offshore areas, shells may be destroyed by other mechanisms, such as predation, scavenging, boring organisms and, to a limited extent (in deep cold water), chemical solution (Section 4.1.6).

Success in interpreting taphonomic history can depend on knowing what clues to look for in specific groups. In echinoids, for example, the presence of irregular fractures cutting *across* plate boundaries indicates that an animal was still alive when it suffered physical impact or was bitten, because in life all the plates are tightly bound by collagen fibres (Section 3.3.2), so that the entire skeleton acts as a single brittle shell. In dead echinoids, however, the fractures follow the plate boundaries because the connecting collagen fibres have decayed.

4.1.6 Chemical attack

By far the most common skeletal material in marine organisms is calcium carbonate ($CaCO_3$). Chemical destruction of this material before burial usually involves dissolution (Figure 4.2d). In open, shallow marine waters, which are usually saturated with $CaCO_3$, dissolution is minimal. Old shells may thus persist for long periods on a shallow sea floor, and survive repeated episodes of burial and exhumation.

In deep oceanic water, chemical changes involving calcareous shells are much more important. Here, a combination of relatively high pressure and low temperature promotes the dissolution of calcium carbonate. The depth at which the rate of dissolution of calcareous debris balances the rate of supply of material sinking down from shallower levels is known as the **carbonate compensation depth** (**CCD**) (Figure 4.4). Below this depth, no carbonate sediment accumulates. The depth of the CCD is greater for calcite than for the more soluble aragonite, and varies from ocean to ocean. For calcite, it is about 4000 m in the Atlantic, whereas in colder waters around Antarctica it is about 500 m. The CCD mainly affects pelagic calcareous organisms that, on death, sink towards the sea floor. Those sinking to depths below the CCD will usually dissolve, but those settling on areas of high relief, such as mid-oceanic ridges, which project above the CCD, will accumulate as calcareous sediments.

In the past, calcium carbonate has sometimes been more prone to dissolution in seawater than today, because seawater chemistry has varied over geological time. In some ancient shallow marine

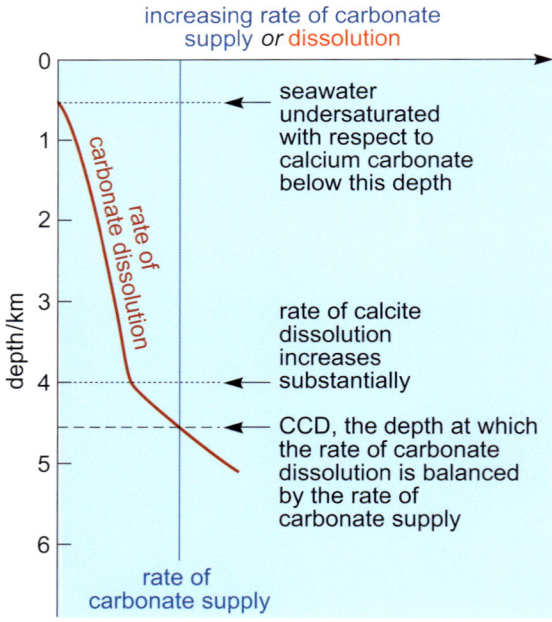

Figure 4.4 Graph to show the position of the carbonate compensation depth (CCD) in relation to the rate of carbonate dissolution and the rate of carbonate supply. Note that the depths shown are typical for tropical regions, but that the absolute depth varies significantly as rates of carbonate dissolution and carbonate supply differ according to latitude and local oceanic conditions.

deposits, aragonite shells show evidence of starting to dissolve even before burial, as they lay on the sea floor, indicating that the CCD for aragonite was much shallower than at present.

The chief mineral component of bone, teeth and fish scales is carbonated calcium hydroxyapatite ($Ca_5(PO_4, CO_3)_3(OH, F, Cl)$), a complex form of calcium phosphate that is less soluble than calcium carbonate. Vertebrate remains therefore sometimes accumulate preferentially in settings where calcareous shells dissolve.

4.1.7 Burial and final burial

So far, we have considered pre-burial factors in isolation. However, some fossils reveal a complicated history before final burial, involving one or more episodes of partial or complete burial, followed by exhumation. This is especially common in shallow marine sediments. In some cases, alteration of the skeleton occurs during preliminary burial, so that on exhumation its resistance to destruction is different. For example, the pore spaces in bones may be infilled by calcium phosphate, increasing their resistance to abrasion if temporarily exhumed. During times of reduced rates of sediment accumulation, a resistant deposit of bone fragments may gradually accumulate to form a *bone bed*, which may be accompanied by inorganically precipitated concretions of calcium phosphate.

4.2 Diagenetic processes following final burial

After a potential fossil is finally buried, it is often subject to processes that alter the microstructure and/or composition of the skeleton. Some of these diagenetic processes may destroy the remains, whilst others can enhance preservation. Diagenetic processes rarely act in isolation, and often grade into each other.

4.2.1 Moulds and casts

The decay of soft parts inside a shell leaves an internal space into which sediment may seep. When lithified, this forms a core of sedimentary rock that replicates in reverse relief the features of the inner surface of the shell.

■ What is the term for this sedimentary infill of a shell? Which specimen in the Home Kit is an example?

☐ It is an *internal mould* (Section 3.1.1). FS D (an ammonite) is an example.

Question 4.3

What features would you expect an internal mould of the bivalve *Circomphalus* (FS I) to show?

The rock surface adjacent to the inside or outside of a fossil shell is called a **mould**. Usually both internal and external moulds are formed, on the inside and outside of the shell, respectively (Figure 4.5). Depending on how the rock breaks in the vicinity of a fossil shell, the shell material itself may have to be removed to see the internal mould within or the external mould outside.

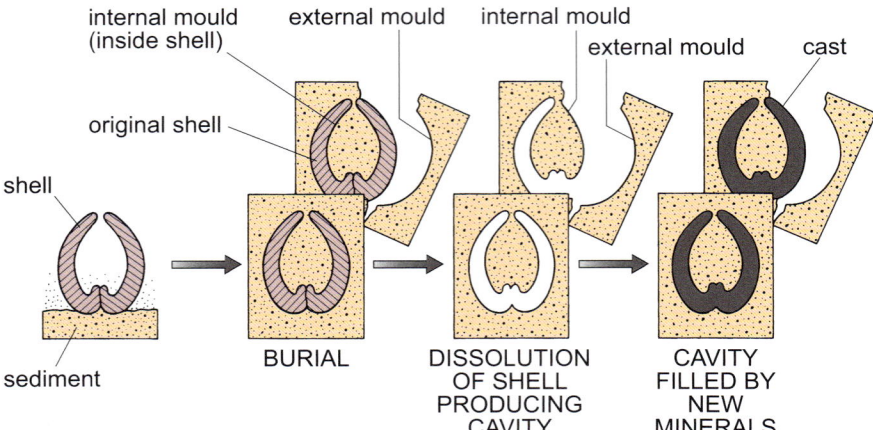

Figure 4.5 The formation of moulds and casts. The diagram shows the sequence of events that may follow burial of a shell, including infilling by sediment, dissolution of the shell and filling of the resulting cavity by cement crystals precipitated from percolating water. The diagrams depicted above and behind show what is revealed if the rock is broken open at different stages.

Often, the internal mould becomes visible because the enclosing skeleton has dissolved away naturally (such as when percolating groundwater becomes acidic after passing through sandstone). In some cases, however, a palaeontologist might deliberately remove skeletal material in order to reveal the internal mould. For example, in fossil vertebrates a mould of the skull interior may give important information on brain size and shape. Recently, though, sophisticated, *non-destructive* medical scanning techniques, such as computed tomography (CT) and magnetic resonance imaging (MRI), can be applied to rare, and even very large, fossils (e.g. dinosaur skulls) to see through matrix and skeletal material, enabling 3D images to be created (Figure 4.6). Remarkable information can now be obtained in this way. For example, scanning of dinosaur skulls has revealed the sizes of various brain lobes and sinuses, indicating the relative importance of senses such as sight, smell and hearing in different species, and thus differences in their way of life.

If at some stage a shell is completely dissolved away, a cavity remains between the internal mould and the external mould. Cavity formation is much more common if shells are originally made wholly or partly of aragonite (e.g. most gastropod and bivalve shells) rather than the less readily soluble calcite (e.g. most brachiopod shells). New minerals may then fill up the cavity, forming a crude **cast** of the shell, which lacks any details of original microstructure (Figure 4.5). In general, casts are rarer than moulds. Commonly, the cavity filling is a

Chapter 4 Fossilisation

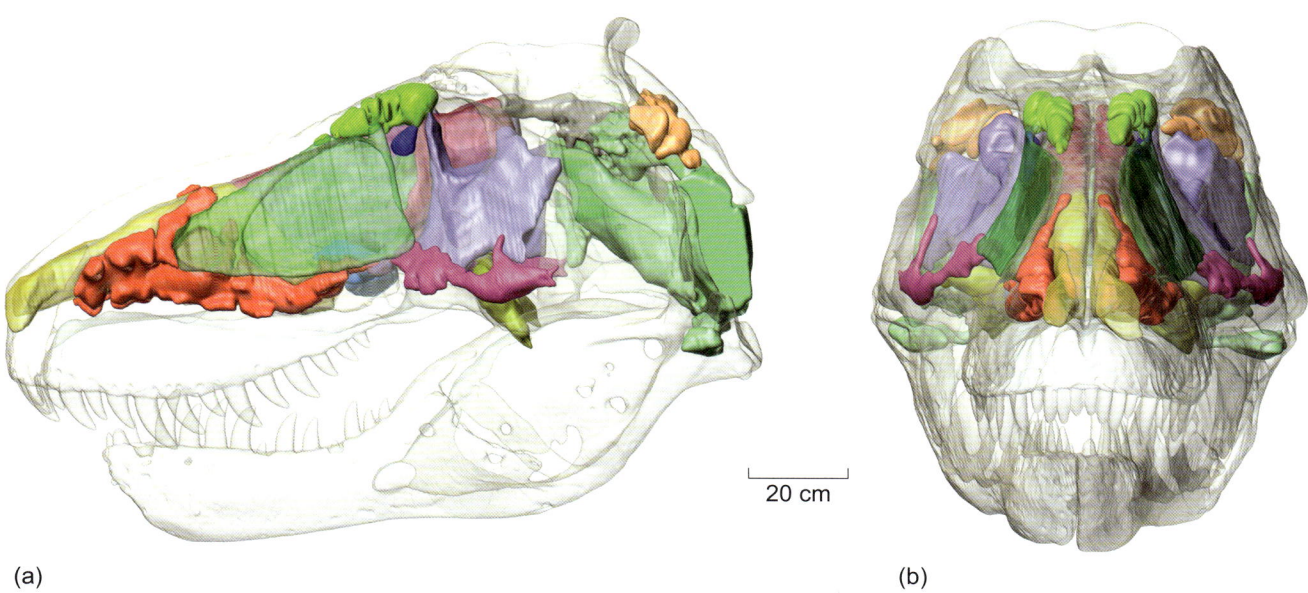

(a) (b)

Figure 4.6 A CT scan and 3D visualisation of a 1.5 metre-long skull of *Tyrannosaurus rex*. The bone is rendered semi-transparent, whilst the sinuses (air spaces) and other components of the head have been coloured. In (a), the cast of the brain can be seen as a small, grey feature near the top of the skull, between areas of purple/blue on the left and green on the right. Digital segmentation images, following CT scanning of this and several other skulls, were combined together to reconstruct details onto this single skull of the specimen nicknamed 'Sue' (after the finder; the sex of the dinosaur is unknown). The head of the living dinosaur weighed about 512 kg (more than four adult male rhinoceros heads). The complex sinuses had many functions, and lightened the head whilst enhancing its strength. Analysis of sinuses and brain lobes show that *T. rex* had the visual systems of a predator and the capability to engage in rapid tracking movements of the eyes, head and neck. The sense of smell was acute and reception of low-frequency airborne sounds was important.

coarse-grained mosaic of cement crystals, precipitated from pore waters. Most casts of originally calcareous shells are themselves calcareous, but infillings may include more exotic minerals, such as opal ($SiO_2.nH_2O$), pyrite (FeS_2), gypsum ($CaSO_4.2H_2O$), hematite (Fe_2O_3), baryte ($BaSO_4$) and cassiterite (SnO_2).

In some cases, as a shell is buried, sediment fails to fill the shell interior entirely. The remaining void may, or may not, be crushed flat during sediment compaction. Minerals may precipitate in any remaining space above the sediment. Incomplete sediment fillings can be very useful for interpreting geological structures in the field, because they usually indicate the horizontal at the time of deposition, like a fossilised spirit level. Not only is the horizontal preserved but, as the sediment fills the lower part first, the original 'way-up' is also preserved, which is especially useful in vertical or steeply dipping beds where 'way-up' is otherwise hard to tell. Examples can often be seen in gastropods, such as in RS 21.

Figure 4.7 A Jurassic brachiopod, broken across to show partial infilling by sediment (3.5 cm long). See text for discussion.

■ Look at Figure 4.7, and imagine the shell was orientated like that in vertical strata, running up and down the length of the page. In which direction do the beds get younger – to the left or to the right?

☐ To the left, as sediment has filled the bottom of the shell first (on the right), and calcite crystals have later precipitated in the void above, on the left.

113

Incidentally, the 'plaster replicas' in the Home Kit are artificial casts taken from rubber impressions, themselves artificially prepared from the original fossil specimens. The preparation of artificial casts (usually of rubber) from natural external moulds is also a valuable technique for studying the fossil content of rocks from which all shell material has been dissolved away, leaving only cavities in the rock.

4.2.2 Neomorphism

Sometimes the structure of a skeleton changes but its chemical composition remains virtually the same. This is called **neomorphism** (meaning 'new form'). Crystalline microstructure can be altered by the growth of new crystals at the direct expense of their neighbours. In other words, ions going into solution from one (dissolving) crystal face almost immediately become incorporated in the lattice of another, growing alongside. Neomorphic alteration to calcite is the usual fate of aragonite if the aragonite has not first been entirely removed by dissolution. Neomorphism can often be detected in thin sections, because original microstructural features marked out by the organic matrix, such as growth lines, tend to be preserved as 'ghosts' within the mosaic of new crystals. This form of preservation can be seen, for example, in some of the shells in the thin section of the gastropod limestone RS 21 using the Virtual Microscope.

Figure 4.8 Specimen of a Jurassic oyster (*Gryphaea dilatata*) that settled as a larva onto a dead ammonite shell, and then grew on it, moulding itself to the ammonite's external surface. The aragonitic shell of the ammonite has completely dissolved away, but evidence of its existence is preserved by what is, in effect, an external mould made by the growing oyster (11 cm across).

As aragonite tends to undergo dissolution and neomorphism, the likelihood of its preservation diminishes with age. Aragonitic fossils are common in the Cenozoic, but relatively uncommon in the Mesozoic and extremely rare in the Palaeozoic. However, preservation of aragonite is more common in some lithologies than in others. Mudstones, for example, tend to provide better protection than sandstones, because their relative impermeability inhibits the flow of dissolving groundwater. Differences in the behaviour of aragonite and calcite lead to biases in the fossil record. In some rocks, e.g. the Chalk (which is highly porous), only fossils with calcite skeletons are usually found, although originally there was also a diverse fauna with aragonite skeletons that failed to survive diagenesis. Evidence for aragonitic organisms exists where a sessile organism with a *calcite* skeleton, such as an oyster, has grown against an aragonite shell (e.g. an ammonite) and so taken an imprint. Figure 4.8 shows a Jurassic example of this.

4.2.3 Permineralisation and replacement

Hard biological materials such as bones, shells and wood usually contain pores (Figure 4.9a). When hard parts are lying buried in sediment, any such pores tend to be filled up with minerals that crystallise out from water seeping through the sediment (Figure 4.9b). If the original spaces in a porous material are impregnated with extra minerals in this way, the material is said

Chapter 4 Fossilisation

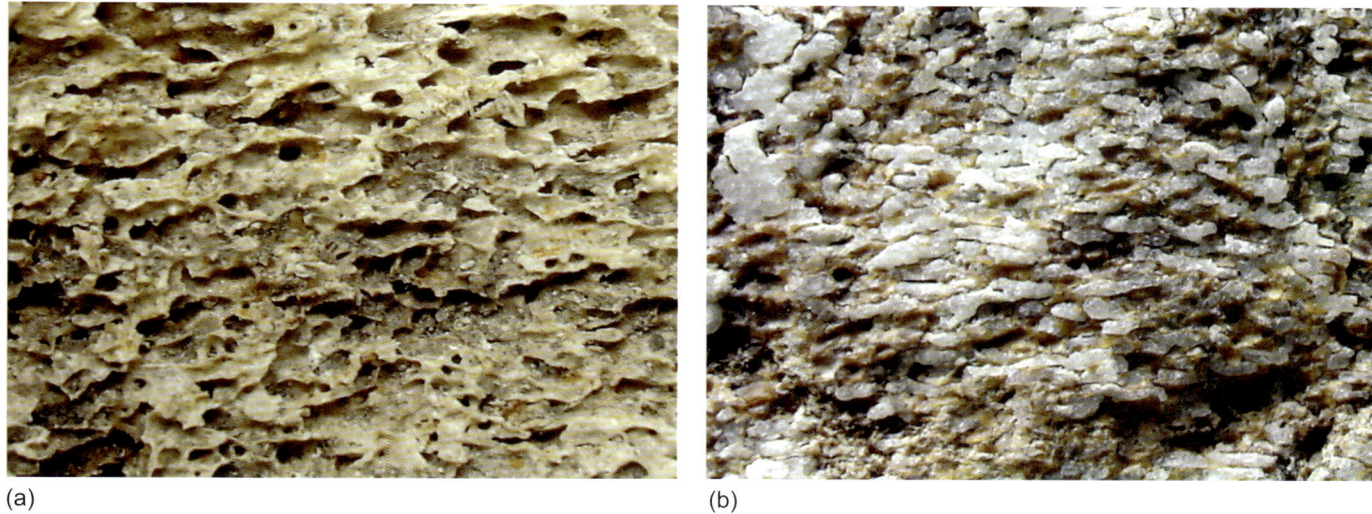

Figure 4.9 (a) Close-up view (2.5 cm across) of a typical piece of Quaternary Ice Age bone about 40 000 years old, showing the spongy texture on a broken surface, revealing little permineralisation as the pores remain mostly unfilled by extra minerals. (b) Close-up view (2.5 cm across) of a typical piece of permineralised Jurassic bone (an ichthyosaur vertebra), showing the spongy texture on a broken surface, with pores completely filled by white calcite.

to be **permineralised**. The growth of new minerals at the expense of any original biological material (such as the cell walls of bone or wood) is called **replacement**. The minerals most commonly involved in permineralisation and replacement include: quartz (when the process is called silicification), calcite, dolomite and pyrite. Replacement may be highly selective, affecting some structures and not others, and is sensitive to local conditions. The commonest minerals that replace calcium carbonate are silica (SiO_2) and dolomite ($CaMg(CO_3)_2$), particularly in limestones. If the limestone matrix of silicified fossils is dissolved, either naturally by rainwater or groundwater, or artificially by immersion in acid, specimens superbly preserved in three dimensions may be retrieved (Figure 4.10). Many of the most spectacularly preserved soft tissues in the fossil record have been replaced by phosphatic minerals, usually calcium phosphate (Section 4.2.5).

Figure 4.10 Silicified Carboniferous brachiopod shell revealed by dissolving away the limestone matrix with acid. The support for the lophophore has been preserved inside the shell, which is about 2 cm across.

Both the filling up of pores (permineralisation) and the replacement of biological materials may occur in a single fossil. Neither of these processes, which together are called **petrifaction** – 'turning into stone' – *has* to occur in fossilisation; sometimes the fossil can still be composed of the original, barely altered shell or bone. A single fossil may show several modes of preservation; most fossil shells, for example, show some degree of permineralisation or replacement, and are normally associated with their internal and external moulds in the adjacent rock.

In some settings, such as fine-grained anoxic sediments, decaying organisms often create local chemical conditions that promote the precipitation of carbonate minerals – especially calcite or siderite (iron carbonate) – in concretions (or nodules) around them (Figure 4.11 overleaf). Such concretions often form early in diagenesis, so that the enclosed fossils (often mineralised themselves) are uncrushed, having been protected from compaction by mineral cements.

Figure 4.11 Concretion containing a Cretaceous fish *Rhacolepis* (length 17 cm). The outer part of the nodule is mainly calcite. The soft parts of the fish have been replaced by calcium phosphate, preserving their morphology in three dimensions (see also Figure 4.17).

4.2.4 Fossilisation of plants

Plant fossils tend to be preserved in a limited number of ways. The outer coat of sporomorphs is sufficiently chemically inert to survive for millions of years, more or less unaltered (Section 3.5.4). Heating, such as during deep burial, alters the sporomorph wall, which turns an increasingly dark colour as temperatures increase. Unless the wall is destroyed, as in extreme cases, the colour changes reveal the burial and heating history of the host rock. This is often used in the petroleum industry to tell whether a source or reservoir rock is likely to yield oil or gas, or if the heating has destroyed the potential for hydrocarbon production.

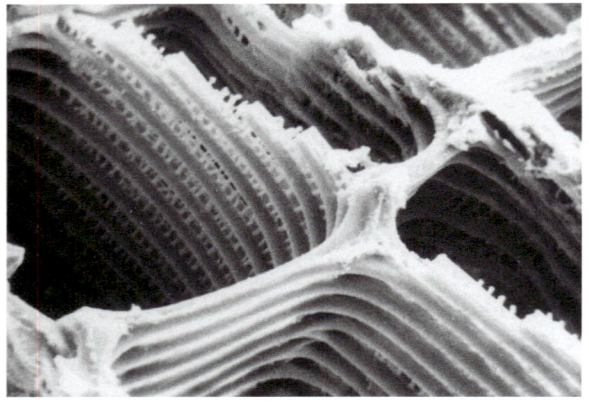

Fossil charcoal is perhaps a surprising example of unusually fine preservation. During forest wildfires, woody materials (but sometimes leaves, flowers and seeds) often become partly burned before burial. Heating in the absence of oxygen alters the original plant material to almost pure carbon, which has a high preservation potential because bacteria, fungi and animals cannot rot it or use it as food. Although often fragile, charcoal contains exquisite detail of the plant's cellular structure (Figure 4.12). The nature of the charcoal depends largely on the temperature to which the plant material is heated, so charcoal can be used to understand the characteristics of ancient wildfires and even to estimate the oxygen content of ancient atmospheres.

Figure 4.12 Charcoal from a Carboniferous tree, showing detail of several cell walls. Field of view 0.3 mm.

In volcanic eruptions, trees can become buried by pyroclastic flows. By studying the charcoal formed as hot ash bakes the wood, it is possible to estimate the temperatures of ancient pyroclastic flows, and hence learn something about their eruption dynamics. This, in turn, can be used to assess potential hazards posed by future eruptions.

Most plant fossils consist of a black, carbon-rich film, representing the remains of ancient tissues flattened onto a bedding plane. Where this film of non-volatile organic residues (usually wholly or mostly carbon) survives, it is called a **compression fossil** (or sometimes a 'coalified compression' fossil) (Figure 4.13, left). Flattening destroys the internal structure of the plant, but preserves its overall shape in two dimensions. Experiments show that virtually no lateral expansion occurs during flattening. Compression of woody stems and trunks is, however, often incomplete, and what was once a cylinder with a circular cross-section develops an oval shape in cross-section. If sandy sediment enters hollow interiors of stems and trunks, compaction may be minimised.

Figure 4.13 *Ginkgo* leaves that grew during a warm phase of the Cretaceous Period, at a palaeolatitude of 75° N. The fossil leaf on the left is a compression fossil, where the organic film representing the original plant tissues is still intact. That on the right has lost most of the organic film, leaving just an impression. Each leaf is about 3 cm across.

Where oxygen is present in the pore waters of a rock during burial, or a compression fossil becomes exposed to air during erosion, the organic film may be destroyed, leaving only an imprint of the original plant material. This is called an **impression fossil** (Figure 4.13, right). Some plant fossil specimens consist partly of an impression and partly of a compression, either due to partial destruction of the organic film or to splitting the rock along the bedding plane in the collection process. In the latter case, there are two compression/impression fossils – a *part* and a matching *counterpart* (terminology that is also used for animal fossils).

As in animal fossilisation, plant tissues can become impregnated with mineral-rich fluids, and minerals may precipitate before compression occurs, preserving the tissues in three dimensions. In permineralisation (Figure 4.14), only the interiors of the plant cells and any voids between the cells are filled with minerals; the original organic cell walls are preserved, albeit often with some chemical alteration. Sometimes the voids are infilled and the organic cell walls are *replaced* by minerals, in which case the fossil has undergone complete petrifaction.

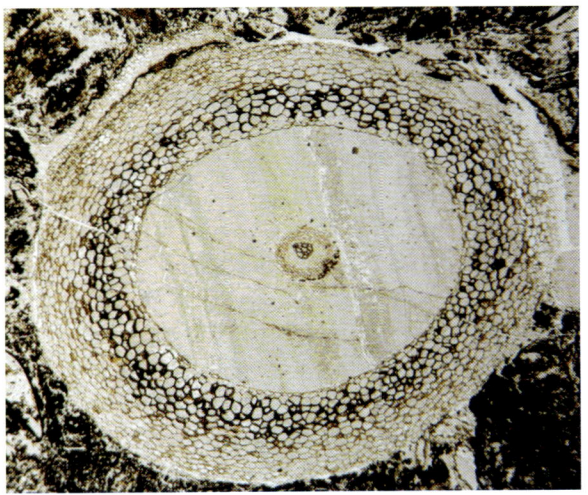

Figure 4.14 Permineralised specimen of a rootlet of the Carboniferous tree *Lepidodendron* (the root system of which is separately called *Stigmaria*, as shown in Figure 3.57). In the very centre of the rootlet are cells that conducted water, surrounded by lighter-coloured protective cells. Surrounding the latter, no cells can be seen, but instead banded layers of calcite fill what was, in life, an air space. Around the outside of the rootlet are large dark-coloured cells and slightly smaller lighter-coloured cells that make up its outer sheath. The brown colour of the cell walls is due to the fact that much of the original organic matter has survived, albeit altered in part to carbon. The rootlet is 5 mm in diameter.

4.2.5 Exceptional preservation of soft tissues

Almost always, soft tissues rot away entirely before they can become subject to processes that might lead to eventual fossilisation. In highly unusual conditions, however, prolonged suppression of bacterial decay, and/or the rapid growth of early diagenetic minerals, may allow such tissues to be preserved. Conditions suppressing decay include deep-freezing, immersion in highly salty brines (effectively, pickling) and desiccation, either directly as mummies in arid settings, or, in the case of small organisms such as insects, through being trapped in resin (forming amber), which withdraws moisture from them (Figure 4.15). Suppression of decay alone provides no guarantee of long-term preservation: most of the fossil examples of delayed decay given above are relatively young (under a million years old), although amber is known as far back as the Carboniferous. More important on geological timescales is the growth of new diagenetic minerals within or around tissues, and here microbes such as bacteria and fungi can make a positive taphonomic contribution, in contrast to their largely destructive role emphasised so far.

(a) (b)

Figure 4.15 Insects preserved in amber, about 25 Ma old, from Mexico: (a) snipe fly (tip of head to tip of abdomen, 4.2 mm long); (b) leafhopper bug, showing original colour banding (6.4 mm long).

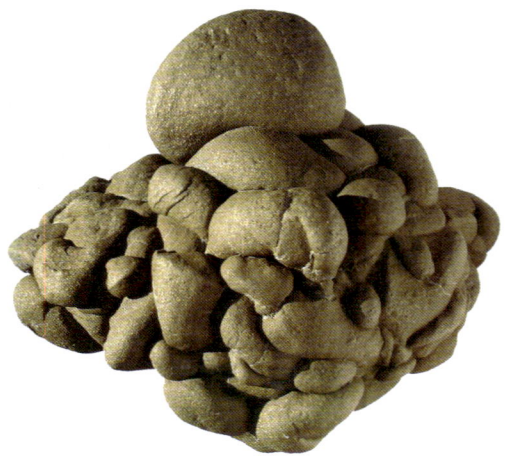

Figure 4.16 Coprolite (fossil dung) preserved in pyrite, from an unknown Jurassic animal. It is 13 cm across.

Most sediments contain organic detritus that is decomposed by microbes. Where there is an adequate oxygen supply, both the detritus and the decomposers are fed on in turn by larger deposit feeders, which make excretory products and faecal material. This can enrich the sediment with nutrients containing elements such as phosphorus, nitrogen and sulfur. The oxygen supply falls off rapidly with depth of burial (Section 4.1.2), and where conditions become anoxic in the sediment, anaerobic bacteria start to reduce compounds, yielding a cocktail of ions, some of which may then combine to form relatively insoluble minerals. For example, sulfide ions (S^{2-}), derived from the reduction of sulfate ions (SO_4^{2-}), are rapidly precipitated in contact with ferrous ions (Fe^{2+}) to form the insoluble compound FeS, which then converts to pyrite (FeS_2). Such new minerals may fill void spaces in biological tissues, or even replace them altogether. Figure 4.16 shows a large **coprolite** – droppings from an unknown marine Jurassic animal – replaced by pyrite. The 3D form of this coprolite is uncompacted and looks as if it might have been excreted by the animal just a few hours ago. Indeed, replacement of the soft faecal material by iron sulfide in an anoxic environment may have occurred almost immediately.

■ How might the nature of the faecal material itself have promoted preservation?

☐ Faecal material usually contains large amounts of anaerobic gut bacteria that, acting together with other bacteria in the vicinity, may well have enhanced sulfide precipitation.

Calcium phosphate (apatite) may sometimes be precipitated in anoxic conditions, and soft tissues can become fossilised with extraordinary fidelity, even showing details of cellular structures (Figure 4.17).

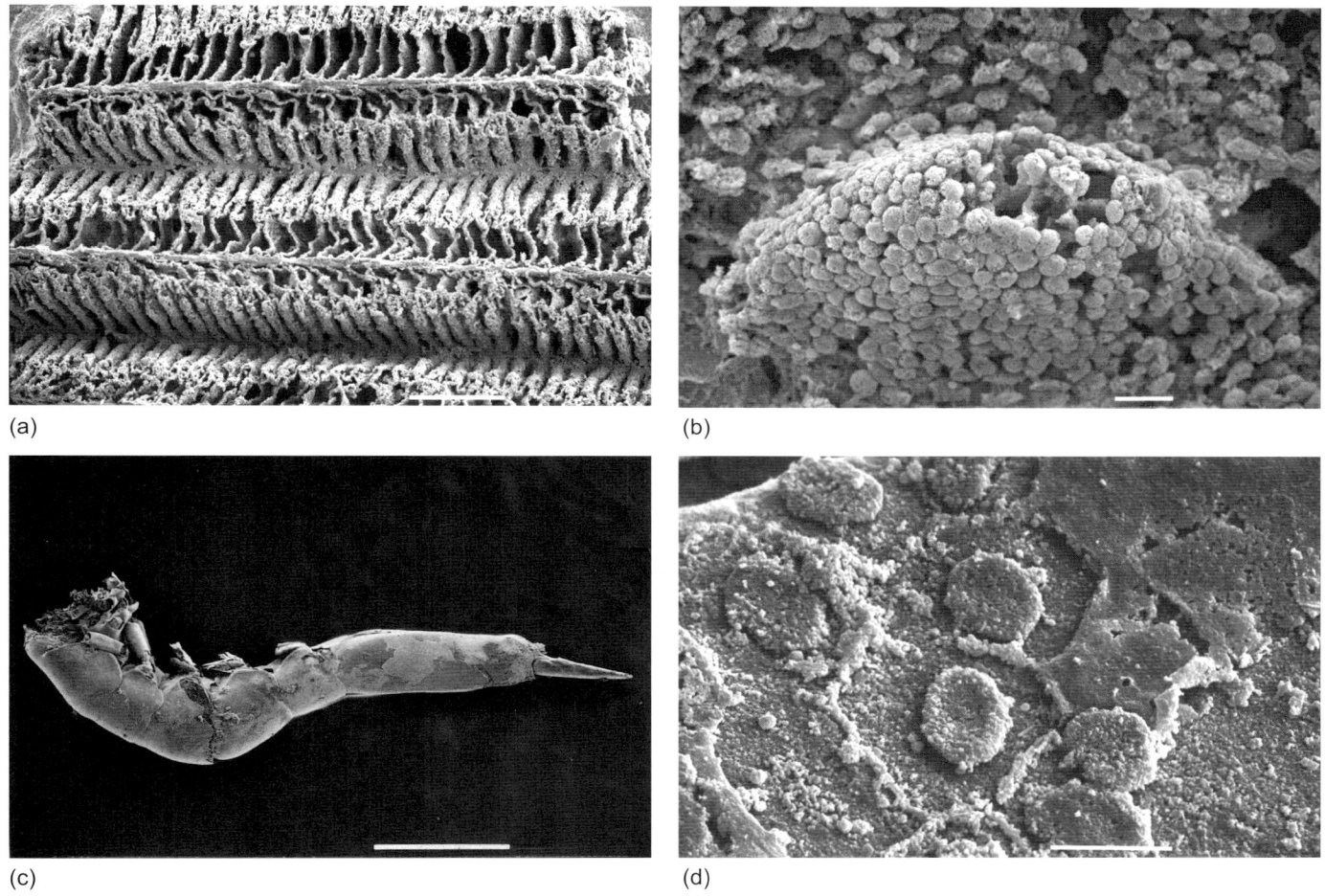

Figure 4.17 Soft parts of Cretaceous fish, including stomach contents, from Brazil, preserved in calcium phosphate (apatite). They are about 110 Ma old. (a) Part of a gill of the fish *Rhacolepis* (scale bar 100 μm). The gill rays run E–W, whilst running N–S are many small blood vessels of the secondary lamellae, giving a large surface area for oxygen absorption. Figure 4.11 illustrated an example of a complete *Rhacolepis* from the same deposit. (b) Cells on a fold on the inside of the gut wall of the fish *Notelops* (scale bar 10 μm). (c) Part of a shrimp found in the stomach of the fish *Rhacolepis* (scale bar 1 mm). (d) Details of the shrimp in (c), showing its cell walls and internal moulds of cell nuclei (scale bar 10 μm).

Figure 4.18 Ink sac of a Jurassic squid-like cephalopod, preserved in calcium phosphate and still containing dark pigment. Note the impressions of blood vessels from the formerly enshrouding muscular bag. The groove at the top is where the duct that led from the ink sac has come away on the counterpart of the fossil (i.e. the other side of the split slab). The ink sac is 3.5 cm long. Reproduced with the permission of the British Geological Survey © NERC. All rights Reserved.

Figure 4.18 shows an exquisitely preserved ink sac of a Jurassic squid-like cephalopod. As with examples of exceptional preservation in pyrite, bacteria are also involved here – in this case in chemical reactions that lead to the precipitation of the mineral apatite. Experimental observation of decay in modern, related, organisms suggests that the replacement by apatite in such fossils must have occurred in a matter of hours or days.

Although very rare, examples of exceptional preservation are still being discovered, adding to a series of spectacular, high-resolution 'snapshots' of life through various stages in its history. Extremely fine details can be fossilised in the right physical and chemical conditions, but there is strictly no such thing as the 'perfect preservation' of an entire organism, though this phrase is often loosely used. DNA itself rapidly degrades, though more slowly in cold and arid environments. Meticulous analysis of fossil insects and other organisms preserved in amber has failed to confirm initial reports that DNA could survive for many millions of years; contamination with genetic material from modern organisms was responsible for the false claims. However, some short sequences of DNA from mammoths, and other Quaternary Ice Age animals and plants, have been retrieved from deeply frozen icy soil or ice up to about 800 000 years old. As the genetic code of many modern species becomes increasingly well known, the former existence of these and related species in very cold areas can be detected *without* the need to find large body fossils such as bones or leaves, because diagnostic DNA fragments can be extracted chemically from frozen soils and ice.

4.2.6 Sedimentary compaction

The effects of compaction on fossils caused by the weight of accumulating sediment are usually most apparent in mudstones and shales, which may shrink to a tenth of their original wet sediment thickness (Section 9.1). Thin-shelled fossils are usually crushed flat, or distorted to give a variety of shapes, depending on their orientation during compaction. If the effects of compaction go unrecognised, the various preservational shapes of a single species may be misidentified as different species.

4.3 Metamorphism of fossils

With increasing temperature and pressure, fossils become distorted, their components recrystallise and undergo change in mineral composition, and so they are eventually destroyed. Fossils may be moderately abundant, albeit highly deformed, in some low-grade metamorphic rocks, such as slates and low-grade marbles (in which the composition of calcium carbonate shells may not change). As discussed in Book 2, Chapter 8, analysis of strain in deformed fossils whose original shape is known can be useful in structural geology. Fossils are extremely rare in medium- and high-grade metamorphic rocks. Uplift, weathering and erosion at the surface may also subsequently destroy any fossil, dispersing widely the chemical elements of which it is made.

Activity 4.1 Fossil preservation

This activity investigates the preservation of various fossils, including some in the Digital Kit.

4.4 Summary of Chapter 4

1. Taphonomy is the study of the various processes that affect the formation of fossils, from the death of organisms onwards (or, in some cases, from the shedding of parts during life). Such processes include those causing death (by biological or physical means), decay (aerobic and anaerobic) and disintegration, as well as diagenetic changes, which are often crucial in the eventual preservation of a fossil. Fossils can be produced in many different ways, and a single fossil may show several modes of preservation. Final burial eventually needs to occur and some degree of permineralisation and/or replacement is often (but not necessarily) involved.

2. The fossilisation potential of organisms varies according to many factors, including their construction (particularly composition and number of skeletal elements), and habitat (especially in relation to sedimentation). The likelihood that a particular species will be found in the fossil record also depends on its total population size and geographic distribution. Skeletal components, the most likely candidates for fossilisation, may be subject to many destructive processes – biological (predation, scavenging and boring), physical (current transport and wave action) and chemical (dissolution). Many kinds of preservation filters (morphological, ecological, sedimentary, diagenetic and metamorphic) stand between a living organism and its long-term preservation as a fossil.

3. Following final burial of an organism, dissolution may leave only moulds of the remains, though new mineral growth in the moulds may produce casts that lack original skeletal microstructure. The simultaneous growth of new crystals of the same chemical composition at the expense of original neighbouring ones (neomorphism) may crudely replicate internal features such as growth lines. Various degrees of petrifaction by permineralisation and replacement are common. In rare cases of exceptional preservation, the growth of minerals such as apatite or pyrite intervenes before the decay of soft tissues has gone to completion. Types of plant preservation include almost unaltered organic material (especially as sporopollenin), charcoal from forest fires, carbon-rich compression fossils, and impression fossils.

4. Fossils may be deformed by sedimentary compaction and by tectonic forces, and during metamorphism fossils may eventually be destroyed by recrystallisation and the growth of new minerals. Uplift, weathering and erosion may disperse the chemical elements of which a fossil is composed, and various processes may move these atoms on around the rock cycle.

4.5 Objectives for Chapter 4

Now you have completed this chapter, you should be able to:

4.1 Outline the various taphonomic processes that can affect the remains of organisms (including manner of death or shedding of parts during life, decay and disintegration, burial history and diagenesis, any subsequent deformation and metamorphism).

4.2 Determine which of these taphonomic processes are likely to have been experienced by given specimens, from observations of their modes of preservation, and so infer the conditions in which they became fossilised and their subsequent preservation history.

4.3 Make a reasoned estimate of the relative fossilisation potential of given organisms, based on their construction, life habits and habitats.

Now try the following questions to test your understanding of Chapter 4.

Question 4.4

Which of the following are taphonomic processes that might affect the fossilisation history of a brachiopod: scavenging, boring by organisms, disarticulation, fracture, abrasion, dissolution, sediment compaction, permineralisation and replacement?

Question 4.5

(a) Briefly describe the state of preservation of FS C (*Pseudodiadema*) from the Home Kit.

(b) In the light of your observations in (a), explain which of the following taphonomic hypotheses is most likely:

 (i) The animal was killed by a sudden influx of fine sediment, which permanently buried it.
 (ii) The animal died by unknown means, became buried a few weeks later, but was re-exposed on the sea floor for a while before final burial.
 (iii) The animal died by unknown means and remained on the sea floor for a few years before being buried.
 (iv) The animal was killed by a predatory fish, quickly discarded without being ingested, and buried a few days later.

Question 4.6

Does the durability and preservation quality of any potential fossil tend to decrease or increase as the grain size of the enclosing sediment decreases? Give your reasons.

Chapter 5   Fossils as evidence of past environments

For this chapter, the only specimen you will need from the Home Kit is FS Q.

The study of how ancient organisms interacted with their environment and with each other is called palaeoecology (Chapter 2). Whilst we can observe at first hand what eats what, or measure birth and death rates among living organisms, the scope of palaeoecology is unavoidably limited by the biased sampling of the fossil record. As an illustration, we will consider the investigation that can be done on a fairly ordinary assemblage of fossils (Figure 5.1).

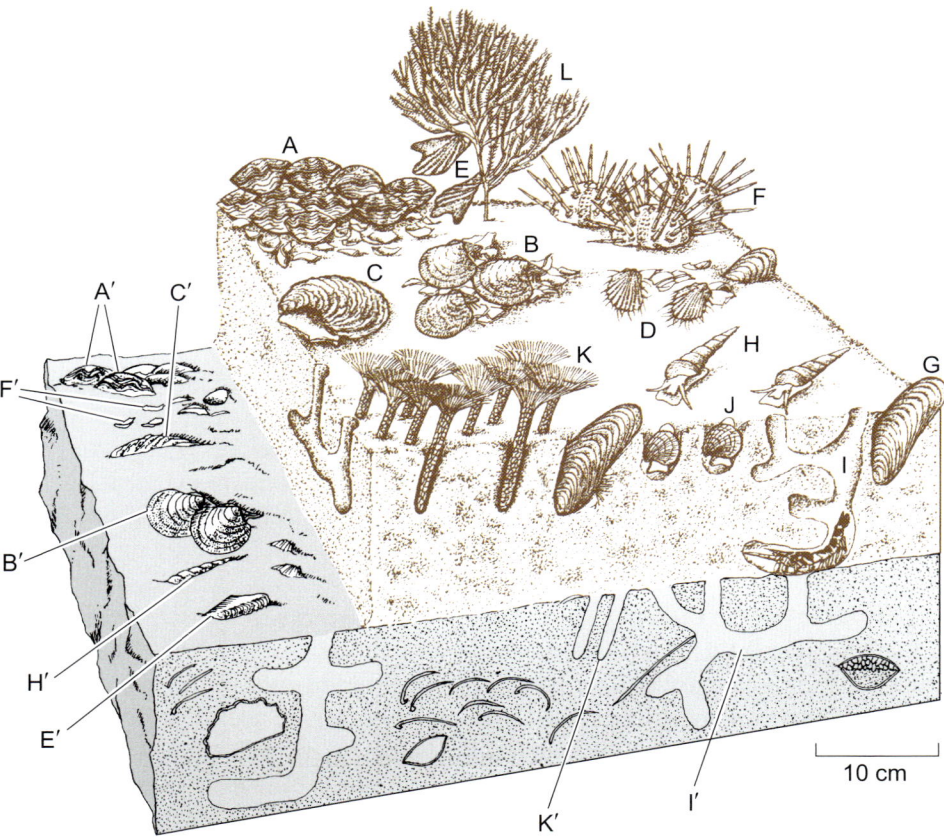

Figure 5.1 Reconstruction of the association of organisms that occupied a Jurassic shelly carbonate mud sea floor. The lower half of the block depicts a slab of the bed of fossiliferous muddy limestone on which the reconstruction is based (though not all the fossils found in the bed are illustrated in the small part that is visible). Part of the upper surface (a bedding plane) of the limestone is shown projecting to the left. The upper half of the figure is the reconstruction of a corresponding part of the original sea floor and the association of organisms that dwelt there. The sediment surface was probably relatively firm, and deposition is likely to have taken place in a shallow but quiet marine bay comparable to modern examples in Florida, USA and the Bahamas. Living organisms and their fossil counterparts (where illustrated in the bed of rock) are given the same letters, the fossils being indicated by a prime mark ('); other features are explained in the text.

■ Which groups seem to be represented in Figure 5.1 by relatively intact body fossils?

□ The upper surface of the limestone block (on the left) shows complete fossil shells of molluscs (e.g. those of the bivalves B' (scallop), C' (oyster) and E', and the gastropod H') and brachiopods (e.g. A'). Note that all of these shells originally consisted of only one or two large components, with a high fossilisation potential (Section 4.1.3).

Now imagine, for example, that you have recovered some fossil shells of the mussel (G). Certain living mussels very similar in shape to this fossil mussel tend to live half-buried, and so the fossilised species might be inferred (by analogy) to have done likewise. This reconstruction might be verified by actually finding whole fossil mussel shells preserved where they lived in their excavations. A fossil retaining the position in which the original organism lived is said to be in **life position**.

Reconstruction of organisms that had a lower intact fossilisation potential tends to depend on lucky finds. The echinoids (F), for example, are represented on the surface of the bed in Figure 5.1 only by fragments of plates – and present quite a jigsaw puzzle for the palaeontologist. In this reconstruction, however, knowledge of the species is based on about 1500 specimens, found elsewhere in a particular quarry, preserved intact with even their spines in place, and in life position.

■ How might the preservation of these intact specimens have come about?

□ Surface-dwelling echinoids such as these would have started to fall apart soon after death (Section 4.1.2), so these specimens were probably buried alive as they are preserved intact (see Figure 3.17 for a similar example). Given the quiet marine bay setting inferred for the deposit, the most likely explanation is the swamping of an echinoid population by a sudden influx of storm-driven sediment.

The reconstruction of the morphology and habits of less readily fossilised species is often more tentative in the absence of unusual preservation. The existence of some organisms (and in particular their life habits) can be inferred only from trace fossils. The branching burrows in the limestone (I') commonly contain sediment-bearing faecal pellets, indicating that the occupant of the burrow was a deposit feeder, at least for some of the time. The forms of the burrows and of the faecal pellets are similar to those associated with certain living infaunal crustaceans. We might, then, infer that the burrows belonged to such creatures (hence I), although, as so often with trace fossils, unequivocally associated body fossils supporting this interpretation are very rare. The presence of the suspension-feeding fan-worms (K) in the reconstruction is again based on distinctive trace fossils (K'), though simple vertical tubes lined with shell debris and fish scales (the details of which are not shown in Figure 5.1) leave some latitude for interpretation.

The scope of palaeoecology at the level of the single organism or species therefore ranges from the precise to the speculative, depending upon the kinds of fossils left behind. But how complete a picture of the original community, or communities, of organisms that dwelt here is this reconstruction? Most of the soft-bodied forms will have left no identifiable evidence. Just attempting to list the constituent species from the incomplete fossil record requires informed speculation. The presence in the reconstruction of the organism labelled L, a sea fan (a sessile colonial cnidarian with very poor fossilisation potential), is hypothetical but is based on the mode of life of the streamlined bivalve species (E). This bivalve is thought to have swung free in the water whilst attached to organisms such as sea fans that were anchored in sediment.

As for the network of feeding relationships between organisms in communities, the evidence is even more patchy. It is not difficult to interpret how various organisms fed, when considered in isolation, as you saw in Chapter 3, but it is quite another matter to reconstruct the intricate web of feeding interactions among the different species. Which, for example, were the main prey species for the predatory or grazing echinoid (F)? Did the suspension-feeding fan-worms (K) compete for food with other suspension feeders, such as the mussel (G) and brachiopod (A); or did they perhaps draw food particles from different levels in the water column, or specialise in food particles of a different size, and so avoid competition with them? And was food (as opposed to, say, space for settlement of larvae) a limiting resource for them, in any case? Often, we can do little more than classify the organisms into broad feeding groups in a **trophic pyramid**, recognising those that synthesised organic matter (e.g. by photosynthesis) – the *producers* – as the basic food source for the *consumers*. Consumers in turn may be divided into those that ate the producers directly and those that ate other consumers.

Some questions are much easier to answer from the available evidence than others. The skill of palaeoecological interpretation often involves deciding which are answerable questions, given the partial nature of the evidence. As many biases intervene between living associations of species and fossil assemblages (Chapter 4), it would be naive to expect to make accurate reconstructions of *entire* ancient communities. Here the focus is on some of the more straightforward palaeoecological inferences that can be drawn from fossils. These insights, together with sedimentological information, can greatly help in interpreting palaeoenvironments.

5.1 Clues to physical and chemical conditions

Imagine you have identified a body fossil in an exposure of sedimentary rock, and you want to know what it can tell you about the original environment of deposition. The next thing to consider is whether or not it lived in or near the place of burial.

If the fossil is preserved in life position, or shows little evidence of transport (Section 4.1.5), then the environmental conditions can be inferred from a knowledge of the organism's biology and environmental preferences (Chapter 3). If, however, it had been transported far, then it might reveal little more about its eventual resting place than any other sedimentary grain. An organism preserved more or less where it lived is said to be **in situ**; it may be either in life position, as described above, or only locally displaced. Sessile organisms that normally lived implanted in the sediment (like bivalve G in Figure 5.1), for example, might have been washed out, but then buried in place as storm currents swept sediment over them. To support the latter interpretation, additional evidence would be needed, such as finding both valves of G intact (implying lack of significant transport), or other specimens nearby in life position. However, if remains have evidently been carried for some unknown distance from their source, by currents, scavengers or whatever, they are described as having been transported.

The possibility of transport can be a serious problem, especially in shallow-water deposits. The best kind of evidence that benthic organisms have been buried where they lived is if they are preserved in life position, such as cemented bivalves still attached to a bedding surface (Figure 5.2a), or burrowing bivalves preserved in their normal burrowing positions (Figure 5.2b). Alternatively, trace fossils may reveal the activities of organisms living at the site (e.g. the circular borings shown by arrows in Figure 5.2a).

(a)

(b)

Figure 5.2 (a) View of the top of a cemented surface (hardground) in limestone showing fossil evidence of both encrusting and boring bivalves. About five encrusting rudist bivalves are visible; in each case, the lower, attached part of one valve can be seen on the surface. Some of the borings, which are roughly circular in cross-section, are arrowed. Field of view about 30 cm across. (b) View perpendicular to bedding showing burrowing bivalves preserved in life position in a bed of limestone. Lens cap gives approximate scale. Both (a) and (b) are from the early Cretaceous.

If, on the other hand, you were to find some intact burrowing bivalves that were not preserved in the normal burrowing position, the problem of how far they might have been transported before being buried alive would remain. So whenever you observe a fossil at an exposure, you should note – preferably with sketches or photographs – its orientation and position in the bed, as well as its state of preservation. (It is best to leave it there for others to see, unless you really need to collect it for further study.)

Following such an assessment, you are then better placed to make inferences about the ambient physical or chemical conditions from what is known about the biology of the original organisms.

Question 5.1

Suppose the only fossils that you could find in a finely laminated, black, pyrite-rich shale were a few ammonites with streamlined shells. What might you infer about the oxygen level (a) at the sea floor, and (b) in the overlying seawater?

One of the most important uses of fossils preserved in aquatic environments is to indicate salinity. As you saw in Chapter 3, many living groups show restriction to marine conditions today and, according to all the available evidence, also in the past. Some living marine groups, such as echinoderms and cephalopods, will tolerate very little salinity variation either side of 'normal marine salinity', which is about 35 grams of dissolved salts (mostly sodium chloride) per kilogramme of seawater (i.e. 35 'parts per thousand', represented by the symbol ‰). Some marine groups may tolerate more variation than others. For example, a very few living species of coral will tolerate salinity variation up to 10‰ above or below normal marine salinity, whereas this is too much for any echinoderm or cephalopod. A few living inarticulate brachiopods such as *Lingula* (Figure 3.36) can briefly survive small salinity variations, but only by closing their valves tightly and waiting for conditions of normal salinity to return. Some groups of organisms, alternatively, may contain species that are restricted to freshwater and are intolerant of even the slightest salinity.

Question 5.2

Which of the following fossils found alone in strata, and in the absence of more detailed identification, cannot be used to infer deposition in marine conditions: a trilobite, bivalve, crinoid, gastropod, ostracod and a belemnite?

Although the ecology of living groups usually helps the interpretation of the palaeoecology of ancient organisms, this approach must be used with care. It is necessary to consider all the evidence provided by the morphology of the fossils themselves, as well as associated *in situ* organisms and features of the enclosing sedimentary rocks. For example, living horseshoe crabs are shoreline inhabitants and tolerant of a wide range of salinities, whilst some fossil horseshoe crabs were restricted to marine conditions and others to brackish water ponds and lagoons. Another example is provided by the corals. The ability of modern forms (scleractinian corals) to build extensive coral reefs has often led to the mistaken assumption that deposits rich in Palaeozoic corals (rugose and tabulate corals) are the remains of similar reefs. In fact, close examination of these Palaeozoic forms shows they were seldom cemented to other living or dead skeletal material, and were mostly adapted to life on soft substrates. Palaeozoic reef builders were mainly fossil sponges and algae; corals played only a subsidiary role.

Fossils can also be used to investigate patterns of sedimentation. Intense bioturbation indicates that the sediment remained unconsolidated near the surface for long enough to become infested with burrowers. This in turn implies not only that conditions at the sea floor supported a thriving benthos, but that net rates of sediment accumulation were sufficiently low to have allowed the burrowers to 'get the upper hand' over current activity, in imposing their imprint on the accumulating sediment (Section 3.6). A sparser record of trace fossils accompanied by abundant current-generated sedimentary structures, by contrast, would suggest that current reworking and/or deposition of sediment 'had the upper hand' in this contest of influences between bioturbation and currents. Where present, the spreiten in trace fossils such as *Diplocraterion* can yield further information about episodes of net sediment gain or loss (Section 3.6.2).

The real value of such observations is that they allow the detection of *changes through time* in sedimentary successions. For example, a shift from mainly bioturbated sediment to mainly wave cross-stratified sediment could be interpreted as reflecting the shallowing of the sea floor to a level where waves routinely affected it in fair weather conditions.

As another example, encrustation of a surface by a fauna like that in Figure 5.2a would indicate that the sediment surface had become hardened before the deposition of overlying beds. Borings in the bedding surface (which cut through sediment grains and shells alike) add further weight to this interpretation.

These examples all illustrate simple inferences that can be made from the mere presence (or absence) of certain kinds of fossils. The approach can be refined by investigating patterns of growth in specimens. 'Abnormal' growth representing a response to adverse conditions may be revealing. A good example is provided by the sea shells of the Baltic Sea; these are markedly smaller than specimens at the same stage of growth found elsewhere on open marine coasts (Figure 5.3). Reciprocally transplanted mussels have been found to grow at rates similar to native mussels at each site, showing that the slow growth rate (and small maximum adult size) of mussels in the Baltic Sea is due to the physiological effects of below-normal salinity. In the fossil record, such stunting of growth might then be taken as signifying abnormal salinity, though care would have to be taken to ascertain that the shells were genuinely stunted with respect to others of their species.

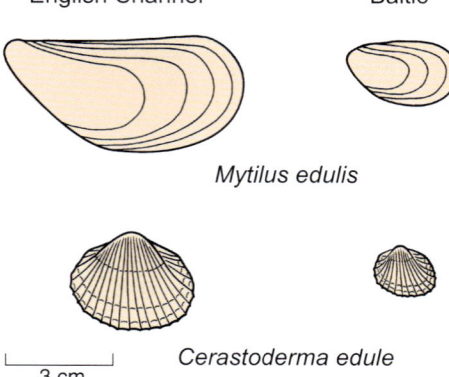

Figure 5.3 Representative shells of the common mussel (*Mytilus edulis*), above, and cockle (*Cerastoderma edule*), below, collected from the English Channel (left) and from the Baltic Sea (right).

■ Apart from stunting due to abnormal salinity, what other possibilities could cause the relatively small size of individuals in a fossil assemblage?

☐ They might be juvenile specimens or a local species characterised by small body size, or alternatively a current-sorted assemblage.

5.2 Palaeoecological relationships between fossil organisms

So far, we have considered inferences from fossils only about the physical and chemical aspects of past environments. For each organism, however, the other organisms around it are just as much part of its environment.

The first task that has to be tackled in determining if species interacted, is whether or not their fossilised representatives actually lived at the same time and place. This may sound simple – just find two fossil species in the same bed that have clearly been buried where they lived – but in practice it is not that easy. Consider the Jurassic Kimmeridge Clay Formation of Dorset, which has a maximum thickness of 540 m, and represents 7.5 Ma.

Question 5.3

What was the average net rate of deposition for the whole Kimmeridge Clay Formation in centimetres per 1000 years?

It is extremely unlikely that sedimentation took place at a regular rate. Instead, deposition at variable rates was probably interspersed with episodes of non-deposition and erosion. This pattern is typical not only of shallow seas, due to storms, tides and so on, but also of deeper areas, as a consequence of fluctuations in both sediment supply and deep currents. Nevertheless, using the average net rate as a basis for calculation, a collection of shells made from one band only (say, a 4 cm-thick layer within the Kimmeridge Clay Formation) could represent a time-span of about 550 years – perhaps one or two orders of magnitude more than the maximum lifespans of most of the shelly organisms present. So a band containing several species of burrowing bivalves could be a record either of successive separate colonisations by each of the different species, or of all or some of the species living there together at the same time. Indeed, a shell-rich bed is quite likely to represent a relatively condensed lag deposit, from which sediment was winnowed away (i.e. a shell bed). In that case, it would represent an even longer period of time than that calculated above. Assemblages of fossils that have accumulated in this way over long periods of time are said to be **time-averaged**. Attempting to unmix the 'cocktail' of organisms that lived at different times can be difficult, if not impossible.

The main exceptions to this problem of demonstrating contemporaneity are assemblages preserved by catastrophic burial (Chapter 4.1.1), which are of special value in palaeoecological reconstruction. In Figure 4.1, for example, we can be relatively certain from their common mode of preservation that the cystoids (stemmed echinoderms) shown there, as well as any associated organisms, did live – and die – together.

Establishing contemporaneity is only the first step. The precise relationships between organisms may still remain obscure. The only direct clues are likely to be traces left by one organism on another and examples of organisms that have clearly grown in intimate contact. Predation and grazing are examples of feeding relationships that might be revealed by trace fossils left on other body fossils. Many predators tackle their prey in a characteristic way, leaving their 'fingerprints' on their victims, such as the distinctive hole of a predatory gastropod in a bivalve (Figure 3.26b).

Figure 5.4 Modern mollusc shells showing characteristic damage inflicted by various predators: (a) cockle shell smashed by a lobster; (b) part of a bivalve's ornament plucked away by the tube feet of a starfish pulling the valves apart; (c) ventral valve rims of a bivalve, ground away by a whelk to allow entry of the gastropod's proboscis; (d) hole drilled in a gastropod shell by an octopus. (a) Actual size; (b–d) slightly magnified.

Figure 5.4 shows some other examples of recognisable damage inflicted on modern molluscan shells by known predators. Examples of prey having survived such attacks, as indicated by repair and continued growth of the shell, show the attacks took place on living prey. Figure 5.5 shows an example of damage to a modern nautilus shell that clearly happened when the animal was alive.

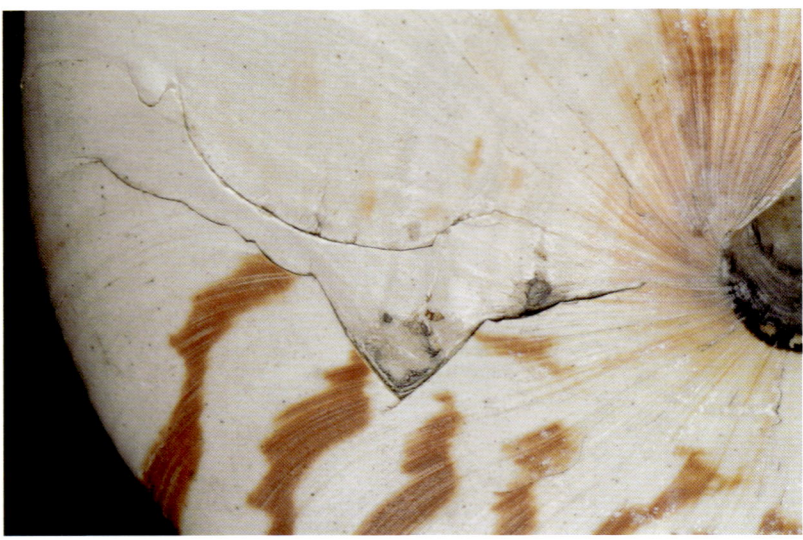

Figure 5.5 Close-up of a modern nautilus shell showing scar tissue and shell repair following two episodes of non-lethal damage, each probably made by a predator such as an octopus or a fish. Note how well the shape of the aperture eventually recovered to its normal outline, as revealed by growth lines after the second trauma (i.e. the one towards the top of the photograph). This shows that the mantle margin, which secretes the shell, cannot have been badly damaged. Field of view 7 cm across.

To illustrate the problems in interpreting closely attached organisms, turn now to FS Q from the Home Kit, which consists of a shell encrusted by a number of sinuous or coiled tubes.

■ To what kind of organism did the large shell belong, and in what way is its shape exceptional for the group?

□ The large shell in FS Q is a brachiopod, called *Torquirhynchia*. It has two unequal valves, the umbonal parts of which, at least, show characteristic brachiopod symmetry (as in Figure 3.1b). The anterior side of the shell (opposite the umbones), however, lacks this symmetry, as the shell margins are obliquely offset from one side to the other.

This highly unusual shape of the shell has nothing to do with the encrusting tubes, because other members of the species, lacking the tubes, also show it. The functional significance of this departure from normal brachiopod symmetry remains unclear.

There are three kinds of encrusting tube on the shell. The largest is in the shape of a question mark near the anterior margin of the smaller, brachial valve of the shell. Nearby, on the margin of the pedicle valve, is a smaller, spirally coiled tube. The third kind is the smallest, and is represented by some long thin tubes running along the radial grooves on the outside of the shell. Three of these form a tight cluster in a single groove between two ribs, about 1 cm down from the spirally coiled tube (you will need the hand lens to distinguish these). All three kinds of tube can be attributed to sessile suspension-feeding worms (like those attached to the oral surface of the echinoid, FS C). From the open end of the tube in similar living examples, the worm projects a fan of tentacles that trap food particles from the surrounding water.

Having established what organisms were involved, we need to decide whether the worms encrusted the shell while the brachiopod was still alive, or after its death.

Question 5.4

In FS Q, judging from the way in which the brachiopod shell itself has been preserved, can you detect any positive evidence that it lay exposed on the sea floor after death and before final burial?

The taphonomic evidence for post-mortem exposure of the shell on the sea floor, identified in Question 5.4, indicates that there was ample opportunity for encrustation by tube-building worms after, as well as during, the life of the brachiopod.

- Can you detect any preferred situation or orientation of the worm tubes with respect to the shell in FS Q?

- ☐ The large sinuous tube and the smaller spiral tube are both situated at the anterior margins of the shell, while the three small tubes near the spiral tube all have their apertures facing towards the anterior margin of the shell (you will need the hand lens to see this). The orientation of the other small tubes is less certain.

As suspension feeders, the tube-dwelling worms probably grew in such a way as to keep their food-trapping tentacles well above the sea floor, to avoid clogging by sediment.

- What, in the light of this assumption, was the likely orientation of the brachiopod shell on the sea floor while it was being encrusted? Does this match the expected life orientation of the brachiopod (Section 3.3.5)?

- ☐ Since the worm tubes are either situated near the anterior valve margins or have their aperture facing towards them, it is likely that the anterior part of the shell was facing upwards from the sea floor while the shell was being encrusted. The umbones would thus have faced downwards. As the pedicle in this species appears to have been reduced, or even lost, it is likely that the animal nestled in sediment with its umbones partially embedded (Section 3.3.5), as reconstructed in Figure 5.6.

Figure 5.6 Reconstruction of the life position of FS Q.

Although the shell was probably encrusted while it was in life position, we still cannot be sure if the brachiopod was actually alive at the time. It is still possible that it had died, but remained in place and was only later encrusted by worm tubes preferentially growing on, or towards, its upper surface. (An example of the interaction of a live organism with a dead one was the oyster that grew on a dead ammonite shown in Figure 4.8). Unless we can see evidence for a direct growth response by the brachiopod to the encrusters (not evident in this case), this is as far as we can go with this one specimen. Encrustation of live brachiopods might, however, be inferred if a survey of worm tubes on a large number of specimens showed them to be concentrated around those parts of the valve margins where feeding currents entered.

The study of FS Q has shown that it may be difficult to establish whether or not attached organisms were contemporaneous, and that single specimens may not yield enough information to settle the issue. Merely finding fossils in close association is not enough. Only when there is secure evidence for live interactions can possibilities such as parasitism or competition, for example, be explored.

For a fine example of palaeoecological detective work with which to end this chapter, study Figure 5.7.

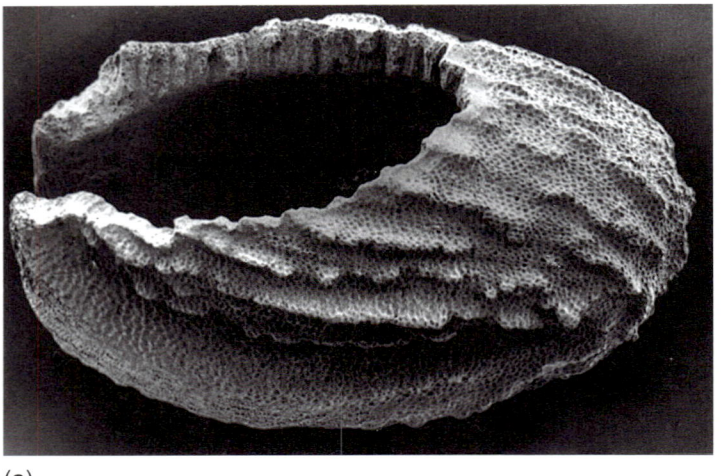
(a)

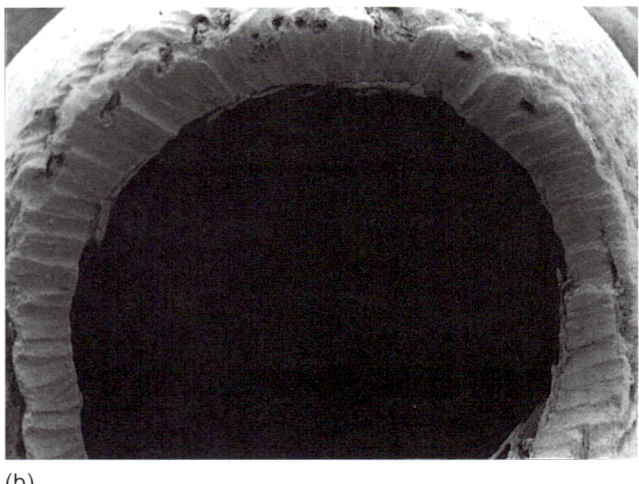
(b)

Figure 5.7 A clear case of animal–plant interaction from the middle Palaeogene, about 35 Ma ago. (a) This fossil seed of the plant *Stratiotes* bears a distinctly shaped hole in its side. The nutritious internal contents of the seed have been removed, leaving the husk. The seed is 5 mm across. (b) A hole (8 mm across) gnawed in a modern seed by a wood mouse, edged with very similar perpendicular grooves to those seen in (a). The grooves are made by the tips of the wood mouse's lower incisors, working from the inside outwards, while it grips the outer wall of the seed with its upper incisors. The strata encompassing the beds with the gnawed fossil seeds yield a diversity of rodents, including two species of the early-evolved dormouse *Glamys* that are of the right size to have caused this damage. One of these dormouse species actually occurs in the same bed as the seed illustrated in (a) and was almost certainly the culprit. All other rodents of this age in UK strata were eliminated on the basis of size, and wood mice had yet to evolve. The ancient dormouse appears to have gnawed the seeds like a modern wood mouse.

Activity 5.1 Fossils as clues to past environments

To round off this and the previous chapters on fossils, you should now watch the video sequence *Fossils as clues to past environments* on DVD 4, which illustrates the links between fossils and sedimentary environments.

5.3 Summary of Chapter 5

1. Palaeoecology and taphonomy together provide evidence for interpreting physical, chemical and biological aspects of past environments. Biases of preservation, however, limit the ecological information available from fossil assemblages. Preservation of a fossil *in situ*, and especially in life position, often allows ambient conditions (such as salinity) to be inferred. The best approach is to combine both palaeontological and sedimentary evidence. For example, changes in the relative frequency and style of both bioturbation and current-generated sedimentary structures may indicate deepening or shallowing of water. Abnormal growth patterns can also be revealing.

2. Reconstructing interactions between fossil organisms is challenging. A key task is to establish whether or not organisms lived at the same time, which can be difficult because of time-averaging. Assemblages preserved by catastrophic burial are thus particularly valuable. Direct evidence for interaction includes recognisable traces on fossils, such as characteristic kinds of damage due to predation, especially if accompanied by examples of repair. Overgrowth (as illustrated by FS Q) still poses the problem of establishing contemporaneity, but this can usually be resolved by studying the distribution of the encrusters in relation to the life habits of the host.

5.4 Objectives for Chapter 5

Now you have completed this chapter, you should be able to:

5.1 Use your understanding of the palaeobiology of organisms to deduce various aspects of the environments in which they lived.

5.2 Outline some of the issues to be resolved when trying to reconstruct ancient communities and interactions between organisms from a study of the fossil record.

Chapter 6   From rocks to sediments

This book has so far mainly looked at various aspects of fossils, including how they form and the morphology, mode of life, habitats and stratigraphic range of major groups of organisms in the fossil record. Many of the observations and inferences obtained from studying fossils and their preservation history can be combined with an understanding of sedimentary processes to produce an overall interpretation of past environments on the Earth's surface. Chapters 6–9 consider sedimentary processes in more detail, and Chapters 10–17 present major sedimentary environments that characterise our planet.

6.1   The weathering process

Most sedimentary deposits generated on land are the product of weathering of rocks that have been uplifted and exposed at the surface.

- What are the deposits called that are derived from the weathering of silicate rocks?
- □ Siliciclastic sediments (see Book 1, Section 7.1).

Agents of weathering range from physical processes such as the freezing of water and thermal expansion and/or contraction, through chemical processes such as dissolution by acid rain, to biological processes such as root penetration. Once broken up, the rock fragments are removed by erosion and accumulate as the familiar sediments of, for example, river systems, beaches and deserts.

6.2   Physical weathering

Physical weathering is a purely mechanical process that can lead to rapid disintegration of rocks. It usually occurs in response to significant temperature changes, particularly in the presence of water. Frost shattering is the most effective process because water, on freezing, undergoes a volume increase of 9%, which can exert a pressure of up to 14 MPa within rock crevices. This pressure is more than enough to shatter granites and other resistant rocks over many cycles of freezing and thawing (Figure 6.1). Another powerful process is thermal

Figure 6.1   A frost-shattered landscape in the Glyder Range of Snowdonia, Wales. The field of view is about 20 m.

Figure 6.2 Large carbonate nodule in a desert showing exfoliation of the outer brown layer. Nodule is approximately 1 m in diameter.

Figure 6.3 Biological weathering: root penetration into the wall of a building, Tung Ping Chau, Hong Kong.

expansion and contraction of rocks in desert regions, where daily temperature differences of up to 60 °C lead to **exfoliation**. This is where thin layers of the rock peel off, much like the outer layers of an onion, hence this process is sometimes also termed onion-skin weathering (Figure 6.2). Exfoliation is aided by the growth of salt crystals in cracks as a result of evaporating overnight dew.

6.3 Biological weathering

Biological weathering can contribute significantly to rock breakdown. Not only can plant roots greatly enlarge cracks and crevices (Figure 6.3), but rock surfaces can also be degraded by lichens – both by physical invasion of fungal strands between mineral grains and by chemical decomposition. Bacteria and fungi play a key role in initiating or speeding up weathering reactions on a nano-scale, as observed by minor changes in feldspar crystal structure when in contact with them. Borers, such as some bivalves, also mechanically weaken the rock structure.

6.4 Chemical weathering

Chemical weathering of rocks most readily occurs when minerals react with water containing dissolved substances, particularly carbon dioxide (Equation 6.1):

$$\underset{\text{carbon dioxide}}{CO_2(g)} + \underset{\text{water}}{H_2O(l)} \rightleftharpoons \underset{\text{hydrogen ions}}{H^+(aq)} + \underset{\text{hydrogen carbonate or bicarbonate ions}}{HCO_3^-(aq)} \quad (6.1)$$

The hydrogen ions, which render the solution slightly acidic, can displace metallic ions within minerals, allowing the ions to be released into solution or to be mobilised in other ways, depending on the water chemistry and climatic factors. Physical disaggregation increases the surface area exposed to such chemical weathering. Plant roots secrete organic acids and also produce carbon dioxide that dissolves in soil water, increasing the soil acidity still further. In lowland areas with a thick soil cover, the predominant form of weathering is likely to be chemical.

6.4.1 Chemical weathering of silicate rocks

Silicate minerals constitute the major components of igneous, metamorphic and siliciclastic sedimentary rocks within the Earth's crust. Their composition and crystal structure control whether they are simply reduced in size or replaced by new minerals, principally clays.

Question 6.1

Briefly summarise the silicate structure of each of the main rock-forming minerals: olivines, pyroxenes, amphiboles, micas, feldspars and quartz.
(*Hint*: Refer to Book 1, Section 4.2 if you are unsure.)

Olivines, pyroxenes and amphiboles have crystal structures that contain metal atoms weakly bonded to their isolated (olivine) or chained (pyroxene and amphibole) silica tetrahedra and are particularly susceptible to chemical weathering. Micas (sheet silicates) and feldspars (framework silicates) are less susceptible, but will degrade nonetheless, albeit more slowly. Iron-rich biotite mica is more susceptible to chemical weathering than iron-free muscovite. Quartz (SiO_2) is highly resistant to chemical weathering because of its strong, interlocking 3D-framework structure and no definite planes of weakness (so no cleavage). The susceptibility to decomposition can also be related to the degree of polymerisation of silica tetrahedra and the order in which silicate minerals crystallise from a melt. Olivine is the least resistant to chemical weathering. It has the lowest degree of polymerisation because none of the silica tetrahedra are linked, and it has the highest temperature of crystallisation. Quartz, by contrast, is one of the most resistant silicate minerals – it has the highest degree of polymerisation (i.e. all the corners of all the silica tetrahedra are linked to other silica tetrahedra), and it crystallises at the lowest temperature (Table 6.1).

Table 6.1 The ratio of tetrahedral (T) sites to oxygen (O) sites for common igneous silicate minerals. See also Book 1, Figure 4.2.

Silicate mineral group	T : O ratio
olivine	1 : 4
pyroxene	1 : 3
amphibole	4 : 11 (1 : 2.75)
mica	2 : 5 (1 : 2.5)
feldspars, quartz	1 : 2

With the exception of quartz, which hardly weathers at all, silicate minerals variously break down to form new clay minerals, insoluble iron oxide, and ions in solution, including silica and metallic ions such as Ca^{2+}, Mg^{2+}, Na^+ and K^+. The clay minerals, such as kaolinite and **illite**, ultimately make up the vast majority of river flood plain and shallow to deep marine inorganic mud deposits and, altogether, account for approximately 60% of the sedimentary rock record.

Question 6.2

Suppose you were examining a sandstone containing 70% quartz and 30% fresh K-feldspar grains, both derived from the same source. What might you infer about the amount of chemical weathering experienced by the source rock, and the resulting sediment?

Due to its resistance to chemical weathering, quartz is the major constituent of sand-grade siliciclastic sediments. Olivine, by contrast, is very rarely found in sedimentary rocks – and then only where there has been little weathering and transport. The susceptibility to chemical weathering provides a useful tool in ascertaining the **compositional maturity** of a sedimentary rock. A rock that consists almost exclusively of quartz (Figure 6.4) would be classed

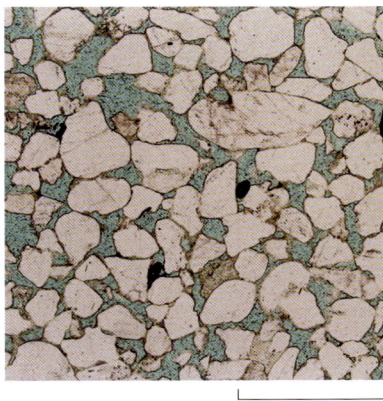

Figure 6.4 Photomicrograph of an Ordovician quartz sandstone from Libya. The spaces between the grains are filled with glue that has been dyed blue.

Figure 6.5 Photomicrograph of an Eocene feldspathic sandstone from East Greenland. The alkali feldspars are artificially stained yellowish-brown.

as compositionally mature, whereas one with a relatively high percentage of feldspar (termed feldspathic) or mafic minerals would be classed as compositionally immature (Figure 6.5).

Activity 6.1 Compositional maturity

In this activity you will assess the compositional maturity of some hand specimens and thin sections, using the Home Kit, Digital Kit and Virtual Microscope.

Mafic minerals (olivines, pyroxenes, amphiboles and biotite mica) contain the reduced form of iron Fe^{2+}, which is oxidised rapidly to Fe^{3+} to form ferric oxide, producing a rusty brown coating on rock surfaces called limonite. In highly oxidising environments (e.g. deserts), ferric oxide may also coat individual quartz grains during or soon after deposition, creating characteristic **red beds** (Figure 6.6). In highly reducing, anoxic (oxygen-deficient) conditions (e.g. waterlogged swamps, or a stagnant seafloor) the sulfide mineral pyrite (FeS_2) may form.

Question 6.3

If basalt, granite and pure quartzite were exposed to weathering for a long period, which of these rocks would be more likely to show a surface coating of iron oxide?

6.4.2 Chemical weathering of carbonate rocks

Limestone ($CaCO_3$), including chalk, is susceptible to chemical weathering by carbonic acid (H_2CO_3), which is a weak acid formed when molecules of CO_2 gas are dissolved in water. Extensive limestone caves and karst systems, formed by dissolving the limestone along joints, are testament to the efficiency of this process (Figure 6.7). The reaction can be expressed as follows:

$$\underset{\text{calcite}}{CaCO_3(s)} + \underset{\text{in rainwater}}{H_2CO_3(aq)} \rightarrow \underset{\text{in solution}}{Ca^{2+}(aq) + 2HCO_3^-(aq)} \quad (6.2)$$

The Ca^{2+} and bicarbonate ions (HCO_3^-) are released into solution and are incorporated via soil water into rivers and, ultimately, the sea.

Figure 6.6 Characteristic red colouration of wind-blown sandstones formed in a Jurassic desert, Red Rock Canyon, Nevada, USA. However, note that red colouration is not exclusive to desert environments.

Figure 6.7 Stalactites and stalagmites in Treak Cliff Cavern, Derbyshire, England. Stalactites form as water drips from the roof of the cave and the carbonate from ions in solution slowly re-precipitates. Stalagmites grow up from the floor of the cavern in a similar manner as drops of water land on the floor of the cavern. Field of view about 5 m.

6.4.3 Rates of chemical weathering

Chemical weathering depends not only on the mineralogy of the parent rock, but also on the rock's texture, climatic conditions, the extent of vegetation cover, and how well jointed or cleaved is the rock. Any features that promote or inhibit the access of aqueous solutions will affect the rate of chemical weathering – so interlocking crystalline textures are usually less susceptible to weathering than fragmental textures, and homogeneous rocks are less susceptible than those with joints, cleavage or fractures. This explains why, in general, crystalline igneous and metamorphic rocks form highland areas long after younger, but less-resistant sedimentary rocks may have been weathered away (Figure 6.8). Chemical weathering is enhanced by warm, wet, tropical conditions with abundant vegetation (which retains moisture and organic acids in the soil).

■ Give two reasons why chemical weathering is enhanced by warm, wet, tropical conditions.

☐ (i) Higher temperatures and abundant water accelerate chemical reactions, and (ii) abundant vegetation increases the availability of organic acids to initiate these reactions.

Activity 6.2 Weathering

In this activity you will consider the mechanisms of weathering and their products.

6.5 Summary of Chapter 6

1. Weathering represents the outcome of interactions between exposed rocks and the atmosphere, water, plants and other organisms. Physical weathering simply fragments rocks, whereas chemical weathering decomposes the minerals within them.

2. Physical weathering is due to mechanical stresses within rocks caused, for example, by water freezing and by plant growth (also part of biological weathering).

3. Chemical weathering requires the presence of hydrogen ions in solution, forming a weak acid. Plant roots can increase this acidity by producing carbon dioxide and secreting organic acids, and microbes can speed up chemical reactions (further biological weathering).

4. The susceptibility of silicate minerals to chemical weathering is related to the crystal structure, the degree of polymerisation and the temperature of crystallisation from a melt. Olivine breaks down most readily and quartz extremely slowly, if at all.

(a)

(b)

Figure 6.8 Different rates of chemical weathering shown by the clarity of the inscriptions on gravestones: (a) negligible chemical weathering of a slate gravestone, dated 1708; (b) relatively rapid chemical weathering of a gravestone made of sandstone with a calcite cement, dated 1878, and 170 years younger than the slate gravestone in (a).

5   The mineralogy of siliciclastic sedimentary rocks is used to determine their compositional maturity, which is a measure of how much chemical weathering they have experienced throughout their history. Quartz-rich sedimentary rocks are classed as compositionally mature, whereas feldspar and mafic-rich sedimentary rocks are classed as compositionally immature.

6   Limestones, including chalk, slowly dissolve in carbonic acid (a weak acid formed from CO_2 dissolved in rainwater), creating karstic terranes, and releasing Ca^{2+} and bicarbonate ions (HCO_3^-) into solution.

7   Rates of weathering are dependent on the mineralogy and texture of the parent rock, climatic conditions, the extent of vegetation cover, and the prevalence of jointing or cleavage.

6.6 Objectives for Chapter 6

Now you have completed this chapter, you should be able to:

6.1 Explain the distinctions between the processes of physical, biological and chemical weathering and recognise examples of each.

6.2 Describe the role of water, carbon dioxide and plants in the weathering of silicate and carbonate rocks.

6.3 Explain how the chemical weathering of silicate minerals takes place and predict, with reasons, the likely chemical weathering products of a rock from its mineralogy.

6.4 Make deductions about the extent of chemical weathering of a siliciclastic sedimentary rock on the basis of its mineralogical composition and degree of compositional maturity.

6.5 Predict the likely rate of weathering of a rock, knowing factors such as its mineralogy, texture, local climatic conditions, and the extent of vegetation cover.

Now try the following questions to test your understanding of Chapter 6.

Question 6.4

Figure 6.1 and Figure 7.19 show different landscapes.

(a) For each, decide whether physical or chemical weathering is the dominant process shaping the landscape, and explain how you have arrived at your conclusions.

(b) Which factors visible in these photographs might enhance the rate at which weathering is taking place?

Question 6.5

Table 6.2 is designed to show the end products of the chemical weathering of the main groups of silicate minerals. The weathering products of olivine are already shown. Complete Table 6.2 to show the weathering products of the remaining groups of minerals.

Table 6.2 The end products of the chemical weathering of the main groups of igneous silicate minerals.

Mineral group	Products in solution	New materials	Residual minerals
olivines	metallic ions, silica	ferric oxide, clays	none
pyroxenes			
amphiboles			
biotite mica			
muscovite mica			
feldspars			
quartz			

Question 6.6

Examine RS 19 (gabbro) and RS 5 (quartzite) in the Home Kit.

(a) Which minerals are present in each of these rocks?

(b) What would be the end products of the chemical weathering of each rock? Make it clear whether the products you describe are in solution, insoluble residues or new materials.

Chapter 7 Sediments on the move

Uplift and subaerial exposure of rocks leads to *in situ* weathering, creating mechanically generated rock fragments, ions in solution and, in the case of silicate rocks, chemically degraded silicate minerals, new clay minerals and residual quartz grains (Figure 7.1). Plant colonisation of the loose debris leads to soil formation – but without this soil cover, the weathered material is more easily dislodged (eroded) and then removed either by fluid flow (which includes water and wind), gravity or ice flow. The continuum of weathering, erosion and transport combine to lower the land surface and to expose yet more rock at the Earth's surface, until a level plain is reached. Most submarine rocks are covered with marine sediment and are therefore protected from erosion, but where the rocks are exposed on the seabed, weathering does take place, albeit at a much slower rate than on land and principally by chemical processes.

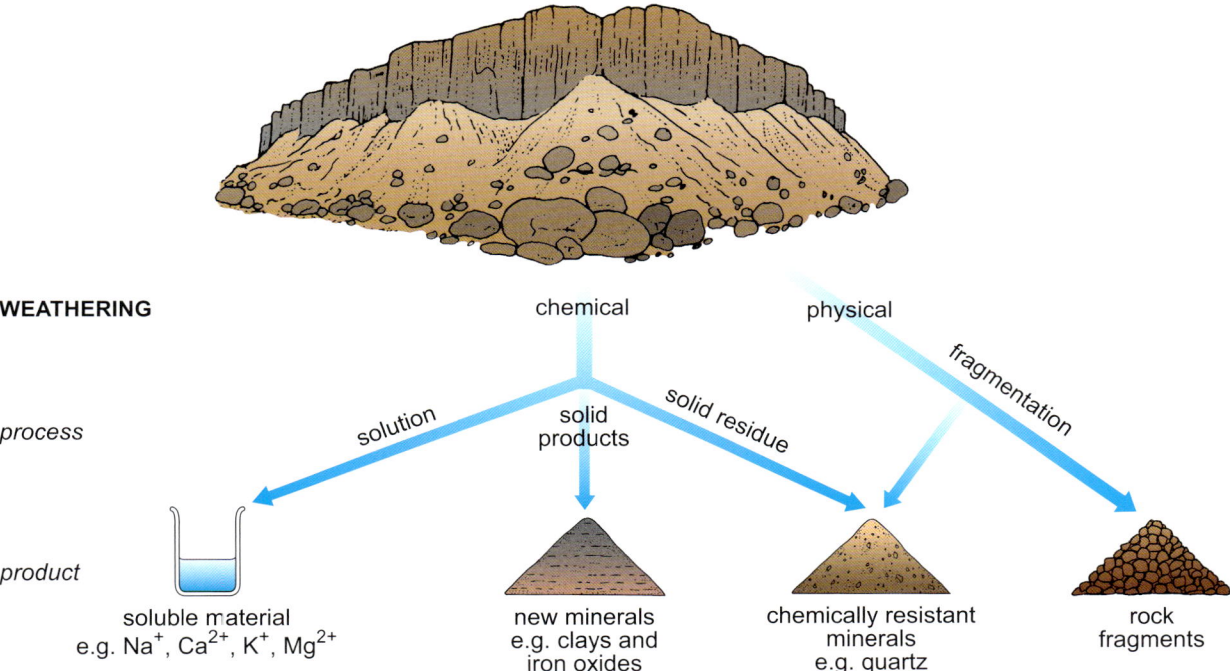

Figure 7.1 A summary of the products of weathering.

7.1 Sediment transport by fluids: water and wind

Most sediment is transported by water or wind. These two fluids possess markedly different viscosities and densities that greatly affect their capacity to transport sediment grains. Water has a density 1000 times greater than air and a viscosity 50 times greater than air, so it is able to transport much coarser grains than the wind – this is why pebbles are common in a river channel or on a beach, and sand predominates in a desert environment (where grains rarely exceed 0.5 mm in diameter).

■ Suggest three factors that control the size of grain that can be moved within a fluid.

□ How fast the fluid is moving, the density contrast between the fluid and the grain, and the shape of the grain.

7.1.1 Laminar versus turbulent flow

The type of fluid flow is another factor that has a particular bearing on *how* the grains move within the fluid:

- **laminar flow** occurs when the fluid moves as a series of parallel layers or laminae (Figure 7.2a)
- **turbulent flow** occurs when the fluid moves in a more chaotic fashion, creating complex eddies (Figure 7.2b).

A familiar analogy would be smoke from a fire in still conditions. The smoke begins to rise in parallel streams (laminar flow) but as it drifts higher into a light breeze, the smoke forms a pattern of swirls (turbulent flow). A similar effect can be seen in flowing water injected with dye (Figure 7.2c).

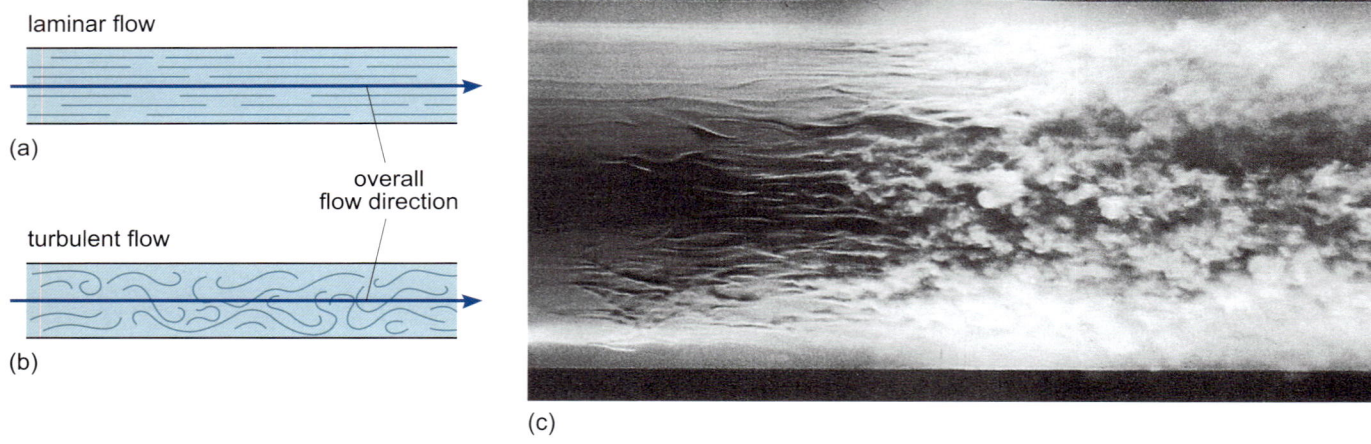

Figure 7.2 The movement of water during (a) laminar and (b) turbulent flow. (c) The transition from laminar to turbulent flow as water, injected with dye, flows over a plate.

Both the speed and viscosity of the fluid determine whether the flow is laminar or turbulent. The likelihood of turbulent flow increases as the speed increases and the viscosity decreases.

■ For a given speed, which of the following fluids is most likely, and which is the least likely, to exhibit turbulent flow: air, rhyolitic lava or water?

□ Air is the most likely to be turbulent because its viscosity is the lowest; rhyolitic lava is the least likely to be turbulent because its viscosity is the highest.

So, for a mass of sediment being transported by a river in which the flow is mostly turbulent, the largest grains will roll or slide along the bottom, whereas smaller grains will bounce along the bottom and ricochet off other grains by a process called **saltation**. All of these grains will form part of the **bedload**. In addition, the smallest grains will be carried along in suspension, caught up in turbulent eddies, rarely touching the bottom, and so forming part of the **suspension load** (Figure 7.3). The actual size of grain that can be carried by these different processes will largely depend on the speed of the current. Clay- and silt-sized grains (<62.5 μm) are almost always carried in suspension, whereas coarser grains will be carried in suspension only at high flow speeds.

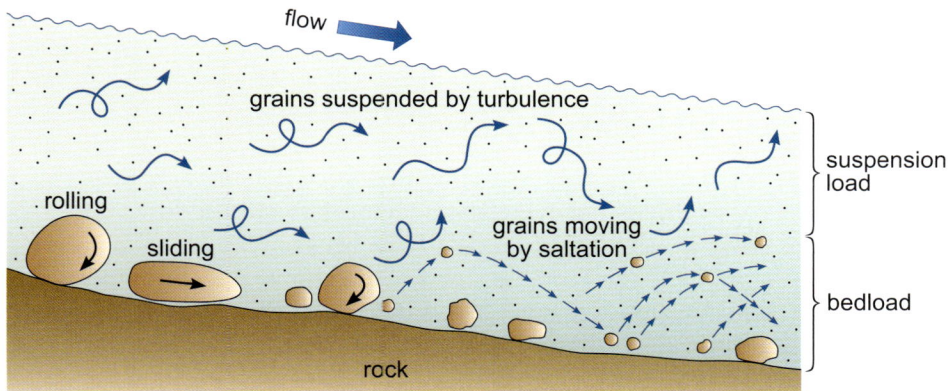

Figure 7.3 Cartoon to illustrate the different forms of particle movement in turbulent water.

Wind-blown (aeolian, pronounced 'ay-owe-lee-an') sediment travels in much the same way as it does in water, but because air has a lower viscosity, and therefore a much lower carrying capacity than water, air transports only significantly finer grains – perhaps just as well for anyone caught in a sand-storm! The lower carrying capacity of air is also partly due to such properties as buoyancy forces and frictional drag, as well as its lower density compared to water. Most of the bedload travels by saltation (pebbles are too large to be moved, even by rolling and sliding) and grains larger than 0.1 mm (100 μm) are rarely carried in permanent suspension.

The process of saltation is more energetic in air than in water as bedload encounters less frictional resistance in air. Significant 'splash up' occurs each time a grain collides with the grain bed, and 'splash down' nudges the grounded grains downwind by a process called **surface creep** (Figure 7.4). Such energetic

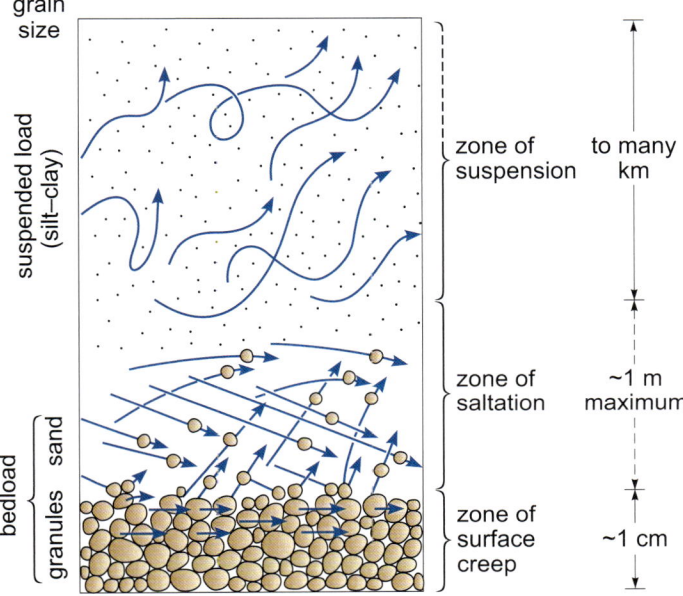

Figure 7.4 Cartoon to illustrate the movement of sand grains in suspension, by saltation and by surface creep during aeolian transport.

(a)

(b)

Figure 7.5 Contrasting surface texture of quartz sand grains. (a) Water-transported quartz sand grains about 1 mm in diameter. The grains are angular to subangular in shape and have a glassy surface appearance. (b) Wind-blown quartz sand grains about 0.5 mm in diameter. The grains are mostly well rounded and have a dull, frosted surface appearance.

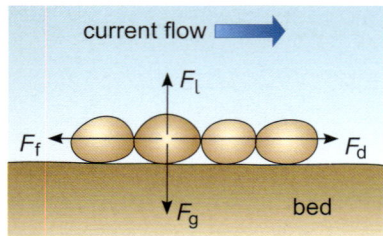

Figure 7.6 The forces acting on a sediment grain at rest on a bed beneath a water current. The force of gravity (F_g) acts to hold the grain down and the frictional force (F_f) impedes motion. However, the forwards push of the water (F_d) and the lift force (F_l) produce a net fluid force that both lifts the grains and propels them forwards.

collisions mean that wind transport is very effective at abrading and rounding both the surface of the sediment grains and any other surfaces that the grains impact. This is in contrast to the cushioning effect of water as grains impact on each other as they move within water. As a result, aqueous environments are often characterised by relatively angular grains with glassy surfaces, whereas aeolian environments are characterised by well-rounded grains with frosted (pitted) surfaces that have a very dull appearance (Figure 7.5).

7.1.2 Initiating sediment movement

Sediment grains begin moving only when the fluid force flowing over the grains (Figure 7.6) is sufficient to overcome both the gravitational force that holds the grains down (F_g), and the frictional force between the grains and their substrate (F_f) (Figure 7.6). The fluid force acts both to move the grains forwards (F_d) and to lift them upwards (F_l), much like an aircraft during take-off. A relatively rough surface and coarse-grained sediment trigger turbulent eddies that can assist this process.

The threshold speed at which grains of various sizes begin moving in water, i.e. when physical erosion begins, can be determined experimentally and plotted using logarithmic rather than linear scales, to reflect the wide range of possible current speeds and grain sizes (Figure 7.7a). Current speed is affected by turbulence, which increases with depth of water, so for a flow 1 m deep, current speeds just above 0.2 m s^{-1} are required to move sand grains 1 mm in diameter (Figure 7.7b). Deeper flows are usually more turbulent and require slightly slower current speeds to move the same grain size, because turbulence increases the amount of lift exerted on individual grains.

- Using Figure 7.7b, what is the finest average grain size that can be transported as bedload by a current of average speed 3 m s^{-1}?

- The average grain size would be approximately 3 mm in diameter, i.e. fine gravel grade.

Activity 7.1 (optional)   Reading logarithmic (log) scales

If you are unfamiliar with log scales you should complete this activity.

Surprisingly, the water current speeds needed to set in motion clay-sized grains (i.e. mostly clay minerals <2 μm across) are comparable to those required to erode pebbles, and significantly higher than those required to erode non-cohesive sand and silt grains (Figure 7.7a). This is because electrostatic forces and surface tension effects cause the clay minerals to clump together so that they behave like larger grains. Sediments with only 5–10% of clay will begin to be cohesive and, in addition, smaller grains can be sheltered by larger grains, making the smaller grains difficult to entrain (pick up) in the current.

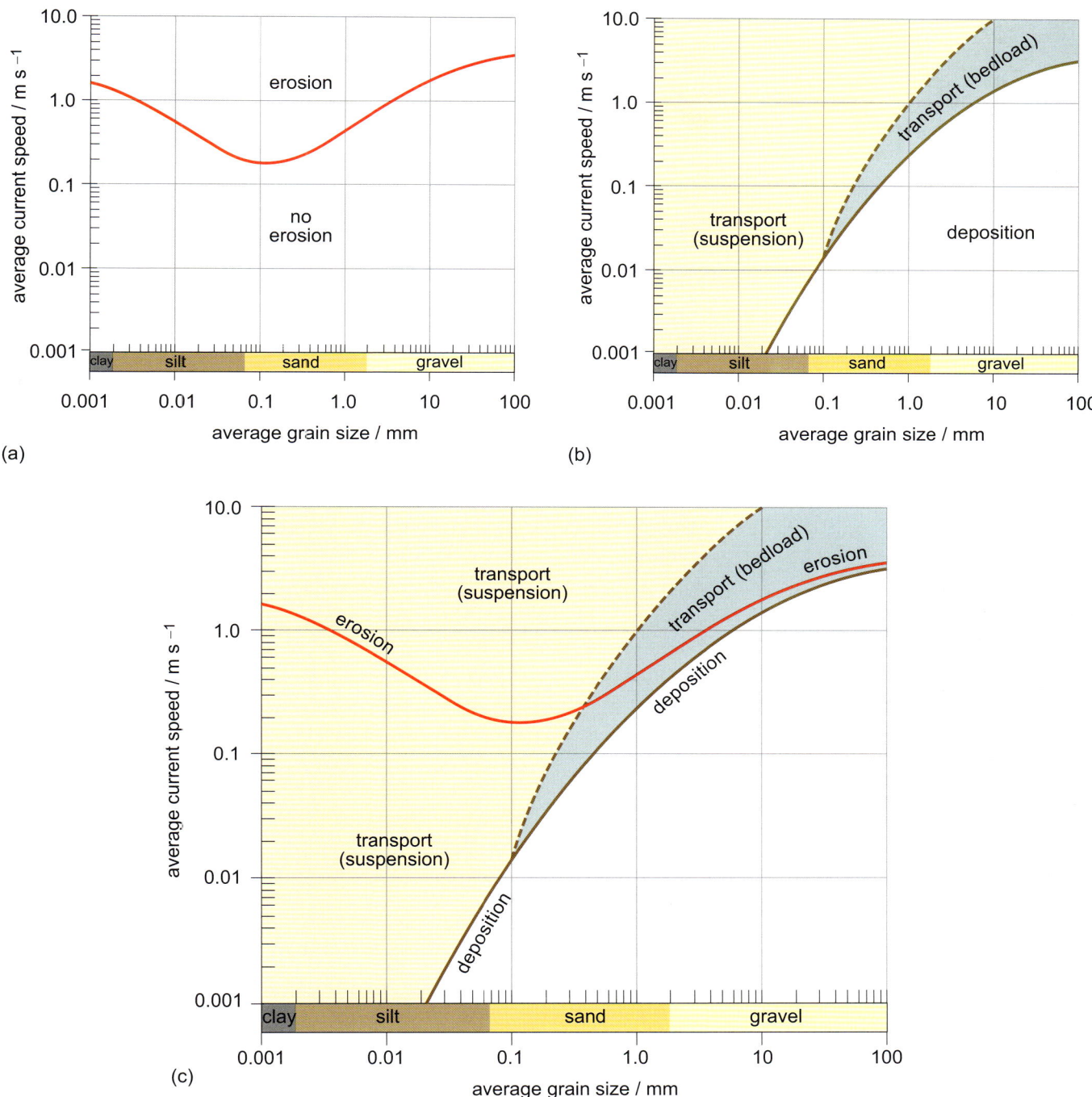

Figure 7.7 (a) Graph showing the range of average current speeds at which sediment grains of different sizes are eroded, i.e. set in motion for a water depth of 1 m. The curve for sediments finer than about 0.1 mm is for relatively uncompacted silts and muds. (b) Graph showing the range of average current speeds at which sediment grains of different sizes are transported in suspension or as bedload, and below which they are deposited. The brown broken line indicates the transition between bedload and suspension load transport. (c) A combination of (a) and (b). Note that grains can be transported below as well as above the red 'erosion' curve and that deposition of fine-grained sediments occurs only at very slow current speeds.

7.1.3 Transporting sediment

Once sediment grains are set in motion, they are transported either as bedload or as suspension load. Sands and gravels move initially as bedload, but they can be lifted into suspension if the water current speed reaches the suspension threshold shown in Figure 7.7b. The boundary between bedload transport and suspension transport is gradational, but the boundary between transport and deposition is sharp, and is triggered when the current velocity falls below the threshold for keeping the grains moving. Only very slow current speeds are required to keep the finest grains in suspension and clays can be transported great distances before being deposited.

Figure 7.7c superimposes the erosion curve (Figure 7.7a) onto the curves for transport and deposition (Figure 7.7b). This combined plot demonstrates that sediment can be transported at speeds below that required to erode it.

■ What is the reason for this?

☐ Sediment requires less energy to keep it in motion than it does to set it in motion.

The disparity between transport and erosional velocities decreases with increasing grain size:

- above 1 mm average grain size, the deposition curve lies just below the erosion curve, whereas below 0.1 mm average grain size, there is a widening separation between the two
- below 0.1 mm average grain size, sediment grains are taken directly into suspension, but above 0.1 mm average grain size, their position (suspension versus bedload) depends on the water current speed.

■ With reference to Figure 7.7a, what is the approximate water current speed required to set in motion grains 2 mm in diameter?

☐ between 0.6 m s^{-1} and 0.7 m s^{-1}.

■ With reference to Figure 7.7b, if the water current speed increases to 10 m s^{-1}, will the transported sediment (average grain size 2 mm) be part of the bedload or part of the suspension load?

☐ Part of the suspension load.

Aeolian transport demonstrates a similar, direct relationship between wind speed and grain size, but requires higher current velocities to move sediment grains because of the lower density and viscosity of air. Gale-force winds reaching 20 m s^{-1} are needed to lift sand grains 1 mm in diameter into suspension. Gravel coarser than 4 mm in diameter would not be moved at all and instead would form a residual concentration or **lag deposit** from which finer grains would be removed by **winnowing**. This inability to entrain coarse grains means that wind-blown sediments are usually better sorted than their aqueous counterparts.

7.1.4 Abrasion during transport

During transport, grains collide and the impacts lead to progressive rounding of grains, as angular corners and sharp edges are worn down. (Note that rounding should not be confused with sphericity, which relates to the overall shape of the grain, and how close it is to a sphere; Book 1, Figure 7.6.) The cushioning effect of water means that abrasion is much less marked in aqueous environments than in aeolian ones, and because higher speeds are needed to move grains of a given size in air compared to water, faster impacts lead to deep pitting or frosting of aeolian quartz grains. These effects give rise to a dull/matt appearance (Figure 7.5b) that is quite distinct from the glassy, much shinier grains typical of aqueous transport (Figure 7.5a). With prolonged transport in either environment, grain rounding becomes more marked: angular or subangular grains become subrounded or rounded during aqueous transport, and well rounded during aeolian transport.

During bedload transport in water, larger sedimentary grains are rounded more quickly than smaller ones because their greater momentum means that they are less easily cushioned by water. Freshly weathered, angular rock fragments can become subrounded boulders and pebbles within a few tens of kilometres of river transport, whereas sand-grade quartz grains would require many hundreds of kilometres of transport to become similarly subrounded. Once the grains reach the sea, wave motion causes the grains in relatively shallow water to be moved to and fro over the equivalent of hundreds of kilometres. Nevertheless, the sand-grade quartz grains are frequently still no more than subrounded or rounded, due to the cushioning effects of water, whereas the beach pebbles are well rounded.

It is worth bearing in mind that inherited characteristics can be misleading when making inferences about the environment based on textural clues. For example, if aeolian sand grains were reworked into a later river system, pitted aeolian grains would not lose their surface frosting, nor change from well-rounded to subrounded grains.

- ■ If aeolian sand grains had been recycled in this way, what other textural clue might be present in the sedimentary rock to enable the correct interpretation of water-transported deposits?

- □ The frosted, well-rounded aeolian grains would be mixed in with more angular, glassy quartz grains that were transported by the water current.

Activity 7.2 Interpreting sedimentary rock textures

In this activity you will interpret sedimentary textures from hand specimens in the Home Kit and thin sections under the Virtual Microscope.

7.2 Erosional structures

Most erosional sedimentary structures are formed either by aqueous and sediment-laden flows or by objects striking the sediment surface during transport. The most common structures are channels and scours, since these occur in nearly all environments (see Section 7.2.1). Flute casts are distinctive erosional structures that occur on the undersides of beds deposited by turbidity currents (turbulent bottom currents) and these are described in Section 7.6.3. Linear ridges found on the underside of sandstone units represent infilled grooves cut into the underlying mudstone by objects carried along by the current. The grooves are generated both by turbidity currents in deep-water settings and by storm currents in shallow marine environments. In glacial settings, erosion caused by moving ice produces characteristic linear grooves called striations (Section 7.8).

7.2.1 Channels and scours

Channels occur on a scale of a metre or more (Figure 7.8), rarely even kilometre scale, whereas scours occur on a centimetre to decimetre scale (Figure 7.9). Both structures show sharp, often irregular contacts with the underlying strata, a concave-up profile and evidence of the removal of part of the underlying material. Channels are generally sites of sediment erosion and transport for appreciable periods of time, whereas scours are usually gouged out during a single high energy event and filled almost immediately. Both channels and scours are usually filled with coarser-grained sediment than the underlying or adjacent strata, and channels frequently display a lag conglomerate at the base. Channels are much longer than they are wide, whereas scours may be only slightly elongate in plan view.

Figure 7.8 The reddish-coloured sedimentary deposits with angular blocks forming the top half to one-third of the cliff is part of a filled Quaternary glacial meltwater channel. The channel has cut down into marine Carboniferous strata, Howick Haven, Northumberland, England. Height of cliff is about 8 m.

Figure 7.9 Sandstone-filled scour cutting down into more muddy sedimentary rocks within the Carboniferous strata exposed near Scremerston, Northumberland, England. Note pencil for scale.

7.2.2 Gutter casts

Gutter casts are distinctive erosional features, comprising V- or U-shaped, straight to slightly sinuous, isolated tubes about 10–30 cm across and almost as deep, and up to several metres in length (Figure 7.10). They are attributed to localised, spiral-shaped eddies generated by storm or fluvial currents and their infills are usually coarser grained than the underlying or adjacent strata.

Figure 7.10 Three gutter casts (indicated by the white arrows) preserved in Middle Jurassic sedimentary rocks of fluvial origin, Burniston, Yorkshire, England. Field of view about 1.5 m.

7.3 Sediment deposition

Sediment is carried along in suspension or as part of the bedload until the current velocity falls below the transport threshold, allowing **deposition** to take place. The depositional curve in Figure 7.7 is based on experiments with glass spheres, but clearly few natural sediment grains are spherical, except perhaps some aeolian grains and marine carbonate grains (ooids – see Book 1, Section 7.2). Instead, grain shapes range from sub-spherical to tabular. Micas in particular form plate-like flakes that offer a lot more resistance to water or air as they begin to settle. As a result, they settle out more slowly than sub-spherical grains of quartz of the same grain size. Mica grains of a particular size will therefore tend to be associated with sediments that are finer-grained. The grain shape, density and transport history of different grain types will thus affect the sorting characteristics of the sediment that is deposited.

Figure 7.11 Normal grading in Palaeogene sandstones, Isle of Wight, England. Note the sharp base and gradual reduction in grain size in the centre portion of the photograph in the direction of the arrows.

Figure 7.12 Inverse grading in Quaternary gravels, Ottawa, Canada. There is a decrease in the grain size in the direction of the arrows.

■ For a flow 1 m deep, a water current speed of around 0.4 m s^{-1} is required to set a sand grain 1.0 mm in diameter in motion (Figure 7.7a). If the current speed increases and then decreases, will the grain be deposited as soon as the speed drops to 0.4 m s^{-1} again?

□ No, it will have to drop *below* this speed, to almost 0.2 m s^{-1} (Figure 7.7b).

A gradual decline in the current velocity will cause the gradual deposition of progressively finer-grained sediment, creating **normal grading** within the depositional unit (Figure 7.11); such a bed can also be described as fining upwards. **Inverse grading** occurs where increasingly coarse-grained sediment overlies finer sediment within the same depositional unit (Figure 7.12), due to buoyancy forces acting within the flow.

After taking into consideration the cohesiveness of sediments and the shape of individual grains, it is clear that grain size and the speed of the transporting fluid are directly related. Thus gravel-sized sediment requires much higher energy conditions to move it than does sand-grade sediment. Broadly, gravel-sized deposits can be equated with high energy conditions, medium to coarse sand-grade sediment with medium energy conditions, and fine sand to mud-grade sediment with low energy conditions. Mud will only settle out of suspension in extremely calm conditions, with virtually no current activity. These relations provide a *qualitative* measure of the **energy of the depositional environment**.

Activity 7.3 Suspension load, bedload and energy of the environment

In this activity you will attempt to interpret transport processes from grain size in hand specimens from the Home Kit.

7.3.1 Grain packing, sorting and textural maturity

Grains that are well sorted, well rounded and of similar grain size, are described as **texturally mature**. This type of sediment will pack together so that the grains are in point contact with each other (like marbles in a jar) to create a grain-supported fabric (Figure 7.13a). Such fabrics possess up to 40% **porosity** (the amount of pore/void space) and high **permeability** (the ability of fluids to move through the pore system, which is related to the degree of connectivity between adjacent pores). Finer grains fill in the spaces (matrix) between larger (framework) grains, leading to closer packing of the framework grains, while retaining a grain-supported fabric. Grains that are angular, poorly sorted, and of differing grain size are described as **texturally immature**. In these types of sediment, packing becomes less predictable. The addition of further depositional matrix can give rise to matrix-supported fabrics, whereby larger grains are no longer touching, but instead are supported by the finer-grained matrix (Figure 7.13b).

Textural maturity reflects the sediment source, the energy of the environment and the conditions during transport – for example wind versus water transport. Prolonged transport tends to round the grains due to abrasion and to increase grain sorting, often by removing the finer grains and leaving behind the coarsest grains. Of course, erosion of a texturally mature sandstone will create a texturally mature sediment, no matter how far the grains are transported.

■ Consider two sandstones, one with a prolonged history of aeolian transport and the other with a rapid history of aqueous transport. Which sandstone should be texturally less mature, and why?

☐ The sandstone transported rapidly in water will be texturally less mature because there has been less time for grain sorting, or for grain rounding, particularly in view of the cushioning effect of the water.

There is considerable variation in textural maturity because of the range of processes operating in different environments and the variety of source material. Wave processes on a beach, for example, lead to a very high degree of sorting, whereas fluvial currents may produce little rounding or sorting.

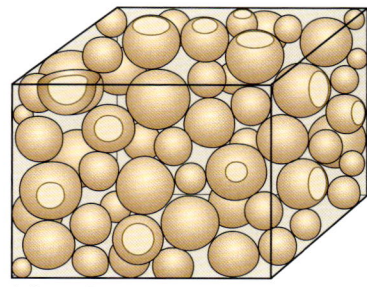
(a) grain supported

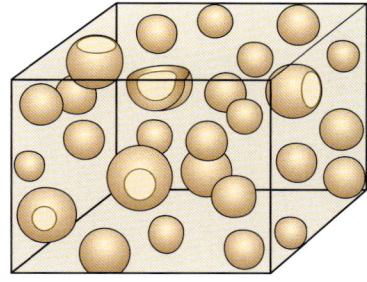
(b) matrix supported

Key
● grains   ☐ matrix
○ projecting grain cut along face of cube

Figure 7.13 (a) Grain-supported fabric and (b) matrix-supported fabric in sediments and sedimentary rocks.

Activity 7.4 Textural maturity

In this activity you will examine textural maturity in hand specimens using the Home Kit.

Whilst compositional maturity (Section 6.4.1) and textural maturity often go hand in hand, this is not always the case. Aeolian sands are invariably well sorted and rounded (i.e. they are texturally mature) but they may contain significant amounts of feldspar, albeit rather degraded (i.e. they are compositionally immature). This disparity is often observed in arid regimes, where the scarce supply of water retards the chemical weathering process.

Activity 7.5 Grain transport and sorting

In this activity, you will answer questions about the mechanisms of sediment sorting during transport.

Activity 7.6 Describing and interpreting siliciclastic sedimentary rocks

In this activity, you will describe the composition and texture of sedimentary rocks and interpret transport and depositional processes, using specimens from the Home Kit, Digital Kit and the Virtual Microscope.

Textural and compositional data can provide clues to the transport and depositional processes and even the environment(s) in which the sediments were deposited. Stepping up from the study of grains to larger-scale features, further evidence of the depositional processes is provided by the nature of the beds themselves and, in particular, the various forms of layering.

7.3.2 Bedding and lamination

Bedding (layers thicker than 1 cm, e.g. Figure 7.12) and **lamination** (layers thinner than 1 cm, e.g. Figure 7.14) are the terms used to describe layering in sediments, whereas **stratification** is the general term used when no specific reference to layer thickness is intended. The same is true for the term **cross-stratification**, which is used for layers of any thickness that are inclined to the principle, and originally horizontal, depositional surface (Figure 7.15), encompassing cross-lamination (layers <1 cm) and cross-bedding (layers >1 cm). In all of these examples, the layering is picked out by changes in grain size, orientation of elongate grains, and grain composition and/or colour. These variations reflect changes in depositional process, sediment source, and environment.

(a)
10 cm

(b)
10 cm

Figure 7.14 (a) Lamination formed by clay minerals and organic matter settling from suspension within Jurassic mudstones near Whitby, Yorkshire, England. (b) Planar stratification (in this case lamination-scale, so it could also be termed planar lamination) formed by traction currents moving the bedload in Carboniferous strata, Northumberland, England (see Figure 7.21 for a block model of planar stratification).

Horizontal, parallel laminations are almost the only structures produced by suspension settling (Figure 7.14a). A similar sedimentary structure of parallel horizontal layers of sediment can build up by deposition of bedload from traction currents moving coarser grains – this is termed **planar stratification** (Figure 7.14b). This structure can be either at a lamination scale or at bedding scale. Deposition from bedload can give rise to a wide range of sedimentary **bedforms** (the 3D shape of the surface of beds), which produce structures such as planar stratification and cross-stratification – where depositional surfaces usually dip downcurrent.

7.3.3 Grain orientation

Grain settling during deposition can generate distinctive fabrics. Platy grains, for example clay minerals and mica flakes, usually settle with the flakes parallel to bedding or lamination, accentuating the depositional fabric. Elongate grains may also be aligned by the current parallel to the transport direction. In matrix-supported sedimentary deposits, gravel-sized clasts can also be aligned parallel to bedding,

Figure 7.15 Cross-stratification in Carboniferous sandstones, Northumberland, England. Compass is 10 cm long.

but in grain-supported gravels transported by fast-flowing rivers, pebbles are often orientated with their long axes dipping *upstream*, creating a fabric known as **imbrication** (Figure 7.16) that provides evidence of flow direction.

7.4 Aqueous bedforms

Depositional bedforms created in water include those produced by unidirectional or bi-directional currents and those produced by fairweather waves (during normal, fairweather conditions) or the larger and more energetic storm waves (produced during storms).

7.4.1 Unidirectional current bedforms

Current-formed ripples form in fine- to medium-grained sands in response to low flow velocities of unidirectional currents. They form in a range of environments, including rivers, estuaries, tidal flats, deltas and shallow marine environments. They are small-scale bedforms with wavelengths of up to a few tens of centimetres and ripple heights of a few centimetres (Figure 7.17). Turbulent eddies in the current sculpt the sediment surface into a series of straight- or sinuous-crested ripples, and sand grains are carried both in the bedload and suspension load up the gently inclined **stoss** side of the ripple,

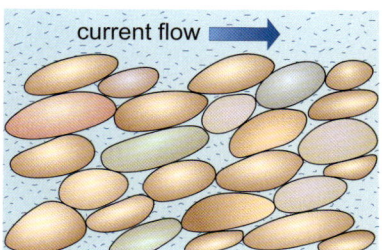

Figure 7.16 Cartoon to show imbrication where the clasts are oriented with their long axes dipping the opposite way to the current direction, i.e. dipping upstream.

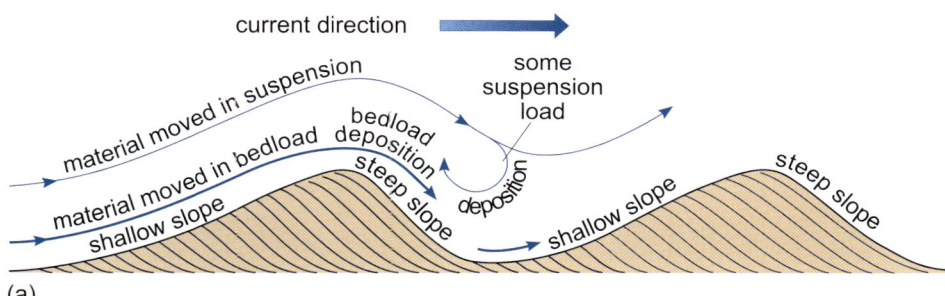

Figure 7.17 (a) The movement and deposition of sediment associated with a current-formed ripple, showing how erosion on the shallow (lee) slope and deposition on the steep (stoss) slope lead to the migration of the ripple in the direction of current flow. (b) Large current-formed ripples (crests oriented bottom left to top right) with smaller ripples (at right angles and horizontal on photograph) superimposed upon the larger ripples.

to avalanche down, and be deposited on the steeper **lee** side of the ripple (Figure 7.17a). The ripples thus move along in the same direction as the current, with the ripple crests at right angles to the flow direction. Current-formed ripples are characteristically asymmetric in cross-section and, as they migrate downcurrent, the front of the ripple is buried beneath successive avalanches of sand grains. This process creates a cross-laminated internal structure and successive laminae mark the former positions of the depositional lee slope at the front of the ripples (Figure 7.17a).

■ How would you explain the two sets of ripples in Figure 7.17b?

☐ The larger set appears to be parallel to the shoreline and is likely to be deposited by the incoming tide (steeper, depositional slopes on the shoreward side). The smaller set is almost at right angles to the larger ripples, and is likely to be the product of wind-driven currents in very shallow water just before low tide, or shore-parallel (longshore) currents operating when the tide is high. Multiple (usually two) ripple sets are often referred to as **interference ripples**.

Figure 7.18 Curved-crested ripples formed by water flowing over a beach, which have then broken up into a series of tongue-shaped or linguoid ripples.

Experiments within flume tanks (artificial channels in a laboratory) demonstrate that with increasing flow velocity, straight-crested ripples become increasingly sinuous until they eventually break up into a series of tongue-shaped or linguoid ripples (Figure 7.18). As the current speed increases still further, the bedforms increase in size, creating subaqueous **dunes** more than a metre high and with wavelengths of several metres (Figure 7.19). These dunes still possess a cross-stratified internal structure, though on a bed-scale rather than laminar-scale.

Figure 7.19 Subaqueous dunes exposed in an estuary at Bantham, South Devon, England.

The size of the bedforms depends not only on the current speed, but also on the depth of water. Still higher flow velocities will eventually flatten the subaqueous ripples or dunes to create a planar surface that builds up as a series of planar laminae or beds. The surfaces of these beds are usually characterised by **current lineations** oriented parallel to the flow direction; these are created by subtle size sorting of coarser sand grains into millimetre-scale ridges (Figure 7.20).

The continuum of bedform size and shape in response to increasing current velocity changes from planar laminae to ripples to dunes to planar beds and culminates in low-relief, undulating mounds. These mounds migrate upcurrent due to erosion on the downcurrent side of the mounds (Figure 7.21). Confusingly, the cross-stratification is inclined in the opposite direction to the current flow, hence the name **antidunes**. These are rarely preserved in the geological record because they are rapidly reworked as the current speed wanes. The presence of antidunes in a modern environment can sometimes be inferred from the development of standing waves in a fast-flowing stream (i.e. waves that appear to stand still within the current) or even waves that move upstream.

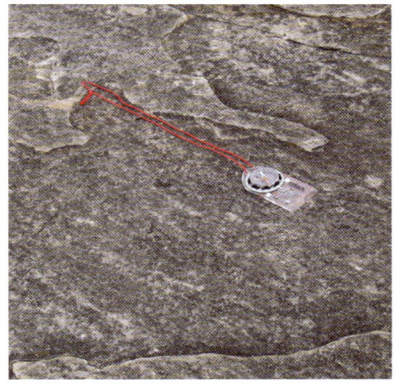

Figure 7.20 Current lineations on the bedding surface of a Carboniferous sandstone bed, Howick, Northumberland, England. The lineations run from bottom right to top left of the photograph parallel to the cord of the compass-clinometer. The compass-clinometer is 10 cm long.

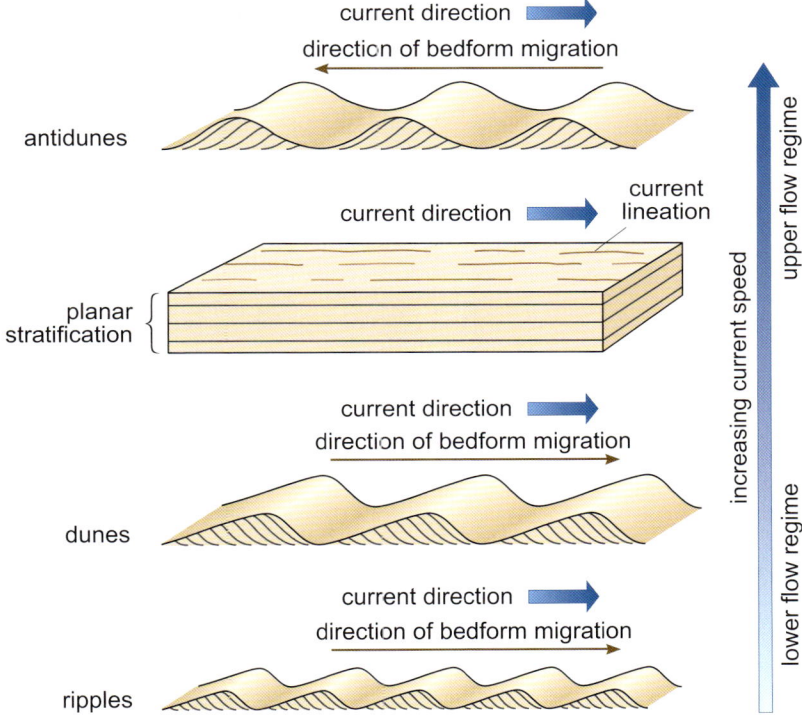

Figure 7.21 The succession of bedforms developed in sand grains 0.5 mm in diameter in response to increasing current speed in a flow around 0.4 m deep. Cross-sections through the bedforms reveal the type of cross-stratification that is preserved, and its direction of inclination in relation to the direction of current flow.

In order to distinguish planar stratification produced at very low current velocities (by suspension settling) from planar stratification produced at much higher current velocities that remove earlier ripples or subaqueous dunes, the terms

lower flow regime and **upper flow regime** are used respectively (Figure 7.22). Only planar stratified beds from the upper flow regime have current lineations (Figure 7.20). The actual bedforms produced are dependent not only on current speed, but also on grain size and water depth. Flume experiments at water depths of about 0.25–0.4 m demonstrate the relation between current velocity and bedform (Figure 7.22).

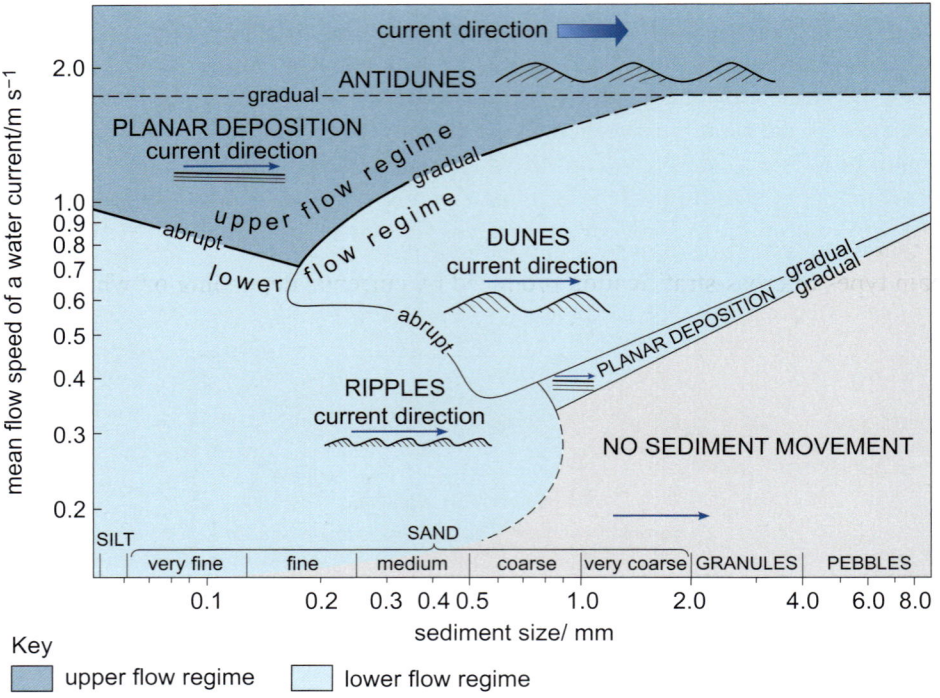

Figure 7.22 The relationship between current speed, grain size and subaqueous bedforms produced beneath a flow of around 0.25–0.4 m depth. The elements of this figure are illustrated by a series of video clips in the 'Processes and environments' section of DVD 5. The area designated upper flow regime is shaded dark blue and the area designated lower flow regime is pale blue; note also the logarithmic scales.

■ For the flow depths shown in Figure 7.22, what is (a) the coarsest grain size from which ripples will develop, and (b) the finest grain size from which subaqueous dunes will form?

☐ (a) Approximately 0.9 mm (coarse sand), and (b) just under 0.2 mm (fine-grained sand).

Figure 7.22 illustrates a number of important points. Ripples can form in silt to coarse sand-grade sediment in water current speeds of up to 1 m s^{-1}, i.e. across the whole of the field delineated by the abrupt boundary with the upper flow regime, dunes and no sediment movement. Coarser-grained sediment gives rise to planar deposition and then to subaqueous dunes at comparable current speeds to those forming ripples in finer-grained sediment. The transition from lower to upper flow regime for silt to fine-grained sand is abrupt, but the same transition for coarser-grained sediment is more gradual. The transition from upper flow regime planar stratification to antidunes is also gradual.

The influence of water depth on the development of bedforms is important. At depths of 0.25–0.4 m (Figure 7.22), subaqueous dunes develop in fine-grained sand or coarser-grained sediment, but the height of these dunes is restricted by the shallowness of the water. In water depths greater than several tens of metres, subaqueous dunes may reach heights of up to 18 m and these are termed **sand waves**. For a given flow speed and grain size, upper flow regime conditions are more likely to prevail in shallow water than in deep water.

The direction of dip of cross-stratification provides evidence of **palaeocurrents**, since the cross-strata dip downcurrent (except in rare antidunes). Large numbers of measurements are required to establish the principal current direction and it is important to understand the exact nature of the bedforms under investigation so that maximum dips rather than apparent dips are measured. There are two main types of cross-stratification produced by currents, depending on whether the ripple or dune has a straight crest or a curved one (Figure 7.23). In order to establish the type of cross-stratification, it is necessary to examine the sedimentary structure in two faces that are roughly at right angles to each other. The migration of straight-crested ripples and dunes produces **planar cross-stratification**. This structure has inclined beds in the faces orientated at right angles to the wave or ripple crest (i.e. parallel to the flow) and horizontal, parallel layers that look similar to planar stratification at right angles to the flow (Figure 7.23b). In contrast, curved-crested ripples or dunes of any shape produce a sedimentary structure that has inclined surfaces parallel to the current flow and nested troughs at right angles to the flow (Figure 7.23a). These troughs give this type of cross-stratification its name, **trough cross-stratification**. The orientation of the bedform needs to be considered when measuring the downcurrent direction to establish the palaeocurrent direction.

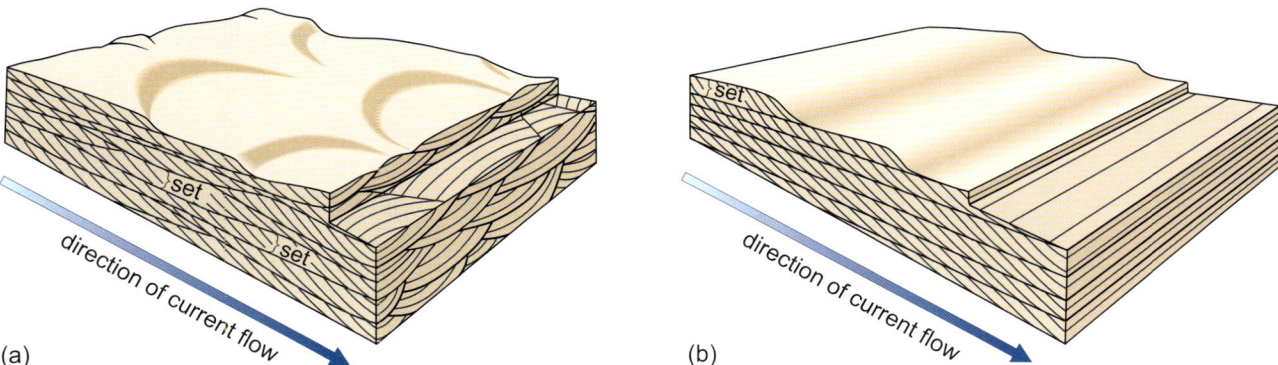

Figure 7.23 Idealised block diagrams showing the two types of cross-stratification. (a) Trough cross-stratification resulting from the preservation of curved-crested (or sinuous) bedforms. Note that the exact geometry of the sedimentary structure will vary depending on continuity and sinuosity of the dune or ripple crest. (b) Planar cross-stratification resulting from the preservation of straight-crested bedforms.

Palaeocurrent data can be plotted as directional rose diagrams, with the greatest concentration of readings corresponding to the principle direction of current flow (Figure 7.24).

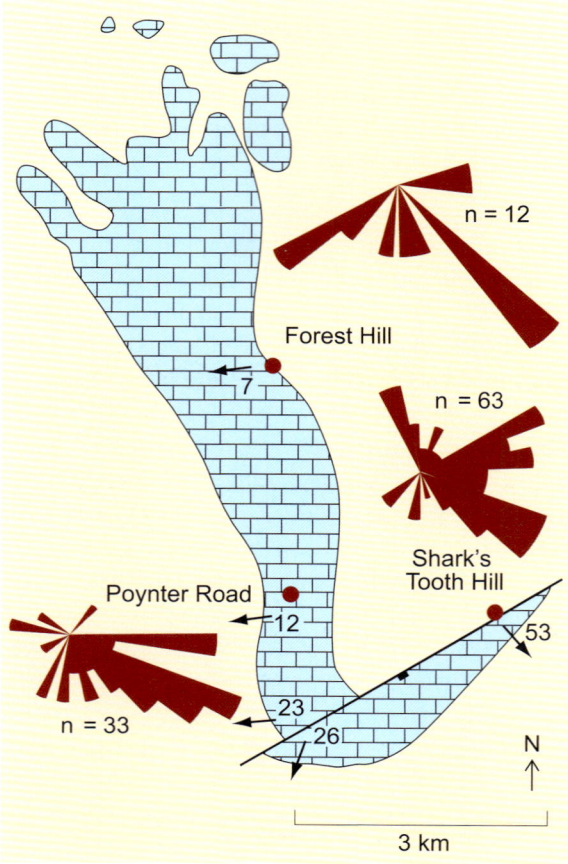

Figure 7.24 Rose diagrams showing palaeocurrent data from cross-stratified bioclastic limestones of Neogene age in Southland, New Zealand. The measurements were taken along the maximum dip of the inclined surfaces and the effect of regional (tectonic) dip has been removed. n = number of measurements.

- In Figure 7.24, what is the principle direction of current flow demonstrated by the three rose diagrams?

☐ To the southeast.

Unidirectional water currents give rise to a succession of stacked sets of asymmetrical ripples or dunes with their crests often eroded by the current depositing the succeeding bedform, and with the sets of strata, or **cross-sets**, all dipping in broadly the same direction. Where the amount of sediment in the flow is very high, and deposition outweighs erosion, sediment is deposited on both the stoss (upcurrent) and lee (downcurrent) sides of the asymmetric ripples. More sediment is deposited on the steep slope than on the shallow slope, causing the ripples to build upwards at an angle to the horizontal, producing **climbing ripples** (Figure 7.25). Both current-formed and wave-formed ripples (Section 7.4.3) can climb.

Chapter 7 Sediments on the move

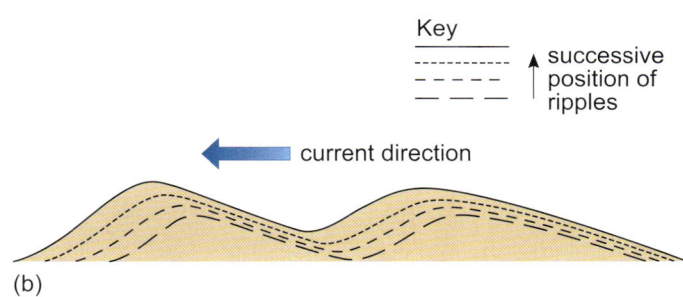

(a) (b)

Figure 7.25 (a) Climbing ripples in Carboniferous sandstones, Howick, Northumberland, England. (b) Sketch to illustrate how climbing ripples build upwards and migrate laterally.

7.4.2 Tidal bedforms

More complex currents occur in tidal regimes, where there are essentially two alternating currents that flow in opposite directions, corresponding to the incoming (flood) and outgoing (ebb) tide. It is quite usual for either the flood or ebb tide to be dominant, so that sediment deposited during the weaker tide is reworked and redeposited during the stronger tide. This gives rise to a succession of cross-sets dipping in one direction, and often indistinguishable from the structures described in Section 7.4.1 for unidirectional currents. Just occasionally, sediment from successive tides of equal strength is preserved, creating a distinctive, **bipolar** pattern termed **herringbone cross-stratification** (Figure 7.26).

Mud drapes represent suspension settling of mud during slack water at high and low tides. The mud forms a thin (sub-millimetre) veneer that drapes over sand ripples and, because it is cohesive, the mud, and the underlying sand, are often preserved through the next tidal cycle.

Figure 7.26 Herringbone cross-stratification in Cretaceous sandstones, Leighton Buzzard, Bedfordshire, England. The hammer is about 30 cm in length.

Tidal currents vary on a daily basis and are affected by the local topography. The largest tidal range of 16 m occurs in the Bay of Fundy, Canada and the second largest range is in the Bristol Channel, England where it is up to 15 m. Both of these ranges are large because of the funnelling effect of the land. In contrast, in the Wash (East Anglia, England) it is about 7 m and in the Mediterranean it is less than 2 m. All parts of the globe are affected by the lunar cycle. During the periods of the 'new Moon' and 'full Moon', when the gravitational effects of the Moon and the Sun are acting in the same direction, high tides are at their highest and low tides are at their lowest. These are the so-called *spring tides* that occur when tidal currents attain their greatest speeds consistent with the largest tidal range. During the period midway between the two spring tides, when the gravitational effects of the Moon and the Sun are at right angles and so do not reinforce each other, high tides are at their lowest and low tides at their highest – the so-called *neap tides*, when tidal currents attain lower speeds consistent with a smaller tidal range.

Cross-stratification that represents deposition in a full lunar cycle may display not only the mud drapes typical of the daily tidal cycle, but also a pattern of increasing and decreasing spacing between the sand–mud couplets, termed **tidal bundles** (Figure 7.27). These correspond to the approaching spring and neap tides respectively, and reflect faster tidal currents in spring tides and slower tidal currents in neap tides (Figure 7.27a). Since there are 29.5 days in each lunar cycle, it is sometimes possible to count 14 tidal bundles in each 14-day spring–neap cycle (Figure 7.27b), though the record can be more complex, particularly since many tides are semi-diurnal (i.e. there are two high tides per day, e.g. around the Atlantic and Indian Oceans) rather than diurnal (i.e. there is one high tide per day, e.g. around the Gulf of Mexico and much of the Pacific). Tidal bundles are the only sedimentary structure wholly diagnostic of tidal processes.

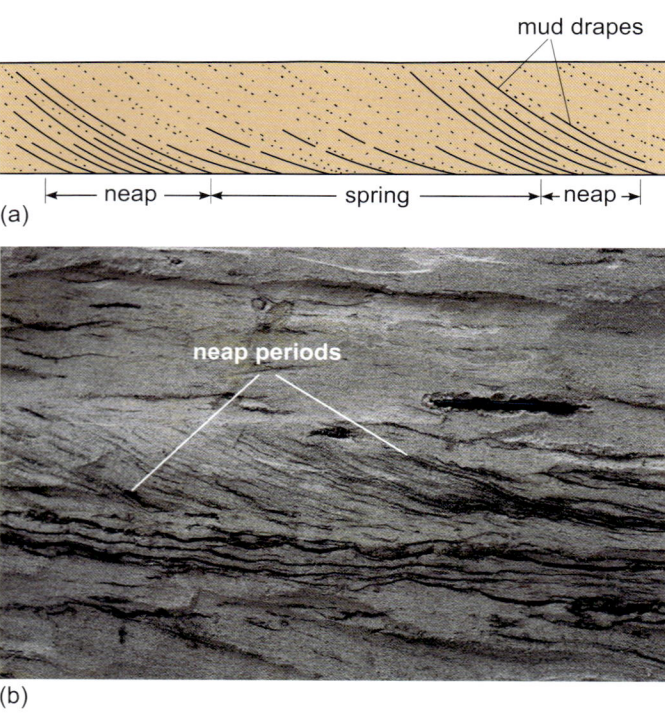

Figure 7.27 (a) Schematic cross-section through tidal bundles formed by variations in current speed with the lunar tidal cycle. The dotted lines represent sand deposition. Each spring–neap cycle represents 14 days. (b) Tidal bundles in Cretaceous sandstones exposed at Leighton Buzzard, Bedfordshire, England. The two areas of closely spaced mud drapes/sandstone couplets represent two periods of neap tides. The more widely spaced mud drapes/sandstone couplets represent spring tides (note pen for scale).

7.4.3 Wave bedforms

Waves are generated by friction as the wind blows over the water's surface. Internally, the water describes an orbital path as the waveform passes, with the highest point of this circular orbit coinciding with the wave crest, and the lowest point coinciding with the wave trough (Figure 7.28). You may have noticed this when swimming off a beach – as each wave passes, your body is not only moved up and then down, but also backwards and forwards a little. With increasing depth, the orbital paths taken by 'parcels' of water decrease in diameter until, at a depth of half the wavelength (i.e. half the distance λ between successive wave crests), orbital motion stops altogether – this location delineates the **wave-base** (Figure 7.28a). In shallow water above wave-base, the wave-generated orbital motion is affected by frictional drag on the sediment surface and motion is more elliptical; immediately above the sediment surface the water moves to and fro as each wave passes (Figure 7.28b).

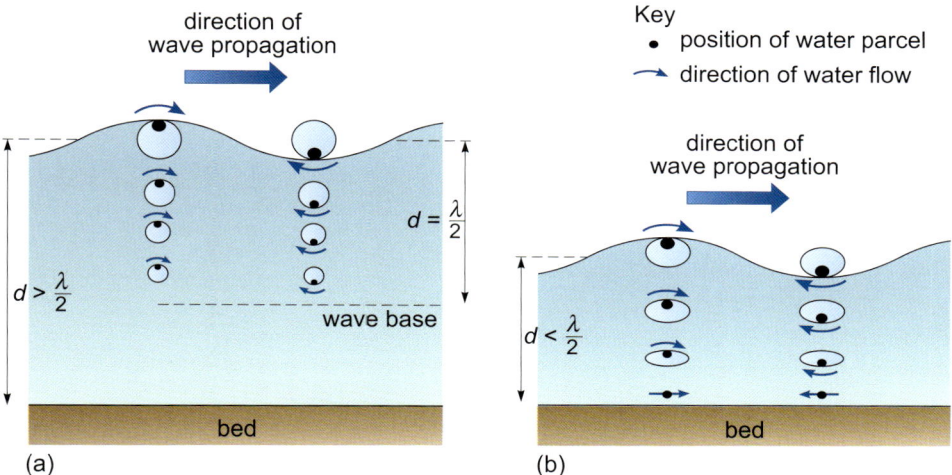

Figure 7.28 The orbital paths of parcels of water in waves associated with: (a) deep water (depth (d) = half the wavelength (λ) or more); (b) shallow water ($d < \frac{\lambda}{2}$).

The current speed increases to a maximum at the wave crest, then decelerates, reverses and increases to a maximum in the opposite direction, coincident with the wave trough. These changes in current speed affect the ability of the flow to transport sediment – the coarser grains are moved during the highest flow velocities and only finer grains are moved in the longer, decelerating flow, often as part of the suspended load (Figure 7.29a–e). These complex and rapidly changing hydrodynamic conditions lead to a high degree of sorting of grains and various internal structures, with forms ranging from chevron-shaped (Figure 7.29e) to undulating (Figure 7.29f). Whatever the internal pattern, **wave ripples** tend to be symmetrical in profile, with pointed or rounded ripple crests that are slightly sinuous in plan view and often bifurcating (i.e. splitting; Figure 7.30a and b).

Figure 7.29 (a)–(d) Cartoon to show the stages in the erosion and deposition of sediment associated with wave-formed ripples as a wave passes. The dashed ellipses show the orbital paths of a parcel of water as a wave passes over a rippled bed. The arrowheads represent the positions of a water parcel at various stages as the wave passes over. (e) Cross-section through a wave-formed ripple to show a typical pattern of chevron-shaped cross-stratification. (f) Wave-formed ripples in Carboniferous sandstone, Scremerston, Northumberland, England. The section is oblique rather than perpendicular to the bedforms, so that ripple crests can be seen in partial plan view as well as in cross-section. Coin 2 cm in diameter near middle of photograph for scale.

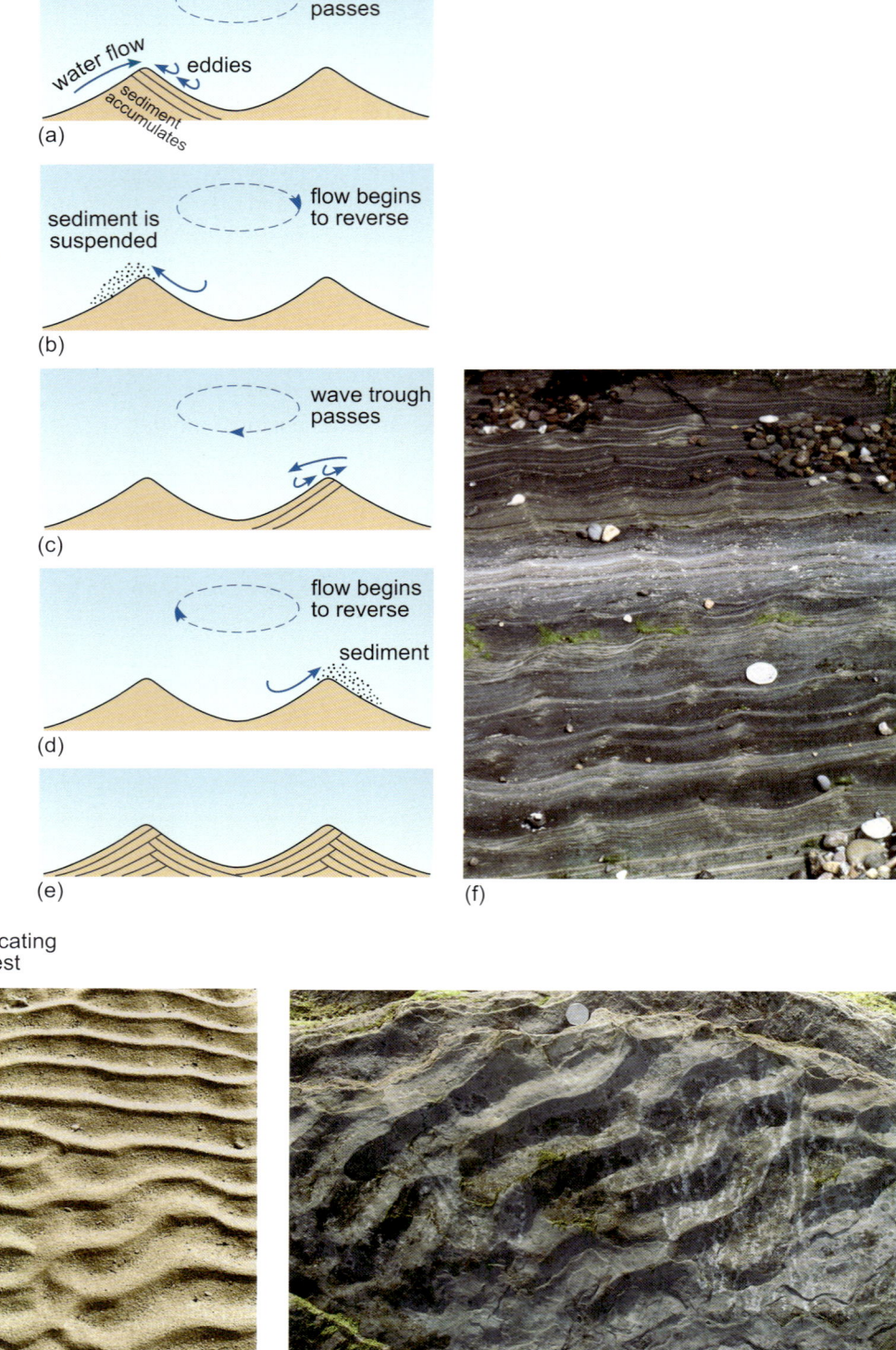

Figure 7.30 (a) Wave-formed ripples on a beach. Note how some of the ripple crests bifurcate (i.e. split into two) when traced from left to right across the photograph. (b) Wave-formed ripples preserved in fine-grained Carboniferous sandstone from Northumberland, England. Coin 2.5 cm in diameter at the top for scale.

■ How do current ripples differ from wave ripples?

☐ Current ripples have an asymmetrical profile and non-bifurcating crests.

As the orbital speed of water increases over the ripples, the wavelength of the ripples also increases, progressively breaking up the ripples into low mounds and ultimately washing them out to create upper flow regime planar beds. The **fairweather wave-base** (**FWWB**) is the maximum depth to which the seabed is influenced by wave motion during normal, fairweather conditions. Higher wave speeds and larger waves are generated during storms penetrating to depths below FWWB, but above the **storm wave-base** (**SWB**), i.e. the maximum depth to which the seabed is influenced by storm waves.

Storm wave-generated oscillatory flows or combined wave and current flows create low domes and basins on the seabed with a relief of 40 cm or more and spaced several decimetres or metres apart. The three-dimensional mounds or **hummocks** and basins or **swales** give rise to **hummocky cross-stratification** (**HCS**) (Figure 7.31a). Where the hummocks are planed off due to shallower water and increased erosion as the waves pass over or are close to FWWB, swales predominate, creating **swaley cross-stratification** (**SCS**) (Figure 7.31b). In cross-section, HCS and SCS are characterised by gently inclined laminations that are mainly concordant with the exterior of the domes and basins, but are occasionally

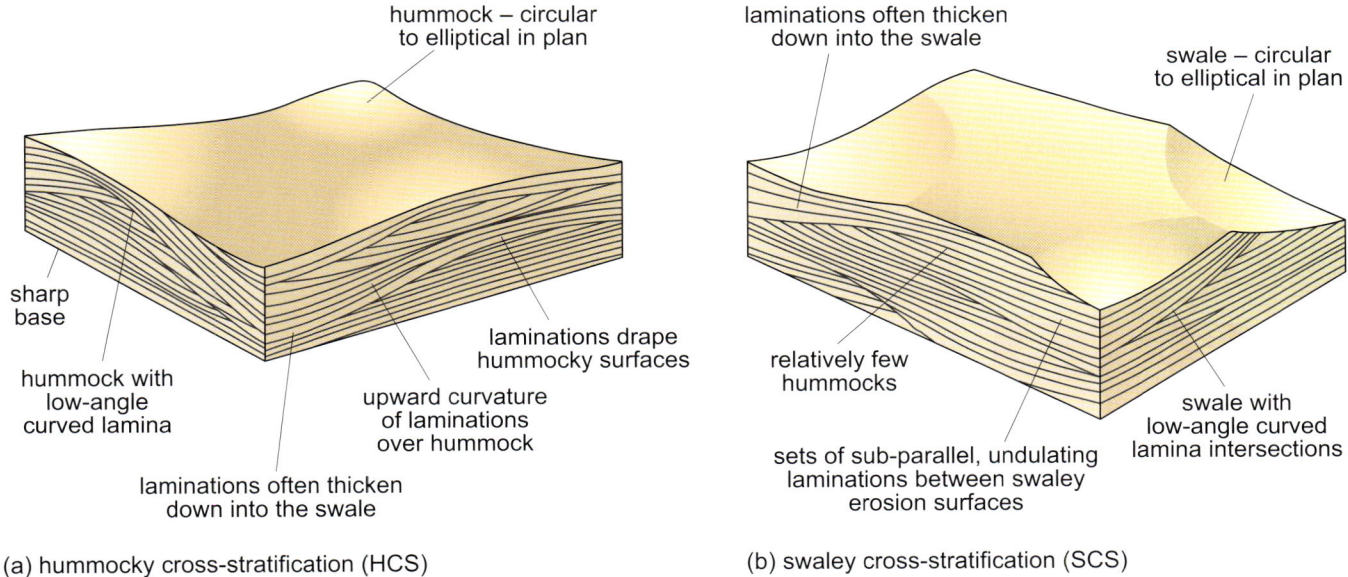

(a) hummocky cross-stratification (HCS) (b) swaley cross-stratification (SCS)

Figure 7.31 Three-dimensional block diagrams of (a) hummocky cross-stratification and (b) swaley cross-stratification, showing both the bedforms and the resulting sedimentary structures.

cut by low-angle erosional surfaces (Figures 7.31 and 7.32). No matter which cross-sectional orientation is examined, the bedding in HCS and SCS will always show the same three-dimensional dome- and basin-like geometry.

Figure 7.32 Hummocky cross-stratification in Carboniferous rocks from Northumberland, England. The penknife is 9 cm long and lies within a low-angle hummock or dome.

■ Is this constancy in shape also found in unidirectional current bedforms?

☐ No. The geometry of trough and planar cross-stratification is dependent on their orientation with respect to the current flow (Figure 7.23).

HCS is commonly preserved close to SWB, where sediment deposition is high enough to preserve the hummocks; SCS tends to dominate in shallower water where the hummocks have been planed off. Both HCS and SCS are reworked during fairweather conditions above FWWB, so they are mostly preserved only between SWB and FWWB.

7.5 Aeolian bedforms

Sediment movement by wind has much in common with water, but because of differences in density and viscosity between the two fluids, there are significant differences in sediment behaviour (Section 7.1). Air flows are vertically much more extensive than most water flows, so air flow depth has much less effect on the bedforms produced. Whereas water ripples start to develop due to slight irregularities in the sediment surface and/or turbulent eddies in the flow, **wind ripples** appear to develop precisely because the wind can only transport a narrow range of grain sizes. Within the bedload, these grains move similar distances by saltation (Figure 7.33). Impacts of numerous grains in the same area lead to small depressions in the sand surface, and further saltation and surface creep create small-scale wind ripples (Figure 7.33c). Slightly coarser grains move only by surface creep, and these migrate up the windward side of the ripples more slowly than the finer grains. This results in coarser-grained sand being concentrated at the ripple crests (the finer grains are effectively winnowed out), in contrast to water ripples where the coarser sand is trapped in the ripple troughs. Wind-generated ripples are asymmetrical, but are slightly flatter than water ripples and their crests are long and relatively straight, with occasional bifurcations.

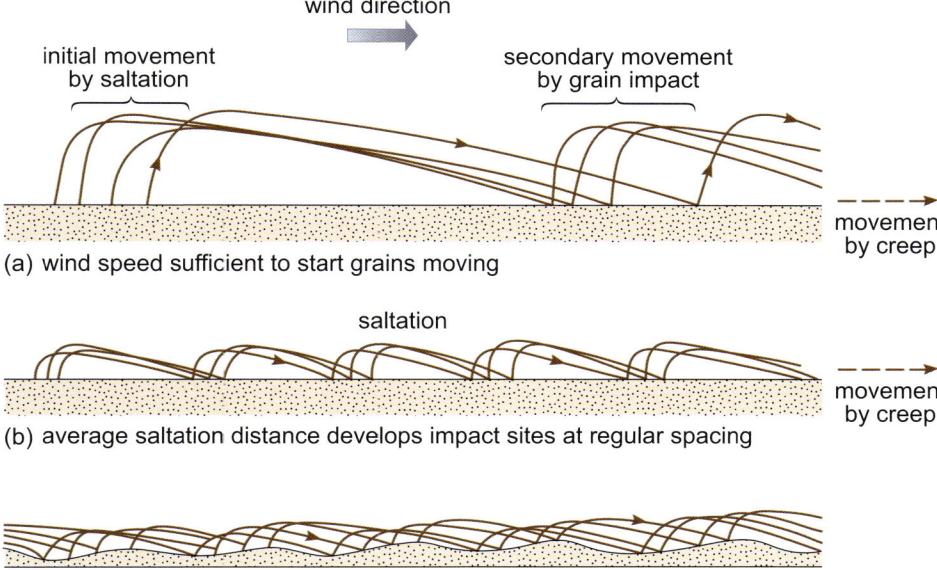

Figure 7.33 Stages in the formation of wind ripples. Note that the scale used to demonstrate initial movement by saltation (a) is considerably larger than that used to demonstrate subsequent ripple formation (b and c).

Whereas subaqueous dunes can develop from water ripples due to increasing flow velocity, increased wind speeds simply destroy aeolian ripples. Instead, some obstacle in the path of the migrating sand, e.g. a clump of vegetation or a rock, creates a 'wind shadow' containing localised eddies that concentrate the drifting sand (Figure 7.34). Once the drift attains a certain size, it becomes independent of the original obstacle and starts migrating as an **aeolian dune**.

Figure 7.34 Wind-blown sand drifts developing round clumps of seaweed and shells on a wind-swept beach in Northumberland, England. Wind direction is from left to right and field of view is about 30 cm.

The shape of the dunes depends on both the variability in wind direction and sediment supply (see Section 12.1). Wind ripples commonly form on the windward surface of the dune and on the **basal apron** in front of the dune (Figure 7.35), but the ripples are continually reworked and/or collapse into the overriding dune so that they are rarely preserved in the geological record.

Figure 7.35 Wind ripples superimposed on a large aeolian dune in Oman. Sand has avalanched down the steep, sheltered, lee slope (to the right) by grain flow processes, creating tongues of sediment on the slipface. The dune is several metres high.

Saltating sand grains bounce their way up the shallow stoss slope of an aeolian dune and periodically avalanche down the steeper, leeward slope as grain flows (Section 7.6.1) that are often interspersed with **grain fall** deposits – thinner, finer sand derived from settling of suspended sediment directly onto the lee slope. The grain fall deposits may be normally graded, as slightly coarser grains settle out first when the wind speed abruptly drops in the lee of the dune. Grain flow and grain fall deposits build up as a series of cross-sets that delineate the former position of the front of the dune as it migrates downwind (Figure 7.36). The tops of the dunes are invariably removed as the sediment gets transported downwind, but most of the **slipface** (sheltered, steeper side of the dune where avalanching of sediment occurs) and also the base of the dune are often preserved as aeolian cross-sets, commonly several metres thick (Figure 7.37), but sometimes reaching 30 m or more.

On average, the angle of the slipface (25–35°) and the thickness of aeolian dunes are both greater than for subaqueous dunes and sand waves.

Chapter 7 Sediments on the move

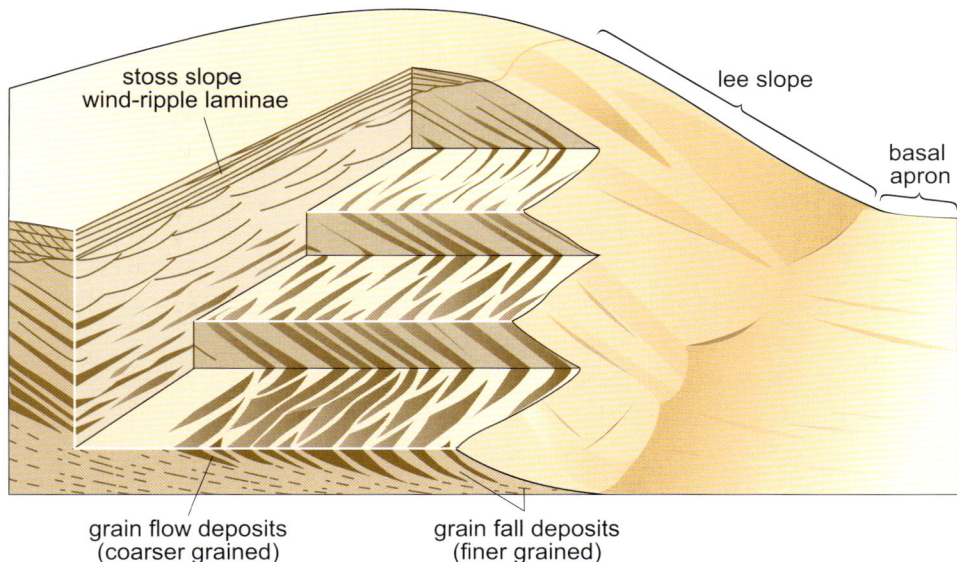

Figure 7.36 Cross-section through an aeolian dune to show interfingering of coarser-grained grain flow and finer-grained grain fall deposits on the lee slope slipface of the dune and wind-ripples on the windward (stoss) slope of the dune. Grain flow deposits result from tongues of sand avalanching down the front of the dune, whereas grain fall deposits result from suspension fall out as the wind speed drops in the lee of the dune.

Figure 7.37 Cross-stratification in Jurassic aeolian sandstones, Utah, USA. The slipface and base of three successive dunes are stacked one on top of the other but the top surfaces of the dunes have been planed off by later dune migration. The exposure shown is ~12 m high.

Question 7.1

What other evidence might you find to confirm that a succession of large-scale, cross-stratified sandstones were aeolian in origin, rather than marine?

169

7.6 Transport and deposition by sediment gravity flows

The discussion of sediment transport has so far concentrated on fluids with little suspended sediment, so that transport depends largely on the properties of the transporting fluid – either water or wind. With increasing amounts of suspended sediment, the viscosity and density of the sediment–fluid mix increases, and the nature of the flow begins to change. Eventually gravity becomes the prime mover, and the sediment itself provides the supporting mechanism, hence the term **sediment gravity flows**.

There is a complete spectrum of gravity flows relating to the degree of consolidation of the material and the proportions of sediment and fluid in the flow. They range from gravity-driven **slides** and **slumps** of weakly lithified material on steep slopes, through unconsolidated sediment-supported grain flows and sediment and fluid-supported debris flows, to fluid-supported turbidity currents (Figure 7.38). Slides and slumps move along a basal slide surface, but whereas slides move as a coherent, largely undeformed mass, slumps display variable amounts of internal folding and brecciation. They may be triggered by events such as earthquakes, heavy rainfall or erosion at the base of a cliff or channel bank.

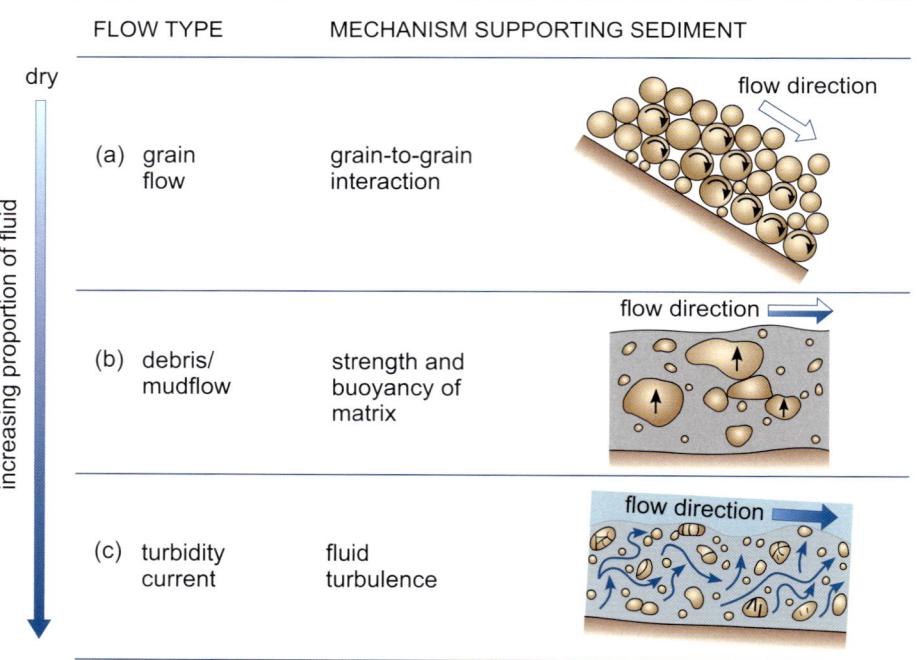

Figure 7.38 Cartoons to illustrate the mechanisms of sediment transport for different types of sediment gravity flows.

7.6.1 Grain flows

Grain flows occur when unconsolidated sediment on a slope exceeds its stable angle of rest, and so begins moving under the influence of gravity. There is almost no fluid involved other than air or water trapped between the grains, which act as a lubricant. Once in motion, the flow is sustained by grains colliding and ricocheting and rolling over each other (Figure 7.38a).

■ Can you think of an example of a grain flow?

☐ The avalanching of sand grains down the steep, down-wind side of an aeolian dune.

There is no grain-size limit for a grain flow, and it may encompass anything from sand grains to boulders (termed rock falls).

7.6.2 Debris flows

Sediment that contains sufficient water to form a viscous mass will flow like plastic. The resultant **debris flows** are supported both by the strength of the matrix and by collisions between the larger grains (Figure 7.38b). The flows consist of slurries of widely varying grain size and sediment concentration that are usually very poorly sorted (Figure 7.39), though some clast-supported flows exhibit preferred orientation of their long axes as the clasts roll or slide within the flow. Amazingly, less than 5% matrix is sufficient to provide a buoyant medium to move large boulders. Matrix-supported debris flows with a high proportion of mud are often termed **mudflows** (Figure 7.40).

Debris flows occur in both subaerial and submarine environments and often funnel down existing channels. Due to the high sediment load and the lubricating effect of watery clay mixtures, they can travel very rapidly, up to 100 km h^{-1} or more. On land they can be extremely destructive, and are usually initiated after heavy rainfall. Debris flows that are sourced from unconsolidated ash on volcanic slopes are called lahars and they can often be more destructive than the eruption itself (Book 2, Section 3.5). Very large debris flows can also result from the partial collapse of a volcano (Book 2, Section 3.6.3).

Figure 7.39 Unsorted, matrix-supported debris flow, Grand Canyon, USA. Field of view is approximately 1.5 m across.

Figure 7.40 Mudflow, Dorset coast, England. The flow is about 5 m across.

7.6.3 Turbidity currents

Turbidity currents are the most important of the sediment gravity flows since they are the principle mechanism for transporting vast quantities of unstable sediment from the edge of the continental shelf and continental slope into the deep ocean. Dense sediment suspensions also create an underflow during snow-melt floods, in river tributaries and in front of deltas. However, turbidity currents are not restricted to water flows since dense suspensions in air may also flow downslope, e.g. powder snow avalanches and pyroclastic flows on the flanks of volcanoes. Failure of unstable sediment may be triggered by a number of mechanisms including earthquakes. As grains begin to move, pore fluids seep between the grains, reducing friction and creating a fluidised turbidity current that flows downslope beneath the lower density fluid, reaching speeds in excess of 50 km h^{-1}.

Turbidity currents can be generated experimentally in a flume (Figure 7.41a). The flow displays well developed 'head', 'body' and 'tail' sections that travel at different speeds. The speed of the head depends on its height, whereas the speed of the body depends on the angle of slope. The body travels faster than the head, so water and sediment are constantly being forced into the head region, increasing the pressure and creating back eddies that keep the sediment in suspension (Figure 7.41b).

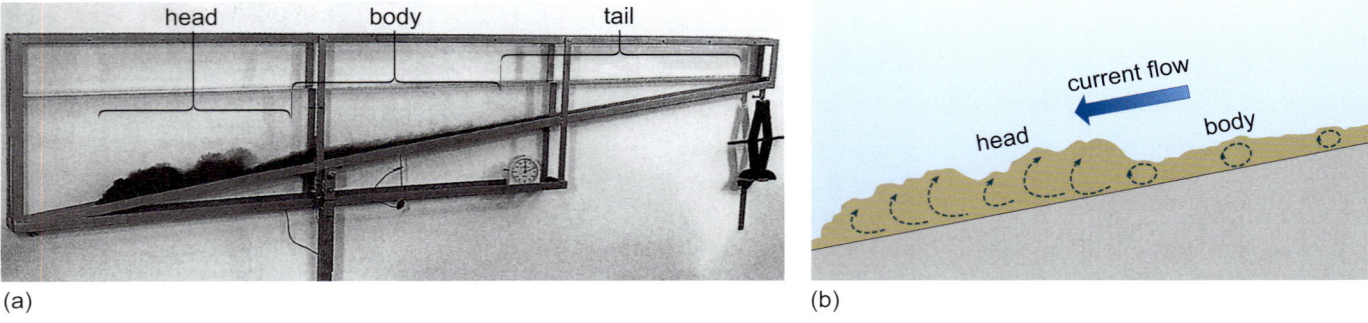

Figure 7.41 (a) The structure of a turbidity current generated when dyed, higher density salt water is allowed to flow down a slope in an experimental tank beneath less-dense tap water. (b) The movement of water and sediment within the head and immediately adjacent body of a turbidity current, shown by dashed lines with arrowheads.

As they gradually slow down and deposit sediment, each turbidity flow produces a characteristic succession of beds known as a **turbidite**. Each bed represents a particular part of the flow, and the size of the clasts and sedimentary structures are diagnostic of particular parts of the flow and the speed of the turbidity current. This characteristic succession of beds was first recognised and described by Arnold Bouma ('bou' is pronounced 'bow' as in a ship's bow) and is thus called a **Bouma Sequence**. Before proceeding, let us first consider the different sedimentary structures that form as the current speed decreases. You may wish to refer back to Figures 7.21 and 7.22.

Question 7.2

(a) What are the three sedimentary bedforms characteristic of the lower flow regime and what sedimentary structure will each produce?

(b) If the current speed was slowing down for a given grain size, which would form first: planar deposition (upper flow regime) or current-formed ripples?

Question 7.3

If a turbidity current transports a mixture of grain sizes including pebbles, coarse-grained sand, clay, fine-grained sand and granules, and the current is slowing down, in which order would the grains be deposited?

As the current speed decreases, a fining-upward succession will form with a predictable pattern of sedimentary structures. This is exactly what is found in a Bouma Sequence (Figure 7.42; see also Section 15.2).

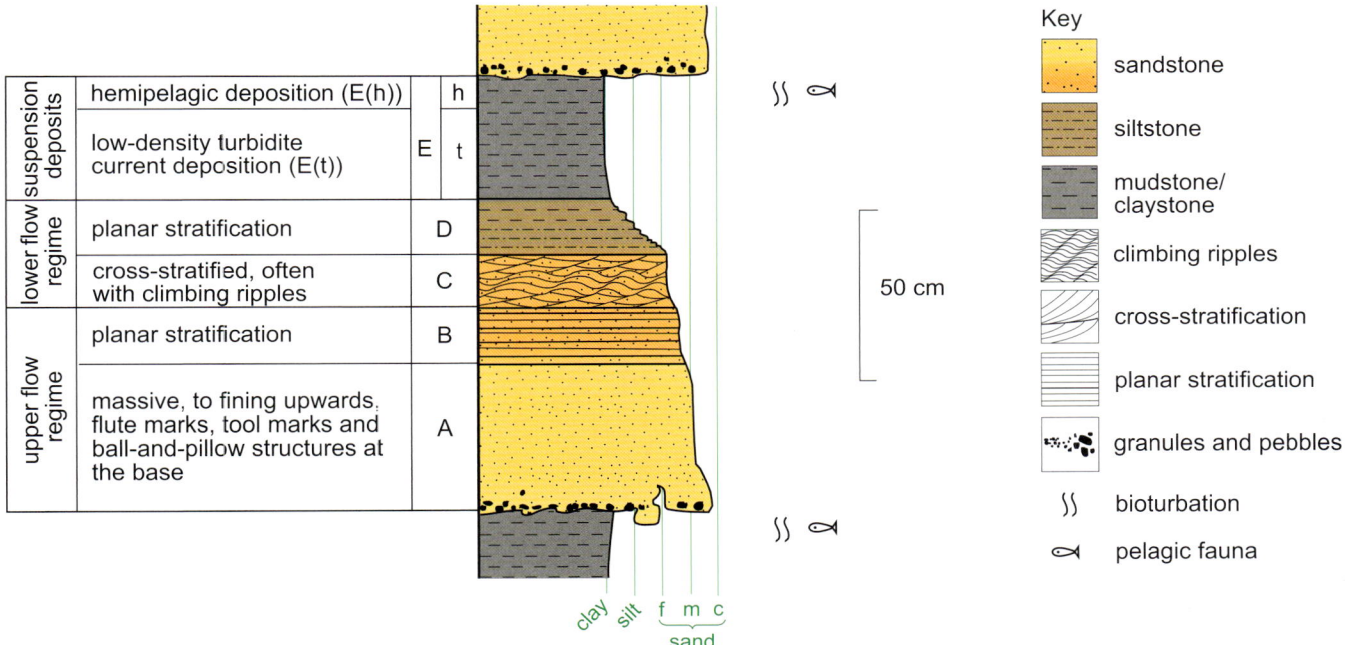

Figure 7.42 A full Bouma Sequence showing the different divisions deposited from a turbidity current, and hemipelagic deposition between intervening flows. See Figure 7.43 and text for explanation. This figure is also shown on the SE Poster. Ball-and-pillow structures are explained in Section 7.7.

Figure 7.43 Cartoons (facing page) to show the various stages in the deposition of a turbidity current resulting in a Bouma Sequence. Overall flow direction of the turbidity current is from right to left across each of the illustrations.

The formation of the individual divisions (A to E) in the Bouma Sequence is explained as a series of time-shot drawings in Figure 7.43. For each time shot, you should read the text and study the illustration. Division E on Figure 7.43 partially consists of **hemipelagic** sediment. Hemipelagic means that it contains at least 25% of sediment derived from the land rather than directly from the sea (e.g. marine microfossils).

Turbidity currents travel at considerable speed and with considerable erosive power, particularly when travelling across unconsolidated sediment. Turbulent eddies carve out tongue-shaped hollows (**flute marks**) in the underlying sediments (Figure 7.44a and b). The eddies cut in more deeply at the upstream end, and as they lose energy, they carve hollows that widen and shallow downstream. The flutes average 5–10 cm across and 10–20 cm in length, and occur in 'swarms', each with a similar orientation, i.e. with the rounded end upstream. The flutes become filled with sediment once the turbidity current begins to slow down and they are best observed as bulbous casts protruding from the base of the overlying unit.

Larger grains carried along by the turbidity current (e.g. shells and bone fragments) can create a range of **tool marks** as they bounce, prod and gouge the underlying surface (Figure 7.44c). These moving objects produce linear **groove marks** that are often parallel to the flow direction.

■ In which direction was the current moving in Figure 7.44b and also in 7.44c?

□ Bottom right to top left in Figure 7.44b, corresponding to the bulbous, upstream end of the flute marks; parallel to the pronounced linear grooves in Figure 7.44c, either bottom left towards top right or vice versa.

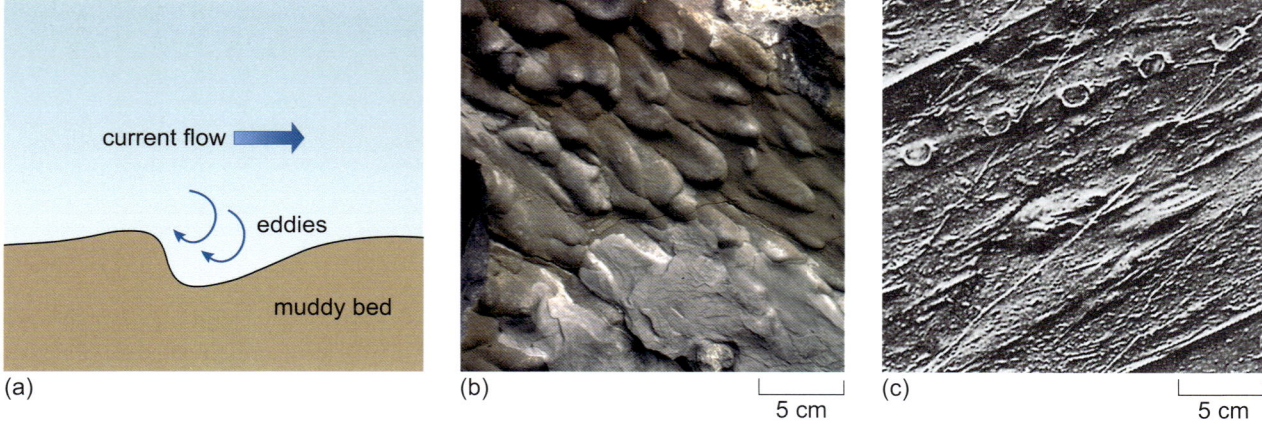

Figure 7.44 (a) Formation of flute marks due to scouring of muds by eddies at the base of a turbidity current. (b) Flute casts on the underside of a bedding plane in Silurian sedimentary rocks from Wales. (c) Various tool marks preserved as casts in fine-grained rocks from Poland. The line of circular marks at the top of the photograph was caused by saltating fish vertebrae.

Chapter 7 Sediments on the move

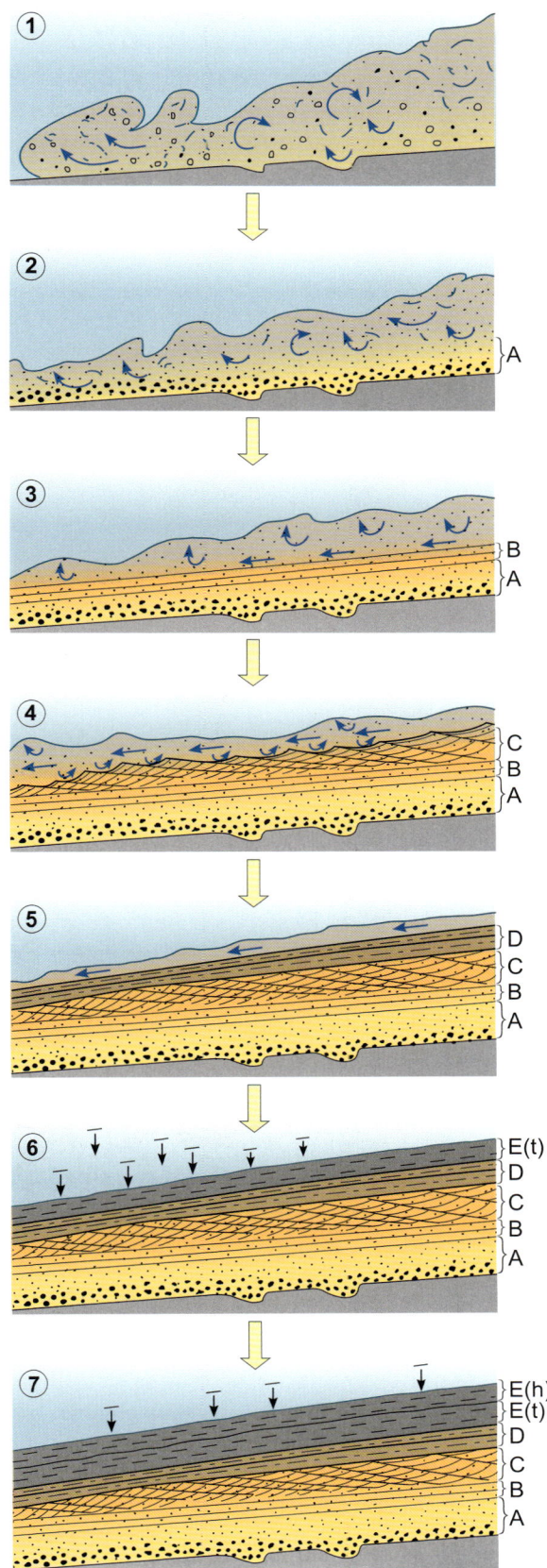

 As a fast-flowing turbidity current passes over a surface, its high energy means that it will usually erode structures into the underlying bed. Because these structures are found as casts on the base of the bed, they are collectively termed **sole structures** or sole casts. They include tool marks, formed as sharp or large clasts scrape and dig into the underlying bed, and flute marks, which form due to small eddy currents scouring the surface. At this stage, the turbulent turbidity current is maintaining all of its sediment load in suspension by fluid turbulence.

 As the speed of the turbidity current begins to decrease, the larger grains settle towards the base of the flow and start to collide with each other, slowing the base of the flow down further. Sole structures will continue to form. Meanwhile, the grains in the upper part of the flow continue moving in the fast turbulent flow. Eventually, the concentration of grains and the slowing down of the base of the turbidity flow will cause the lowermost layer to be deposited. This forms division A; it contains the coarsest grains, and is generally massively bedded although it may show some fining upwards.

 As the current slows down further, the medium-sized particles start to settle out. The current is still quite fast moving (upper flow regime), and the sediments are planar stratified. This forms division B.

 With further slowing of the current into the lower flow regime, current-formed ripples will form and so create cross-stratification; if there is a lot of sediment present, these will be climbing ripples. This forms division C.

 Continued deceleration in the lower flow regime allows both fine- and medium-grained sediment to be deposited in laminae forming planar stratification. This forms division D.

 As the flow comes to a stop, the final stage is to deposit the finest-grained sediment still in suspension; this forms division E (turbiditic, or E(t) for short).

Division E (hemipelagic; or E(h) for short) is the 'background' sedimentation of fine-grained hemipelagic sediment deposited from suspension and which has not been carried by the turbidity currents.

175

7.7 Post-depositional structures

There are a number of post-depositional processes other than sliding and slumping that operate during sedimentation or immediately afterwards. Three of the most common post-depositional structures are load casts, desiccation cracks and bioturbation. Together with cross-stratification, they can be very useful for determining the way-up of sedimentary rocks that have been structurally deformed.

Load casts are generated by storms and turbidity currents that rapidly 'dump' sand onto underlying muds, resulting in squeezing up of the mud to form **flame structures** (Figure 7.45a), and sinking of lobes of sand into the mud. Detached sediment lobes that form ball or pillow shapes are termed **ball-and-pillow structures** (Figure 7.45b).

Figure 7.45 (a) Flame structures in red mudstone created by rapid loading of sand onto unconsolidated mud, Devonian, Dingle Peninsula, Ireland. Note hand-lens near left margin for scale. (b) Ball-and-pillow structures created by rapid loading of sand onto unconsolidated mud, Carboniferous, Northumberland, England. Note the hammer (about 30 long) for scale. (c) Desiccation cracks and mud polygons curled into flakes formed in a dried up wadi pool, Oman. (d) *Chondrites* burrows within Silurian siltstones, Dingle Peninsula, Ireland. Note coin for scale.

Desiccation cracks usually form in drying muds due to shrinkage and they provide evidence of subaerial exposure. The cracks are polygonal in outline and often the mud polygons curl into flakes (Figure 7.45c) that may be reworked in intertidal environments to create mud-flake conglomerates.

Bioturbation is the name given to evidence of sediment disturbance by organisms (Section 3.6). It ranges from churning up of the sediment, which shows little or no discernible internal structure, to discrete burrows (Figure 7.45d), though only rarely are such trace fossils attributable to a particular type of animal. In every case, the original sedimentary structures are disrupted or destroyed, and changes in sediment fabric are often accentuated during post-depositional compaction and cementation.

7.8 Glacial processes

Moving ice can be considered as an extremely viscous fluid in which the high viscosity supports clasts up to several metres in diameter. Ice is a crystalline solid that moves by bulk sliding and/or gravitational deformation in response to its own weight and influenced by gravity. A mound-like ice sheet will flow radially outwards, irrespective of the underlying topography, so in places it commonly flows uphill. Valley glaciers, on the other hand, move downhill due to gravity.

Ice movement is largely possible because melting caused by pressure at the base of the glacier enables sliding on a fluid-rich and highly porous sediment, lubricated by glacial meltwater. These glaciers are often referred to as 'warm' and wet. 'Cold' and dry ice beds are extremely rare because basal ice is usually above its melting point. The speed of movement is usually measured in metres per year ($m\ y^{-1}$) but increased volumes of glacial meltwater lead to surges in ice flow at greater speeds.

Glacial ice contains debris derived from the top, sides and/or the base of the glacier. Freeze–thaw action creates large volumes of angular clasts, forming screes that avalanche onto the glacier, often falling into crevasses within the ice. Subglacial erosion occurs by plucking, abrasion, crushing and fracturing, creating abundant rock 'flour' that dominates the suspended load of glacial outwash streams. The erosive power of sediment-laden ice is enormous and has moulded many upland landscapes, with their characteristic U-shaped valleys (Figure 7.46a), such as the Lake District, Scottish Highlands and the European Alps. Exposed bedrock that has been scoured by glacier ice is characterised by extensive linear grooves called **glacial striations** (Figure 7.46b).

Glacial drift constitutes all the rock material transported by glacier ice; **till** is the term applied to unsorted, unstratified glacial sediments that are deposited by glaciers and not reworked by later flowing water. The term **moraine** is used for till forming directly at the margins of a glacier and, in particular, includes two distinctive tills:

- **melt-out till** is derived from the slow release of glacial debris from melting ice (Figure 7.46c)
- **lodgement till** forms either by plastering of glacial debris from the sliding base of a moving glacier by pressure melting, or by continual shearing of soft sediment moving en masse beneath the ice (Figure 7.46d).

Book 3 Fossils and Sedimentary Rocks

Figure 7.46 (a) Glaciated U-shaped valley, Arran, Scotland. (b) Glacial striations, Dingle Peninsula, Ireland. (c) Melt-out till from melting of Brúarjökull glacier, Iceland. The till has formed a series of hummocky moraines. (d) Lodgement till, Sólheimajökull glacier, Iceland.

Activity 7.7 Physical processes in action

This activity is based on the *Book 3 Resources* DVD and the *Coastal Processes* video sequence.

7.9 Summary of Chapter 7

1 Most sediment is transported by water or wind and their differing viscosities and densities greatly affect their capacity to transport sediment grains: air has a much lower carrying capacity than water. Movement is by laminar flow or turbulent flow and sediment forms either part of the bedload or suspension load.

2. Grain surface textures are affected by the transporting fluid. The cushioning effect of water gives rise to subangular grains with a glassy surface, whereas abrasion during wind transport leads to well-rounded grains with a pitted (frosted) surface.

3. The threshold speed at which grains begin to move is greater than the speed necessary to keep them in motion. Clay minerals are cohesive and require more energy to start them moving than silt or fine sand grains. The lower carrying capacity of the wind leads to a narrower spread of grain sizes and better sorting than in water.

4. Most erosional structures are formed by aqueous and sediment-laden flows prior to deposition and by objects striking the sediment surface during transport. Channels and scours (including gutter casts) are the most common structures, and flute and groove marks created by turbidity currents are the most distinctive. Glaciers are extremely erosive, giving rise to U-shaped valleys and glacial striations.

5. Grain size and the speed of the transporting fluid (excluding ice) are directly related. Gravel-sized sediments equate to high energy conditions, sand-sized sediments equate to medium energy conditions and mud-grade sediments equate to low energy conditions.

6. Textural maturity reflects the energy of the environment and the conditions of transport. Prolonged transport rounds the grains (wind is more effective than water) and increases grain sorting, often by removing fine-grained sediment and leaving behind the coarsest grains.

7. The terms bedding (thicker than 1 cm), lamination (thinner than 1 cm) and stratification refer to layered sediments. Suspension settling gives rise to parallel laminations, whereas bedload layering may be parallel (planar stratification) or inclined (cross-stratification).

8. Unidirectional, aqueous bedforms range from asymmetrical current ripples to subaqueous dunes, depending on the current speed and water depth. With increasing speed, current ripples and subaqueous dunes produced in the lower flow regime are washed out and replaced by planar stratification in the upper flow regime. Internal cross-stratification represents the former positions of the front of the ripples or dunes and provides evidence of palaeocurrent direction, with the current pointing down dip.

9. Tidal currents create cross-stratified sand–mud couplets that reflect the daily tidal cycle, with mud drapes settling out of suspension at slack water. A pattern of increasing and decreasing spacing between these couplets, termed tidal bundles, corresponds to the approaching spring and neap tides respectively. Usually, either the flood tide or the ebb tide is dominant, and cross-stratification dips in one direction only. In rare cases, however, sediments from both the incoming and outgoing tides are preserved as herringbone cross-stratification, and these, together with tidal bundles, are diagnostic of tidal processes.

10 Symmetrical wave ripples are generated by oscillatory flow above fairweather wave-base, whereas larger mounds or basins termed hummocky cross-stratification (HCS) and swaley cross-stratification (SCS) are created by storm waves above storm wave-base. They are usually preserved only between fairweather wave-base and storm wave-base.

11 Aeolian bedforms include impact (wind) ripples and aeolian dunes. Sand avalanches down the lee side of the dunes as grain flows, or falls out of suspension as grain fall deposits, creating cross-sets that represent the former positions of the front of the dune. Aeolian dunes are often steeper and thicker than subaqueous dunes, but other evidence such as grain textures should be used to confirm their origin.

12 There is a complete spectrum of sediment gravity flows relating to the degree of consolidation of the material and the proportions of sediment and fluid in the flow. They range from gravity-driven slides and slumps of weakly lithified material on steep slopes, through unconsolidated, sediment-supported grain flows, to sediment- and fluid-supported debris flows, and thence to fluid-supported turbidity currents. The latter produce characteristic successions of beds and bedforms termed Bouma Sequences, often with flutes and grooves at the base.

13 Post-depositional structures disrupt primary sedimentary structures, and range from slumps and slides, to load casts, desiccation cracks and bioturbation, the latter created by organisms churning up or burrowing into the soft sediment.

14 Ice behaves like an extremely viscous fluid and it moves by bulk sliding and/or gravitational deformation. Ice sheets can move uphill and valley glaciers can carve out distinctive U-shaped profiles aided by sediment-charged ice at the base of the flow. Glacial striations are common on exposed bedrock surfaces.

15 Glacial drift constitutes all the rock material transported by glacier ice. Till is the term applied to unsorted, unstratified glacial sediments that are deposited by glaciers and are not reworked by later flowing water.

7.10 Objectives for Chapter 7

Now you have completed this chapter, you should be able to:

7.1 Explain how the mechanism of fluid flow depends on the properties of the fluid and describe how transport by fluids takes place, distinguishing between transport by water currents, waves and wind.

7.2 Explain how the speed of a moving fluid is related to the sediment grain size that it is capable of eroding, transporting and depositing, and interpret graphs that illustrate these relations.

7.3 Estimate qualitatively the energy of a depositional environment from the grain size of the sedimentary deposits.

7.4 Describe the textural characteristics of siliciclastic sediments in hand specimen and thin section, including grain shape, rounding, sorting and proportion of matrix, and estimate the degree of textural maturity of the rock.

7.5 Distinguish between bedforms produced by water currents, waves and wind, and briefly explain how each bedform is created.

7.6 Explain how the process of sediment gravity flow depends on gravity and the properties of the sediment–fluid mix, and how the proportion of fluid in the flow determines its character.

7.7 Describe, using sketch diagrams, the structure of a turbidity current and the elements that make up an idealised Bouma Sequence.

7.8 Describe the mechanisms responsible for ice flow and the associated transport and erosional processes.

Activity 7.5 gave you an opportunity to test your understanding of Objectives 7.3 and 7.4. Now try the following questions to test your understanding of some of the remaining objectives.

Question 7.4

With reference to Figure 7.7:

(a) Given a flow depth of 1 m, what is the water-current speed required to begin eroding a sand grain 0.6 mm in diameter?

(b) When the grain begins moving, how will it be transported: as bedload or in suspension?

(c) How will the same grain be transported if current speed increases to 1 m s^{-1}?

(d) To what speed must the current decrease again before the grain can be deposited, and how will the grain be deposited – from the bedload or from suspension?

Question 7.5

With reference to Figure 7.22, a water current flows at a speed of 0.4 m s^{-1} over a bed of gravel composed of grains 3 mm and more in diameter. The depth of the flow is around 0.3 m. What is the succession of bedforms you would expect to develop if the current speed increased to 2 m s^{-1}? How rapid would the transitions be from one type of bedform to the next?

Question 7.6

Look carefully at the sandstone exposure in Figure 7.15.

(a) Assuming the exposure is a section parallel to the current flow that deposited the sandstone, and that the cross-stratification was produced by bedforms formed in the lower flow regime, which way was the current flowing?

(b) Explain whether the water current varied in speed or was fairly constant.

Book 3 Fossils and Sedimentary Rocks

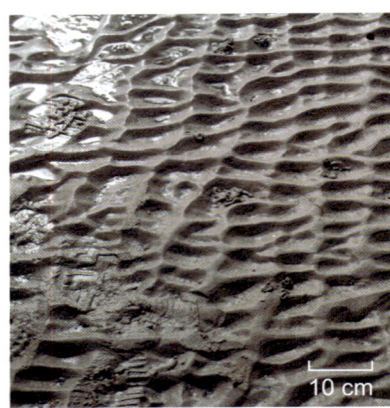

Figure 7.47 Two sets of interference (ladder) ripples exposed at low tide. (For use with Question 7.7.)

Question 7.7

Figure 7.47 shows two sets of ripples that formed in the intertidal zone. The ripples have a distinctive pattern, often referred to as ladder ripples (a type of interference ripple).

(a) Identify the process that has formed each of these ripple sets, providing evidence to support your identification.

(b) Which ripple set formed first and why?

(c) What evidence is available to determine the direction of one or both currents?

(d) Two features have disrupted the sedimentary structures. Describe these features and provide names that encompass both.

Question 7.8

Figure 7.48 shows three different sedimentary structures, labelled A, B and C. Identify each of the three structures and provide evidence to justify your choice in each case.

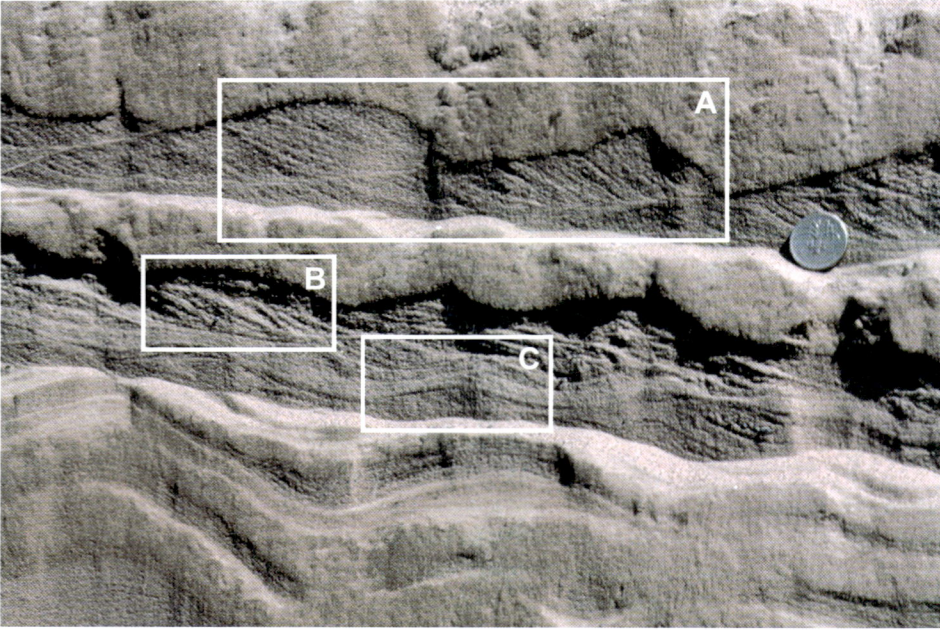

Figure 7.48 Sedimentary structures displayed in Triassic sandstones from Cheshire, England. Note coin for scale. (For use with Question 7.8.)

Chapter 8 Sediments from solution

The preceding chapter has explored the mechanisms of fluid transport and deposition of sediment created by weathering and erosion of exposed rock. Weathering processes also result in solution and the dissolved ions are transported to the sea via percolating groundwaters and rivers. Some of the dissolved materials in freshwater and seawater are incorporated into the shells of organisms by biological processes to form **biogenic sediments**, or are precipitated directly to produce inorganic, **chemical sediments**, including evaporite minerals such as halite and gypsum (see Book 1, Section 7.3.1).

■ What is the main textural difference between such biogenic and chemical sediments?

□ Biogenic sediments have a fragmental texture whereas chemical sediments have a crystalline texture.

Most shelly material consists of calcium carbonate – hence the term carbonate sediment – but some organisms secrete silica skeletons, giving rise to siliceous sediments. Wherever the soft parts of animals and plants are preserved from decay, then organic, carbon-rich (carbonaceous) sediments can also accumulate.

8.1 Carbonate sediments

Carbonate sediments are the most common of the biogenic and chemical sediments, giving rise to limestones ($CaCO_3$) and dolomites ($CaMg(CO_3)_2$). The majority of limestones are marine in origin, and they accumulate wherever siliciclastic input is minimal and the seawater is saturated with respect to calcium carbonate. In contrast to siliciclastic sediments that are transported into the depositional site, carbonate sediments are produced within the basin, and are rarely transported very far. In very deep water, beneath a zone called the carbonate compensation depth (CCD, around 3–4 km deep), the water is undersaturated with respect to calcium carbonate, and carbonate skeletons begin to dissolve.

There are three polymorphs of calcium carbonate (i.e. minerals with the same chemical composition but different crystal structures):

- aragonite
- high magnesian calcite (high-Mg calcite, in which small but significant amounts of Mg substitute for Ca in the crystal lattice)
- low magnesian calcite (low-Mg calcite).

Aragonite and high-Mg calcite are stable in seawater and low-Mg calcite is stable in both seawater and freshwater. Many bioclasts, for example gastropods and many bivalves, are made up of aragonite, and these are very susceptible to dissolution when exposed to freshwater. The altered aragonitic shells rapidly adopt a soft, 'chalky' texture. High-Mg skeletons, such as echinoderm plates, remain largely unchanged when exposed to freshwater, despite a loss of magnesium from the crystal lattice. Thus high-Mg and low-Mg calcite have much greater preservation potential than aragonite.

8.1.1 Biogenic carbonate grains

Calcium ions are released during chemical weathering of silicate minerals, such as plagioclase feldspar and some amphiboles, and also when limestone is dissolved in rainwater charged with CO_2 (see Section 6.4.2). Equation 8.1 describes the precipitation (forward reaction) or dissolution (back reaction) of $CaCO_3$:

$$\underset{\text{Ca ions in solution}}{Ca^{2+}(aq)} + \underset{\text{bicarbonate ions}}{2HCO_3^-(aq)} \rightleftharpoons \underset{\text{calcium carbonate}}{CaCO_3(s)} + \underset{\text{water}}{H_2O(l)} + \underset{\text{carbon dioxide}}{CO_2(g)} \quad (8.1)$$

Like all gases, CO_2 is *less* soluble in warm water than in cold water and, by the same token, $CaCO_3$ is also less soluble and therefore more prone to precipitation in warm water. As a result, carbonate production is more prolific in warm, tropical waters (where evaporation is also highest) than in cool waters at higher latitudes.

Calcareous organisms extract Ca^{2+} and bicarbonate ions from water to secrete their hard parts. After the deaths of these organisms, which include everything from microscopic calcareous phytoplankton (e.g. unicellular algae) and zooplankton (animals), to larger animals such as fish, clams, mussels, sea urchins, corals and bryozoans (colonial animals much smaller than corals), their hard parts accumulate on the sea floor (Figure 8.1), eventually to form a limestone (Figure 8.1b). Chalk is a very fine-grained limestone consisting predominantly of the skeletons of marine algae called coccolithophores that are only a few microns in diameter. Each coccolithophore is made up of individual coccolith plates, arranged in a series of interlocking discs to create a coccosphere (Figure 8.2).

(a)

(b)

Figure 8.1 (a) Bryozoan colonies collected *in situ* from the Otago Shelf, New Zealand (field of view 10 cm). (b) Bryozoan limestone, Tertiary, Southland, New Zealand. Key is 5 cm in length.

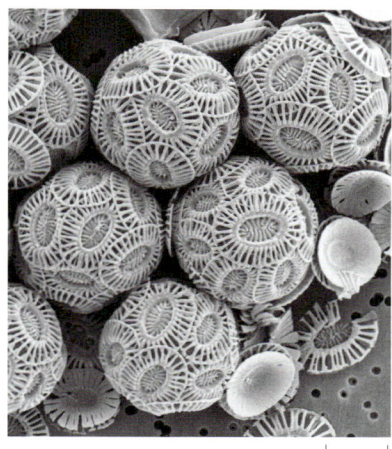

Figure 8.2 Scanning electron microscope image of Recent coccolithophores, *Emiliania huxleyi*. Each coccolithophore has a calcareous skeleton made up of coccolith plates, arranged in a series of interlocking discs to create a coccosphere.

2 µm

Bioclasts are not the only grains of biogenic origin. Since almost everything (including the substrate) is eaten by something else, and passes through some organism's gut, faecal pellets are common. **Peloids** are fine to medium sand-grade, ovoid grains made of structureless carbonate mud that are conspicuous if they rapidly harden (Figure 8.3), but are almost indistinguishable from carbonate mud if they are squashed together soon after deposition. Most peloids are of faecal origin, but some are probably altered bioclasts, though these may be slightly more irregular in shape. Marine algae (including coccolithophores) are eaten by numerous organisms, especially tiny crustaceans such as copepods and shrimps, so that the White Cliffs of Dover are little more than a pile of shrimp droppings! (This also explains why such fine-grained deposits as chalk are able to settle out of suspension when there is evidence of current activity. By aggregating the coccolith plates that are individually only a few microns across, the much larger faecal pellets can settle out of suspension.)

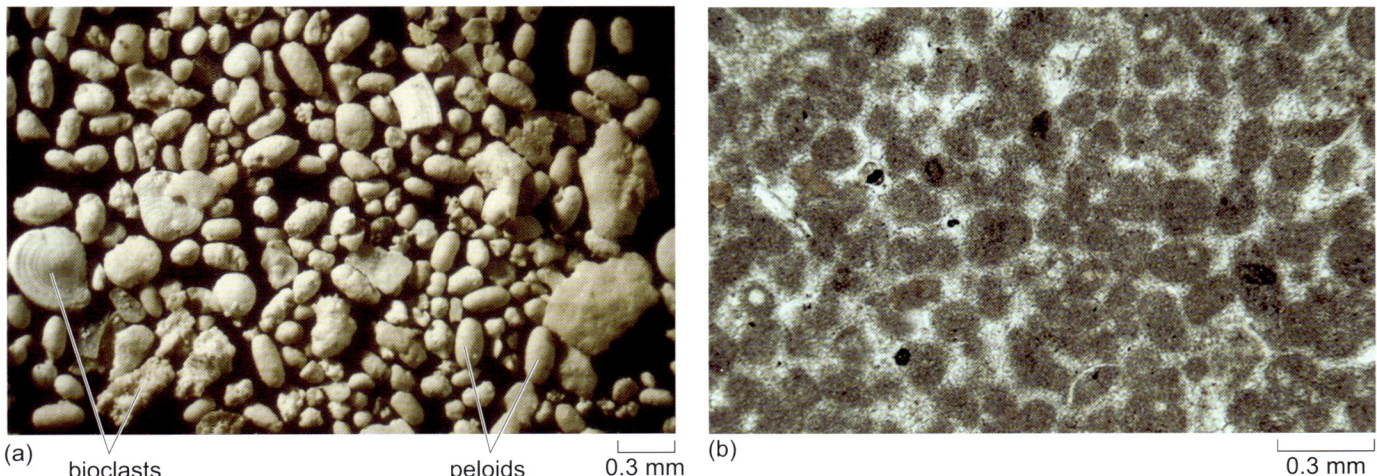

Figure 8.3 (a) Peloids and bioclasts extracted from modern marine sediments. (b) A thin section of calcite-cemented peloids in Jurassic limestone from Dorset, England (viewed in plane-polarised light).

8.1.2 Carbonate ooids

There is a very distinctive type of carbonate grain, typically about 0.2–0.5 mm in diameter that today forms in warm, shallow waters supersaturated with $CaCO_3$ in places such as the Bahamas. Each grain is characterised by a smooth, sub-spherical shape (similar to fish roe, hence the name ooid, meaning egg-like; Figure 8.4a). The internal structure comprises concentric laminae of needle-shaped crystals of aragonite arranged in layers around a nucleus such as a quartz grain or shell fragment (Figure 8.4b). Ooids typically form in agitated waters where they are frequently moved to and fro by tidal and storm currents and by wave action. Single layers of aragonite needles grow randomly on the ooid surface, but are flattened against the grain as they are rolled around. The ooids rarely exceed 0.5 mm in diameter, and this may be because above this size abrasion becomes a more important process than accretion. During burial, the aragonite is altered to calcite (it is calcitised). Accumulations of ooids ultimately form oolitic limestones, such as the Jurassic Great Oolite of southern England.

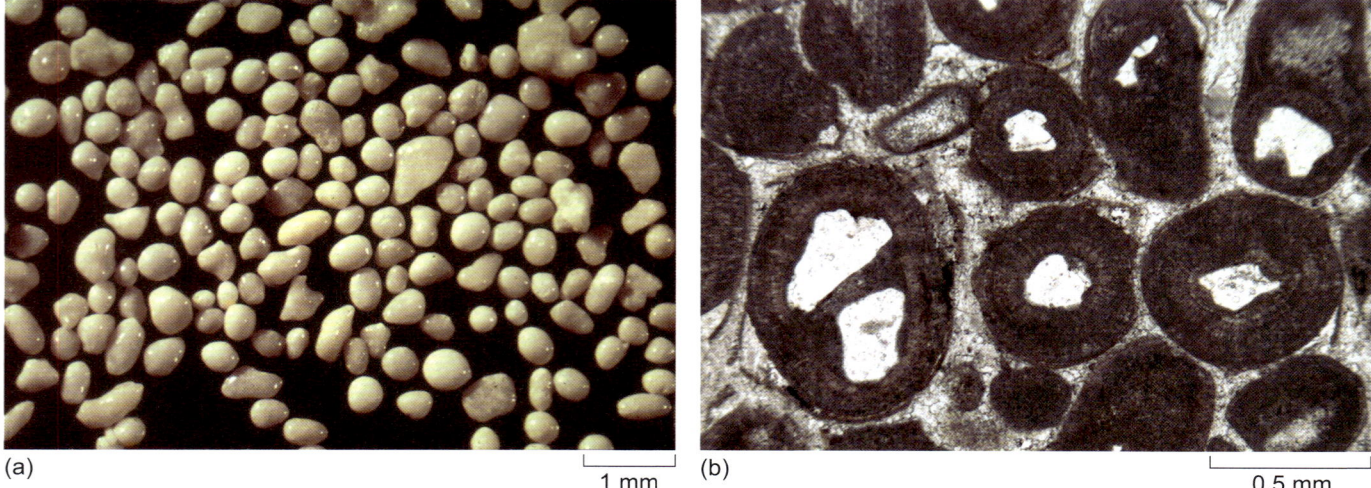

Figure 8.4 (a) Modern ooids from marine sediments. (b) Ancient ooids viewed in thin section (plane-polarised light); many of the ooids here possess a quartz grain nucleus. The original aragonite crystals have been replaced by calcite.

Ooids are undoubtedly the product of chemical precipitation, but a bacterial influence is also likely, borne out by the presence of organic films within the ooid structure. Coarser, rather more irregular grains with an irregular concentric structure form in environments above high tide, including within soils.

8.1.3 Carbonate mud

Many limestones contain a matrix of carbonate (lime) mud that consists of fine-grained calcite crystals commonly less than 0.004 mm (4 μm) in diameter called **micrite** – *micr*ocrystalline cal*cite* – that forms in a number of ways. Some originates as a chemical precipitate forming directly in the water, but clouds of aragonitic needles known as whitings, which occur both in the Arabian Gulf and on the Bahama Banks, may instead be due to shoals of bottom-feeding fish stirring up the mud. In lakes and marginal marine environments, huge quantities of micrite are produced biochemically by cyanobacteria and it is likely that most of the extensive marine, Bahamian-type muds are produced largely by the disintegration of calcareous green algae. Most of the remaining carbonate mud is the product of bioerosion of carbonate skeletons by organisms such as boring sponges, sea urchins and parrot fish.

Whatever the origin, whether depositional or a chemical precipitate, most carbonate mud undergoes recrystallisation to more equidimensional micrite crystals during burial (Figure 8.5).

■ What does the presence of micrite matrix in a limestone suggest about the energy of the environment in which the limestone was deposited?

☐ Micritic matrix implies low energy conditions, though the trapping effect of marine grasses can result in the accumulation of carbonate mud under higher energy conditions.

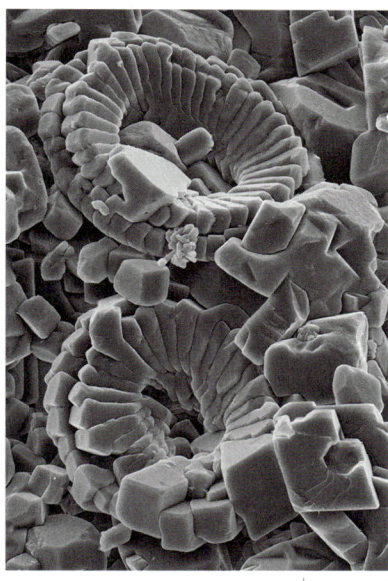

Figure 8.5 Scanning electron microscope view of a micrite-cemented coccolith limestone from the Cretaceous Chalk of North Yorkshire, England. The coccolith plates are arranged in shallow discs that are cemented by calcite crystals, some of which are rhombic in shape.

8.1.4 Carbonate textures

As you saw in Section 7.3.1, textural maturity of siliciclastic sediments is ascertained from grain shape and sorting – well-rounded, well-sorted grains are texturally mature, whereas angular grains of variable size are texturally immature. Coarse-grained siliciclastic sediments are the products of high energy processes, whereas fine-grained sediments are the products of low energy processes. Textural maturity of carbonate sediments on the other hand is more difficult to assess, as many carbonate grains such as ooids are inherently rounded and well sorted, whilst others, such as corals, are very variable both in shape and size and may be preserved where they grew. In addition, the size of the carbonate grain depends on the organism represented, rather than the energy of the transporting medium. It is therefore inappropriate to infer the degree of textural maturity in carbonate sediments in the absence of evidence relating to breakage, sorting and transport.

Activity 8.1 Describing and identifying limestones

In this activity you will be guided through the description and interpretation of limestone grains, matrix and cement, using hand specimens, the Digital Kit and the Virtual Microscope.

Figure 8.6 Rippled carbonate sands formed on a tidal flat at Hamelin Pool, Western Australia. The large, rounded masses are carbonate stromatolites created by sediment-binding blue–green algae. Field of view is approximately 8 m across.

8.1.5 Carbonate bedforms

Bedforms such as planar laminations, ripples, sand waves, and sediment gravity flows are not confined to siliciclastic sediments. They are equally well represented in carbonate sediments, but since carbonate grains are often highly porous as well as very variable in shape, they possess slightly different hydrodynamic properties compared to siliciclastic grains of a similar size. Carbonate ripples are common in shallow marine carbonate sediments (Figure 8.6).

8.2 Siliceous sediments

Some organisms secrete skeletons made of opaline silica ($SiO_2 \cdot nH_2O$). These include radiolarians (marine zooplankton; Figure 8.7a), diatoms (marine or freshwater phytoplankton; Figure 8.7b) and siliceous sponges (marine and freshwater). Both freshwater and seawater are undersaturated

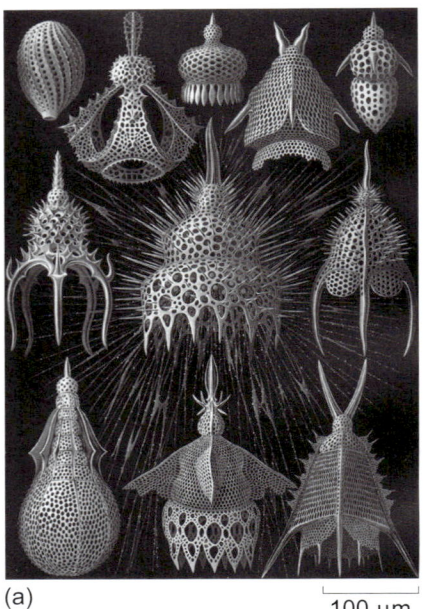
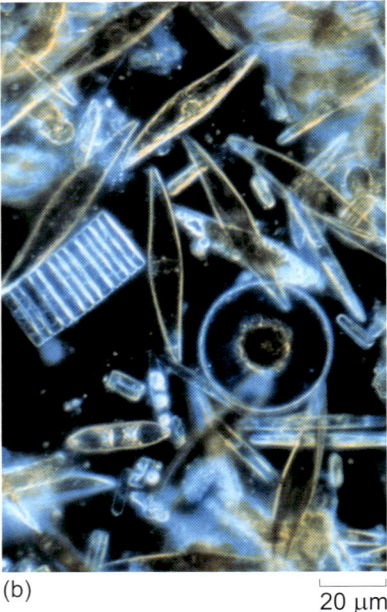

(a) 100 μm (b) 20 μm

Figure 8.7 Modern siliceous planktonic organisms. (a) A selection of radiolarians, which are marine zooplankton. (b) Marine diatoms (phytoplankton) found living between crystals of annual sea ice in McMurdo Sound, Antarctica.

with respect to silica and it is estimated that 95% of opaline silica dissolves either as it sinks through the water column or at the sediment–water interface. If it comes out of solution, it may form **bedded chert**, which may be laterally extensive (Figure 8.8). Bedded chert may also originate from dissolved and reprecipitated silica from volcanic sources.

Figure 8.8 Neogene age pale grey, bedded chert layers within mudstones exposed in California, USA. In this case, the cherts are composed of diatoms. Field of view is approximately 5 m across.

Figure 8.9 Nodular chert (flint) within the Chalk, West Runton, Norfolk, England. The chert nodules surround a large burrow-fill. Note boot on the left-hand side for scale.

Some limestones contain **nodular chert** (or flint), often concentrated along particular bedding planes, commonly nucleated within burrow-fills, and formed by the dissolution and reprecipitation of biogenic silica skeletons within the carbonate sediment (Figure 8.9).

8.3 Carbonaceous sediments

Organic-rich, carbonaceous sediments (not strictly derived from solution) accumulate in a number of environments.

- ■ What are the conditions necessary to preserve organic material?
- □ Anoxic conditions with reduced bacterial activity to prevent or slow down decomposition.

These conditions are found in bogs and swamps, including tropical mangrove swamps. Decaying vegetation, particularly trees and leaves, accumulates as peat (even after transport along the coast). With progressive burial, the peat is converted to lignite (brown coal) and then to bituminous coal and finally anthracite (Book 1, Section 7.3.3). In stagnant lakes and in stratified marine waters with restricted circulation, decaying phytoplankton (algae) and plant debris settle out of suspension and become incorporated into the fine-grained sediments and can be preserved as organic-rich mudstones. The organic matter is first converted to **kerogen** (Figure 8.10) and, if optimum burial conditions prevail for sufficient time, to oil (temperature = 70–100 °C; depth = 2–3.5 km) or gas (above 150 °C).

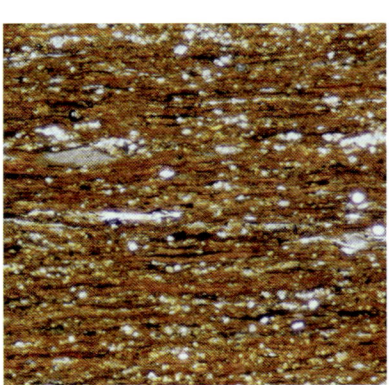

1.0 mm

Figure 8.10 Kerogen-rich laminae (brown) within Jurassic mudstone, Kimmeridge, Dorset, England.

8.4 Evaporites

Sediments called evaporites can be formed by direct precipitation from solution, usually due to evaporation under arid conditions. Halite **pseudomorphs** result from the precipitation of halite crystals on or immediately beneath the sediment surface, dissolution of the halite during rare rainfall, and filling in of the cube-shaped holes with more sediment (Figure 8.11). Evaporites range from anhydrite nodules ($CaSO_4$) (Figure 8.12) and cubic halite crystals (NaCl) that form within arid coastal sediments above high tide (known as **sabkhas**; see Section 14.4.1), to extensive bedded gypsum ($CaSO_4.2H_2O$) and halite deposits within evaporating, marine basins with a restricted circulation such as the Mediterranean during the Miocene (Figure 8.13). As it requires 1 km of seawater to generate just 13 m thickness of halite, clearly a basin has to be replenished many times over to enable thick successions of halite to develop, such as in the Permian succession of the North Sea (which is >1 km thick).

Figure 8.11 Halite pseudomorphs in Jurassic sabkha deposits, Dorset, England.

Figure 8.12 White anhydrite bands within supratidal (above high tide) sediment. The anhydrite has replaced the sediment as the crystals grew, Abu Dhabi.

Figure 8.13 Bedded, slightly contorted gypsum deposits, Miocene, Sicily. 2.5 cm diameter coin for scale.

8.5 Summary of Chapter 8

1. Weathering processes create ions in solution that are transported to the sea and are either precipitated directly in seawater to produce crystalline, chemical sediments, or processed organically, to create biogenic sediments. Calcium carbonate and silica give rise to carbonate and siliceous sediments; organic, carbon-rich material gives rise to carbonaceous sediments.

2. Calcareous marine organisms extract Ca^{2+} and bicarbonate ions from seawater to secrete their hard parts, which can be made of aragonite or calcite (high- or low-Mg calcite). These bioclasts will eventually be buried to form limestone above the carbonate compensation depth, and in the case of microscopic marine algae (coccolithophores) to form fine-grained limestone termed chalk.

3. Bioclasts and peloids (mostly faecal pellets) are of biogenic origin, whereas ooids are chemical precipitates, though probably influenced by bacterial activity.

4   Carbonate mud or micrite forms both a matrix and a cement. Carbonate mud is produced as a biochemical precipitate, by mechanical disintegration of calcareous green algae, or by bioerosion. The cement forms after deposition and most of it is recrystallised during burial.

5   Textural maturity and high versus low energy processes can only be measured in carbonates when there is clear evidence relating to breakage, sorting and transport. Some grains such as ooids are naturally rounded and well sorted, whilst others such as corals are very variable in shape and size.

6   Carbonate bedforms are often identical to those produced in siliciclastic sediments. They include current ripples, wave ripples, sand waves and sediment gravity flows.

7   Some organisms secrete opaline silica skeletons, which are very soluble. Those that are buried before they can be dissolved form either bedded chert or, where they are dispersed within carbonate sediments, may be dissolved and reprecipitated as nodular chert.

8   Carbonaceous sediments accumulate in anoxic conditions with reduced bacterial activity. Vegetation accumulating in swamps and bogs, or even after being transported along the coast, will form peat. Algal and plant debris settling out in stagnant lakes and stratified marine waters and incorporated into the fine-grained sediments will form organic-rich mudstones. With progressive burial, peat is converted to coal, and organic matter in marine sediments to kerogen and then to oil and gas.

9   Evaporite minerals such as anhydrite, gypsum and halite are precipitated directly due to evaporation under arid conditions. Repeated replenishment of the restricted marine waters is required for thick evaporite deposits to accumulate.

8.6   Objectives for Chapter 8

Now you have completed this chapter, you should be able to:

8.1 Describe the components of carbonate, siliceous, carbonaceous and evaporitic sediments.

8.2 Outline the differences between biogenic and non-biogenic carbonate grains.

8.3 Outline the origin of carbonate ooids and carbonate mud.

8.4 Explain the fundamental difference between the textures of siliciclastic sediments and carbonate sediments.

8.5 Explain the significance of carbonate and silica undersaturation.

8.6 Distinguish between biogenic and chemical limestones on the basis of their dominant grain composition.

8.7 Outline the biological and chemical processes that lead to the formation of chert, coal, oil and gas.

8.8 Outline the conditions under which evaporites form and explain how the textures of evaporites can be related to their mode of formation.

Now try the following questions to test your understanding of Chapter 8.

Question 8.1

Extensive deposits of halite (rock salt) occur within Triassic sandstones of Cheshire in NW England. What does this suggest about the geographic conditions (including climate) in Cheshire during the Triassic Period, and why?

Question 8.2

At the present day, we find that carbonate sediments increase in abundance towards low-latitude tropical and equatorial regions. Suggest a reason for this observed pattern.

Question 8.3

Oolitic limestones are locally abundant among Jurassic rocks in Britain and they sometimes exhibit large-scale cross-stratification. What can you infer about the energy of the environment and the climatic conditions in which they were deposited, and why?

Chapter 9 From sediments to rocks

Contrary to popular understanding, very few sediments become rocks simply by being squashed. Rocks do indeed undergo minor to appreciable compaction during burial (mudstones are compressed by as much as 90%) but they also require the addition of a natural cement to transform them into solid (lithified) rock. Some sediments also undergo dissolution and recrystallisation; the term diagenesis is used to include all changes that occur after deposition and before any metamorphism takes place (which starts at temperatures above 150–200 °C). The nature of these changes depends on sediment composition, the chemistry of the water in the pore spaces between the grains, cement availability and diagenetic environment – whether subaerial or submarine, shallow or deeper burial. The resultant changes that take place during burial and uplift can be both variable and complex. Early cementation provides a rigid framework that resists compaction during burial, whereas late cementation takes place too late to prevent very close packing of grains.

9.1 Compaction

Unconsolidated sediments are highly porous – sands initially contain about 40–50% intergranular pore space, depending on grain shape and degree of sorting (fine grains more easily fill the spaces between coarser grains) and muds contain about 70–90% water. Below the **water table**, all pore space is filled with fluids, usually water, and during **compaction**, grains are packed more closely together and much of the pore water is squeezed out (Figures 9.1 and 9.2).

■ Which two distinctive structures are produced by very rapid dewatering of mud when sand is dumped on top of it (either by turbidity currents or by storm action)?

☐ Flame structures and ball-and-pillow structures, where mud is squeezed up into the overlying sand (see Figure 7.45a and b).

Mudstones contain approximately 30% water at 1 km depth, and significant further dewatering requires temperatures of ~100 °C at depths of 2–4 km.

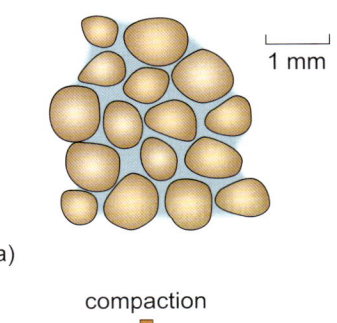

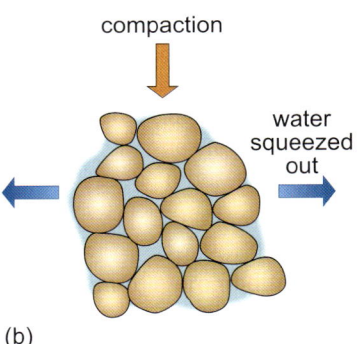

Figure 9.1 Cross-section through a sand-grade sediment (with well-rounded, well-sorted grains): (a) at deposition: (b) after compaction.

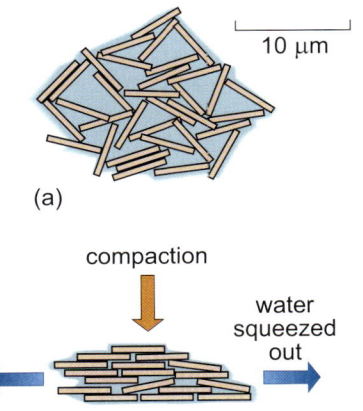

Figure 9.2 Cross-section to show the compaction of a claystone: (a) clay flakes soon after deposition; (b) clay flakes during compaction.

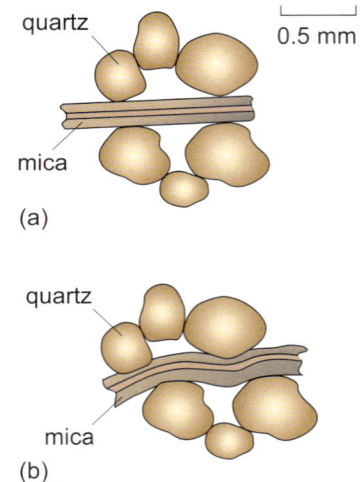

Figure 9.3 Deformation of a mica flake where it is bent around quartz grains (a) before, and (b) after compaction.

Initially, grain-supported fabrics are characterised by point contacts between adjacent grains, i.e. the grains are just touching (though in a 2D image, some grains are in contact out of the plane of the section; Figure 9.1a). Within a few metres depth or even less, elongate grains such as clay minerals and micas are rotated to a more stable, horizontal orientation (Figure 9.2b) and all grains are in closer contact.

With progressive burial, soft grains such as micas become distorted around more resistant grains, usually quartz (Figure 9.3), and as the pressure of the overburden increases, grains in contact with each other are dissolved. This **pressure dissolution** gives rise to concavo-convex grain contacts where only one of the adjacent grains dissolves, or to sutured contacts where both grains have suffered dissolution (Figure 9.4). More pervasive pressure dissolution produce sutured planes called **stylolites**, which are particularly common within limestones (Figure 9.5).

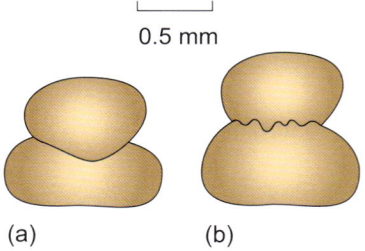

Figure 9.4 Sketch to show (a) a concavo-convex grain contact, and (b) a sutured grain contact caused by pressure dissolution.

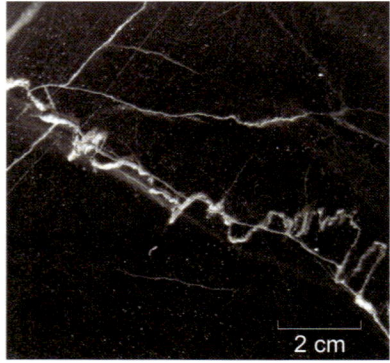

Figure 9.5 Fine-grained, pelagic limestone of Triassic age from Slovakia, with veins and a stylolite seam, along which significant dissolution has taken place. In this example, the serrated stylolite is sub-parallel to the bedding.

9.2 Cementation in sandstones

Sandstones are commonly cemented by quartz, carbonate, clay minerals and/or iron oxide. Many of these cement phases are derived from the chemical breakdown of unstable mineral phases such as feldspar, olivine, pyroxene and amphibole.

■ Recalling what you know about the processes that take place during compaction, suggest an additional source of cement.

□ The silica released by pressure dissolution during burial.

Cement is derived either locally or from an external source, such as deeper parts of the succession where temperatures are higher and grain solubilities are greater.

9.2.1 Quartz cement

Quartz grains often possess quartz **overgrowths** that are in optical continuity with the host grain so that both grain and cement go into extinction at the same time (Figure 9.6b). The grain–cement contact may only be apparent in thin section when hematitic (iron oxide) dust coats the host grain (Figure 9.6a). In hand specimen, the overgrowths often mask grain surfaces and give rise to a crystalline fabric similar to a quartzite (metamorphosed quartz sandstone).

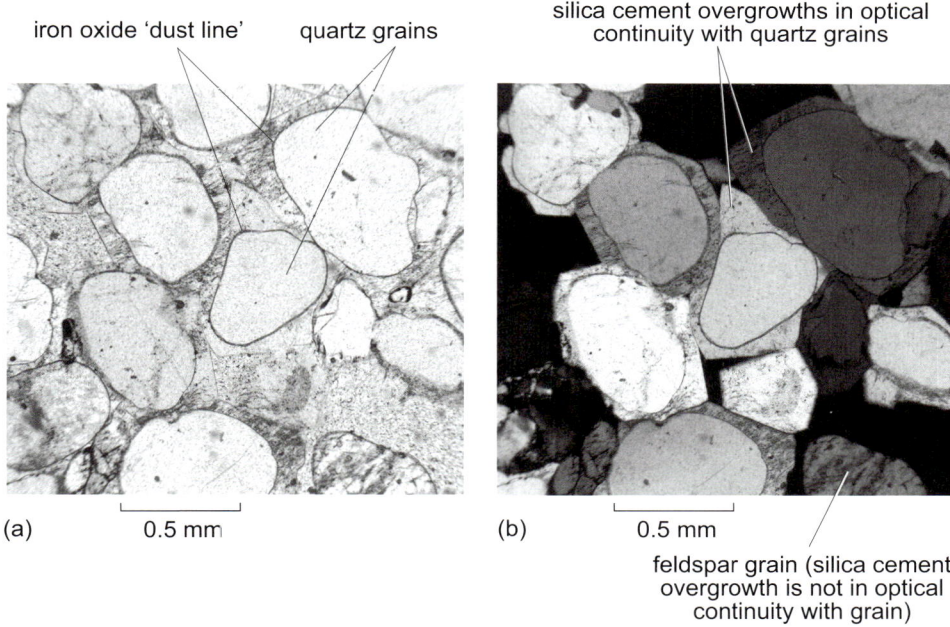

Figure 9.6 Thin section of a sandstone with a quartz cement. (a) Viewed in plane-polarised light. The quartz grains can be distinguished from their overgrowths only by iron oxide-rich dust coatings on the grain surfaces. (b) Viewed between crossed polars, grains and overgrowths can be seen to be in optical continuity, since they demonstrate the same degree of extinction.

■ What evidence would you look for in a thin section of a quartz-cemented sandstone to suggest that pressure dissolution might have contributed some of the cement?

□ Sutured or concavo-convex contacts between grains.

9.2.2 Carbonate cement

Calcite ($CaCO_3$) is the most common carbonate cement in sandstones, but dolomite ($CaMg(CO_3)_2$) and **siderite** ($FeCO_3$) may be locally important. Calcite (and sometimes siderite) is often the first-formed cement, precipitating from alkaline pore waters in near-surface sediments. In environments exposed to the air, such as river plains, deserts and in soils, especially in arid and semi-arid climates, coarse sediments often become cemented by calcite to form

Figure 9.7 Cemented beachrock exposed by storm action (darker, left-hand side of photograph) and loose pebbles (paler, right-hand side of photograph), Olu Deniz, Turkey.

calcrete. In these settings, calcite is precipitated as the Ca-rich pore fluids continually evaporate and are recharged, but in submarine environments, the $CaCO_3$ is probably sourced mainly from dissolving bioclasts. Beach sands and gravels may also be cemented by calcite or aragonite to form a **beachrock**, usually formed a few centimetres beneath the surface, and then later exposed by storm action (Figure 9.7). Later carbonate cements that are precipitated during burial often post-date quartz overgrowths and may precipitate in response to an increase in pH and/or temperature. Water percolating through siliciclastic sandstones may become acidic, inhibiting the precipitation of carbonate cements.

Fine-grained dolomite cement is often the product of near-surface evaporation. Coarse dolomite may form later when Ca^{2+} and Mg^{2+} concentrations are increased, often by leaching of silicates and alteration of clay minerals (Figure 9.8).

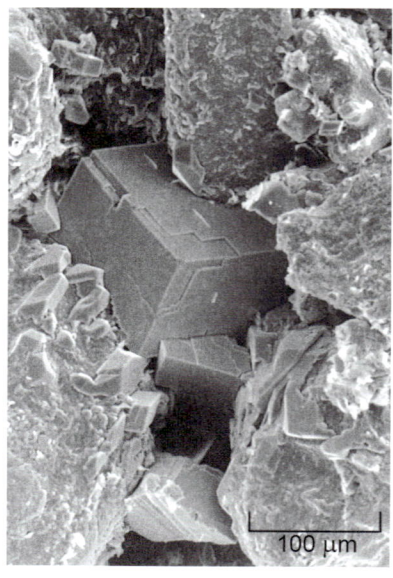

Figure 9.8 Scanning electron microscope photograph of dolomite cement (rhombic shape) within a sandstone. Core sample, Permian, southern North Sea, UK.

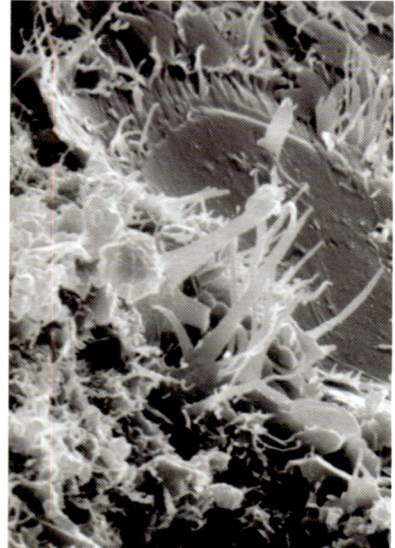

Figure 9.9 Scanning electron microscope photograph of fibrous to ribbon-like illite that fills much of the pore space between adjacent grains. Core sample, Permian, southern North Sea, UK.

9.2.3 Clay minerals

Clay minerals are precipitated in the shallow subsurface or during deeper burial from pore waters moving through the sediment. Illite and kaolinite are the most common clay cements, but chlorite is an important early cement in marine environments, as are smectitic clays in desert environments. Many of the clay minerals form open networks of plates and fibres that slightly reduce intergranular porosity, but can drastically reduce permeability – the ease with which a fluid can move through the rock – by blocking the narrow channels between adjacent pores (Figure 9.9).

9.2.4 Iron oxide cement

Many continental siliciclastic rocks are red in colour due to the presence of surprisingly small quantities (less than 1%) of hematite (Fe_2O_3). Some of the hematite may be detrital in origin but much of it is formed *in situ* from weathering of mafic minerals. The iron released is precipitated as a yellowish-brown, hydrated iron oxide called **limonite**. This slowly converts to hematite in a strongly oxidising environment, particularly deserts, creating so-called 'red beds' (Figure 9.6; see also Section 6.4.1). Red sedimentary rocks are often associated with a continental rather than a marine environment. (If reducing conditions prevail, the iron is present in the more soluble ferrous state and when incorporated into clays, it will impart a green colour to the sediments. The green colour may revert to red on weathering.)

9.3 Cementation in limestones

Limestones are very susceptible to dissolution and cementation due to their inherent solubility. This is clearly illustrated by the development of extensive cave systems that are created as limestone is gradually dissolved by slightly acidic rainwater, and then reprecipitated as crystalline calcite to form stalactites and stalagmites (Figure 9.10).

Aragonite is the principle building block of marine skeletons such as corals, gastropods, infaunal bivalves, and some bryozoans, but aragonite is stable only in seawater, so aragonitic skeletons are prone to dissolution if exposed to freshwater. Carbonate cementation and/or dissolution starts on the sea floor, often at the same time as sedimentation, and continues through burial and/or uplift. Sea-floor cementation gives rise to **hardgrounds** that provide substrates for various encrusting and boring organisms (Figure 9.11).

Figure 9.10 Stalactites in Treak Cliff Cavern, Derbyshire, England. See also Figure 6.7.

Figure 9.11 Oysters and serpulid worms encrusting a Miocene hardground, Southland, New Zealand.

Calcite cementation continues during burial, either as clear **sparite** >4 μm across, though often appreciably larger, or as clay-grade micrite (<4 μm across; see Section 8.1.3). In thin section, the micrite appears dark grey because 7 or 8 crystals would be stacked one on top of the other in a 30 μm thick section (Figure 9.12). Sparite is usually visible in hand specimen, and is characterised by highly reflective cleavage planes; it is precipitated in the pore spaces after

Figure 9.12 Thin section from the Miocene submarine hardground illustrated in Figure 9.11. Notice the dark-grey depositional micrite matrix (e.g. top right) and cavity in the centre lined with inclusion-rich calcite (replacing submarine aragonite), then coated with dark-red **goethite** (a hydrated iron oxide), and later completely filled with clear sparite (containing bubbles in the mounting resin).

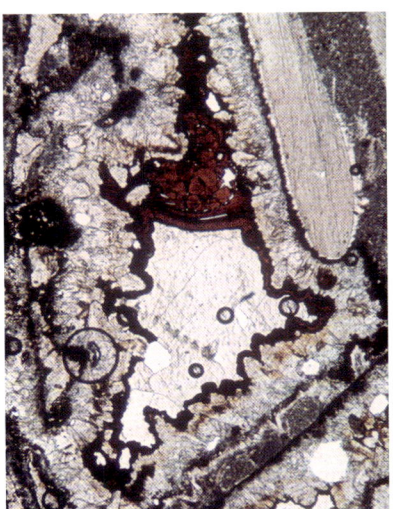

deposition of the surrounding sediment. Micrite forms a dense, white to grey material in hand specimen, representing either a post-depositional cement or a depositional matrix.

■ In Figure 9.12, what do you notice about the position of the dark-red goethite?

☐ It lines and partly bridges the calcite-lined cavity. This pore-bridging aspect is typical of cement precipitated in only partly saturated pores (i.e. the sediment was exposed at the surface above the water table) and goethite precipitated as meniscus bridges (i.e. water in the pore adhered to the pore margin as a meniscus and cement was precipitated just within this water-wet zone).

Dolomite may precipitate in strongly evaporating conditions at the surface or during burial, where it forms either a cement or a replacement for limestone (Figure 9.13). Dolomite typically forms well-formed rhombic crystals, but so too can calcite, and the two minerals can only be reliably distinguished in thin section if a chemical stain is applied. (In the field, dolomitic and calcitic rocks can be distinguished because calcite reacts vigorously with dilute hydrochloric acid, whereas dolomite does not.)

Figure 9.13 Thin section of a Permian dolomite, Abu Dhabi, United Arab Emirates, showing dolomite rhombs that have replaced the original limestone fabric. The pore spaces are stained blue.

9.4 Cementation in mudstones

Compaction is the main diagenetic process to affect mudstones but there can also be important changes in clay mineralogy as the temperature rises at depth – kaolinite and smectite are replaced by K-rich illite and chlorite, and some clays are recrystallised.

9.5 Nodules

Nodules (often called concretions) are common diagenetic structures found in sedimentary rocks. The nodules consist of ovoid or irregular lumps of cemented carbonate, anhydrite or silica that are much harder than the surrounding sediment and often protrude during weathering. They range in size from about 1 cm to 2 m or more in diameter and some nodules coalesce to form almost continuous tabular sheets (Figure 9.14).

Figure 9.14 Calcite nodules that range from a few millimetres to several centimetres in diameter are aptly named 'cannonballs', Roker Rocks, Sunderland, England. Coin about 3 cm in diameter for scale.

Calcite nodules tend to occur more often in sandstones and calcareous mudstones, siderite nodules in iron-rich mudstones, and silica (chert or flint) nodules in fine-grained limestones. Most nodules are found along particular bedding planes, reflecting either a change in the rock chemistry or the pore water chemistry, or a particular biological feature (e.g. a burrow or bed rich in body fossils), which has provided a nucleation site for the nodules (Figure 8.9). Calcite nodules are commonly derived from dissolution and reprecipitation of calcite or aragonite. **Septarian nodules** are distinctive carbonate nodules with prominent shrinkage cracks filled with calcite, barite or other minerals (Figure 9.15). Chert nodules are usually sourced from siliceous bioclasts (Section 8.2). The nodules usually form during early burial, often nucleating on a fossil or burrow-fill, which are then protected from subsequent compaction. Anhydrite nodules occur within supratidal (above high tide) evaporites, forming very soon after deposition of the host sediment, and displacing it as they grow (Figure 8.12). Nodules can also form during late burial.

Figure 9.15 Calcite-filled shrinkage cracks in part of a septarian nodule of Jurassic age, Dorset, England.

Activity 9.1 Recognising diagenetic features

In this activity you will use the Virtual Microscope to identify some diagenetic features.

9.6 Classifying sedimentary rocks

Rocks exposed at the Earth's surface are weathered and eroded, and the products transported via rivers into marine basins, where sediments are deposited, and altered, during progressive burial.

■ What are the two ways in which siliciclastic sedimentary rocks are classified?

□ Grain size and mineralogical composition (see Book 1, Section 7.1).

Grain size and composition broadly reflect the origin and transport history of the sediment, and these properties provide clues to their depositional environment. Post-depositional changes provide evidence of their burial history. It is therefore no accident that grain size and composition figure largely in the standard classification of sedimentary rocks.

9.6.1 Siliciclastic rocks

The grain size of siliciclastic rocks provides valuable information concerning the energy of the environment in which the sediments were deposited, since it relates to the speed of the transporting fluid (Section 7.1). Muds, which include clay and silt, are defined as <62.5 μm in diameter, sands are 62.5 μm–2 mm, and gravels, which encompass granules, pebbles, cobbles and boulders, are >2 mm in diameter. Detailed subdivisions are given in Figure 9.16 and these are rigorously applied to each sedimentary rock description. It is also important to measure the range of grain sizes since this provides a measure of grain sorting. Overall grain size of the sediment or rock is based on the most volumetrically abundant grain size. Such detailed measurements can be achieved using a grain-size scale.

Grain size (most volumetrically abundant grains)[a]	**Sediment name**		**Sedimentary rock name**	
>256 mm	boulders	gravel	conglomerate (rounded fragments) or breccia (angular fragments)	
64–256 mm	cobbles	gravel	conglomerate (rounded fragments) or breccia (angular fragments)	
4–64 mm	pebbles	gravel	conglomerate (rounded fragments) or breccia (angular fragments)	
2–4 mm	granules	gravel	conglomerate (rounded fragments) or breccia (angular fragments)	
62.5 μm–2 mm	sand		sandstone	
4–62.5 μm	silt	mud	siltstone	mudstone (shale)[c]
<4 μm	clay[b]	mud	claystone	mudstone (shale)[c]

[a] μm = micrometre = 10^{-6} m.

[b] Clay can refer to material with a *grain size* of less than 4 μm; or to a certain type of sheet silicate mineral.

[c] Shale, if fissile.

Figure 9.16 Grain-size scale for siliciclastic sedimentary rocks. Note that it is the volumetrically most abundant grains in a sedimentary rock that determine its classification.

Activity 9.2 Classifying siliciclastic rocks

In this activity you will use hand specimens, the Digital Kit and the Virtual Microscope to apply the classification of siliciclastic rocks.

A siliciclastic rock with predominantly sand-grade grains is a sandstone, but this name does not convey *any* compositional information. In Book 1, Figure 7.15 you saw how sandstones, mudstones and limestones can be informally named according to how much quartz, clay or calcite they contain. Names such as 'muddy sandstone', 'sandy limestone' or 'calcareous mudstone' give you a good idea not only of the grain size of a rock and the minerals that it contains, but also the relative proportions of the three main minerals.

Question 9.1

What are the main minerals you would expect the following rocks to contain: (a) a muddy sandstone, (b) a sandy limestone and (c) a calcareous mudstone? In each case, state which mineral you would expect to form the highest proportion.

Rocks such as RS 12 in the Home Kit contain not only sand- to gravel-sized quartz grains and rock fragments, but also some feldspar and significant amounts of mud-grade matrix. Classifying these rocks using only grain size, or even based on the amount of quartz, calcite or mud, omits valuable information that could be used to aid the interpretation of the rock's origin or depositional environment.

■ What feature related to chemical weathering will aid this interpretation?

☐ Compositional maturity (Section 6.4.1).

The four main grain types: quartz, feldspar, rock or lithic fragments, and clay minerals each convey information on the compositional maturity of the rock, reflecting the composition of the parent rocks, the length of exposure at the surface, intensity of weathering and duration of transport. The first three grain types form the main detrital constituents of sandstones, whereas clay minerals are the main constituent of claystones (fine-grained mudstones) and occur as a matrix within siltstones, sandstones and conglomerates.

Sandstones with less than 15% mud-grade matrix are classified as **arenites** and those with between 15% and 75% mud-grade material as **wackes**. The abundance of clay minerals in wackes makes many of them grey hence the old term greywackes for poorly sorted, muddy sandstones. Mudstones contain more than 75% mud.

The sandstones can be usefully subdivided, based on their quartz, feldspar and rock (lithic) fragment content. This subdivision provides a powerful tool in reconstructing the origin and history of the sediments. In each case, the amount of quartz, feldspar and rock fragments (ignoring any clay content) are recalculated to 100% and plotted on a ternary, QFR (Q = quartz, F = feldspar, R = rock fragments) diagram (Figure 9.17).

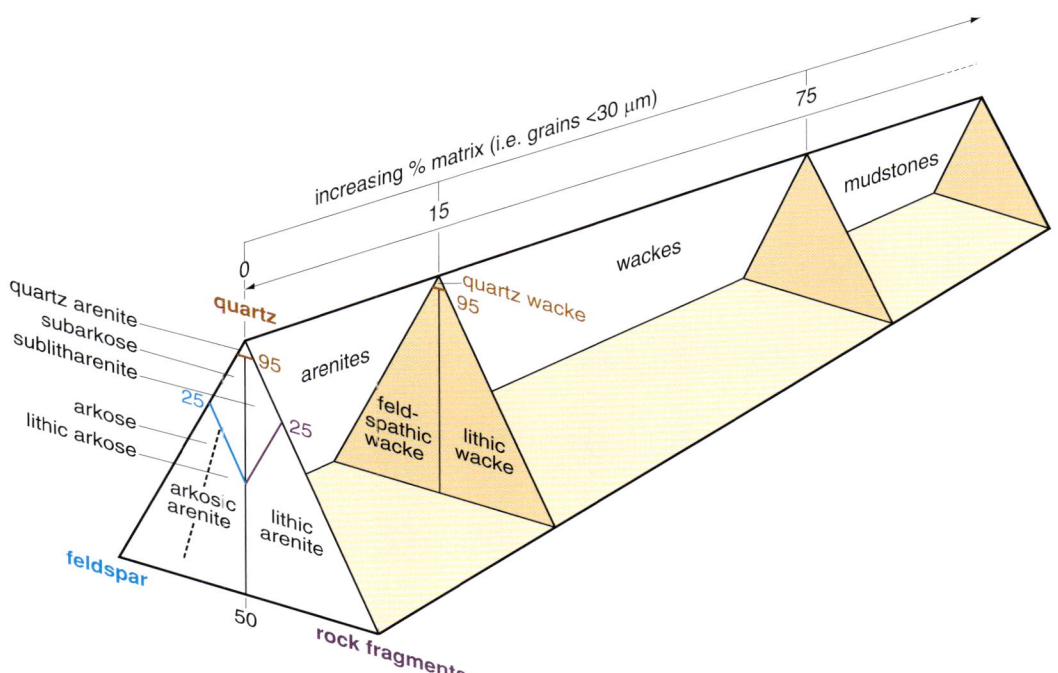

Figure 9.17 Sandstone classification using a QFR diagram (Q = quartz, F = feldspar, R = rock fragments). Note that on the end face, the components and the field boundaries are colour-coded.

The arenites are classified as follows:

- quartz dominant: **quartz arenite** >95% quartz
- feldspar dominant: **arkosic arenite** >25% feldspar; **subarkose** <25% feldspar
- rock fragments dominant: **lithic arenite** >25% rock fragments; **sublitharenite** <25% rock fragments.

Arkosic arenites can be further subdivided into **arkose** and **lithic arkose** (Figure 9.17). The wackes (with more than 15% mud) are similarly classified:

- quartz dominant: **quartz wacke** >95% quartz
- feldspar dominant: **feldspathic wacke** >5% feldspar
- rock fragments dominant: **lithic wacke** >5% rock fragments.

Question 9.2

(a) Using the sandstone classification in Figure 9.17, classify the following arenites (with <15% mud-grade matrix). (Hint: if you cannot remember how to use a ternary diagram, refer back to Box 4.1 in Book 1.)
  (i)   80% quartz, 15% feldspar, 5% rock fragments
  (ii) 40% quartz, 10% feldspar, 50% rock fragments
  (iii) 55% quartz, 35% feldspar, 10% rock fragments.

(b) Considering each of the three arenites, which would be the most compositionally mature, and why?

(c) Classify the following wackes (with >15% mud-grade matrix):
  (i) 97% quartz, 3% feldspar, 0% rock fragments
  (ii) 50% quartz, 40% feldspar, 10% rock fragments.

Activity 9.3   Diagenesis and classification of siliciclastic rocks

In this activity you will complete your diagenetic observations and classification of siliciclastic rocks using hand specimens, the Digital Kit and the Virtual Microscope.

9.6.2   Carbonate rocks

A rock with more than 50% calcite is a limestone and a rock with more than 50% of the mineral dolomite is, somewhat confusingly, called dolomite (or more logically a dolostone). A dolomitic limestone contains a significant amount of the mineral dolomite but more than 50% calcite.

Question 9.3

In carbonate rocks, what are (a) the three main types of carbonate grain, and (b) the two common types of intergranular carbonate material?

Limestones can be further subdivided in a number of ways. Two classifications are described here to reflect the grain composition of limestones and to provide some information related to their energy of deposition. The first classification provides compositional qualifiers for the dominant grain types:

- bioclastic limestone
- oolitic limestone
- peloidal limestone.

The second classification (Table 9.1) reflects both the dominant grain type and the nature of the interstitial (in between) material, whether it is principally a sparite cement or a micritic matrix (though micritic cements also occur; see Section 8.1.3). In most cases, the presence of micrite suggests low energy conditions, whereas the absence of micrite implies higher energy conditions that winnowed out any depositional lime mud.

- a bioclastic limestone cemented by sparite, and with little or no micritic matrix would be a **biosparite** (Figure 9.18a)
- an oolitic limestone cemented by sparite, and with little or no micritic matrix would be an **oosparite** (Figure 9.18b; though strictly speaking this is a bio-oosparite – see footnote to Table 9.1)
- a peloidal limestone cemented by sparite would be a **pelsparite**.

Using the same logic, a bioclastic limestone with a micritic matrix would be a **biomicrite** (Figure 9.18c). An additional category termed **biolithite** is used for those limestones that have grown *in situ*, such as limestones with framework-building corals or sediment-binding stromatolites (Figure 9.18d).

■ What would be the terms for an oolitic limestone with a micritic matrix, and a peloidal limestone with a micritic matrix?

□ **Oomicrite** and **pelmicrite**, respectively.

Table 9.1 Classification of limestones based on composition of the grains and intergranular material.

Principal grain type	Limestone type*	
	cemented by sparite (little or no micrite matrix)	micrite matrix
bioclasts (*bio-*)	biosparite	biomicrite
ooids (*oo-*)	oosparite	oomicrite
peloids (*pel-*)	pelsparite	pelmicrite

* Note that 'hybrid' limestones that contain significant amounts of more than one type of grain or intergranular material are common. The descriptive classifications of such limestones should reflect this mix. For example, a limestone containing bioclasts, together with a significant proportion of ooids (but subordinate to bioclasts) and a sparite cement, would be classified as an oobiosparite. A limestone containing mostly ooids and subordinate bioclasts in a sparite cement would be a bio-oosparite. A bioclastic limestone with significant amounts of micrite matrix and sparite cement would be a biomicsparite.

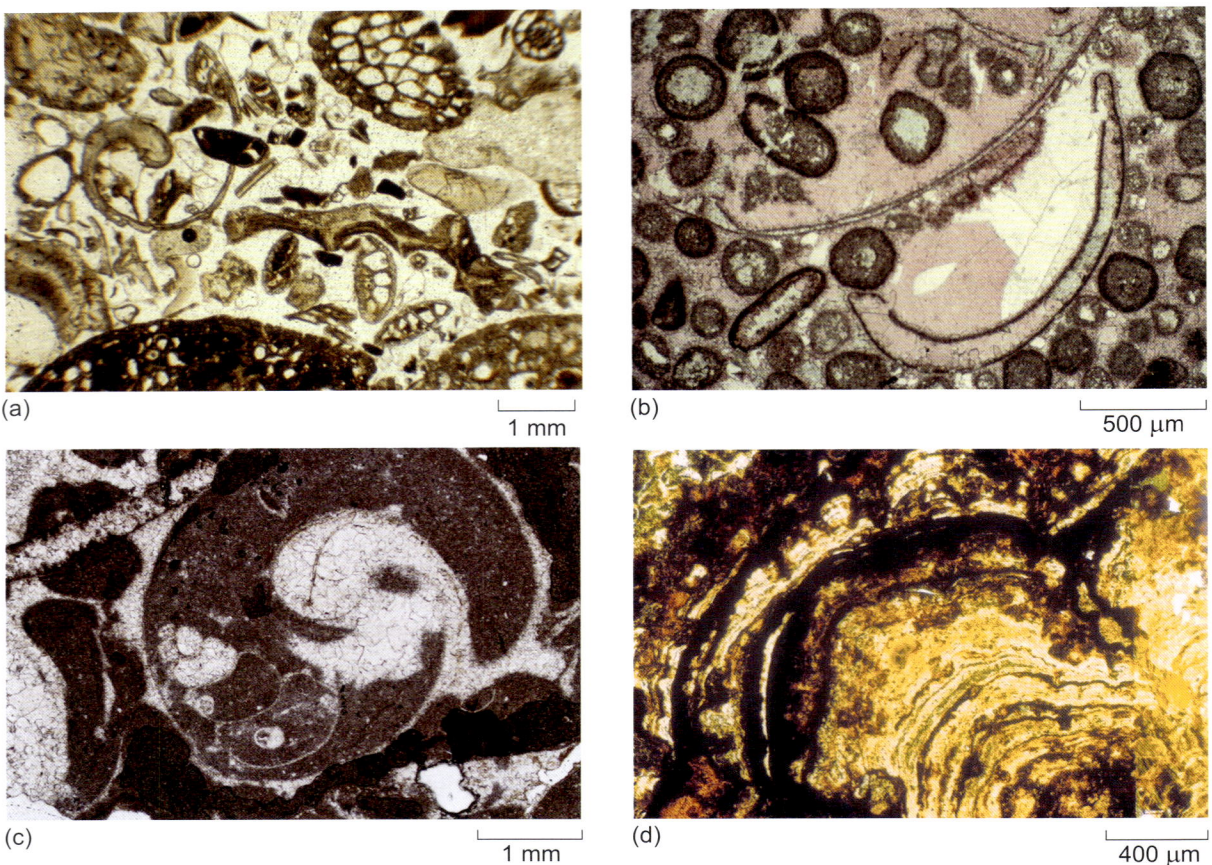

Figure 9.18 (a) Thin section of a bryozoan biosparite in the making (not yet cemented), Otago Shelf, New Zealand. (b) Thin section of a bio-oosparite from Abu Dhabi, United Arab Emirates. The ooids and bioclasts (preserved by thin micritic coatings) are cemented by calcite (artificially stained pink). (c) Thin section of a biomicrite from the Bembridge Limestone, Isle of Wight, England. Coiled gastropods are filled with grey micrite and the aragonitic shells have been replaced by calcite. (d) Biolithite of Tertiary age from New Zealand. The fine crinkly laminations of these algal stromatolites are picked out by a black manganese mineral.

Activity 9.4 Classifying limestones

In this activity you will classify some limestones using hand specimens, the Digital Kit and Virtual Microscope.

9.7 Summary of Chapter 9

1. Most sediments undergo compaction and cementation and often some degree of dissolution, recrystallisation and replacement during lithification. Diagenesis encompasses all those changes that take place after deposition and before the onset of metamorphism.

2   Recently deposited sands contain 40–50% intergranular pore space, and muds contain up to 90% water. During compaction, grains are packed closer together and the water is progressively squeezed out. Increasing burial leads to pressure dissolution at grain contacts, creating sutured grains and, especially in limestones, stylolites. The dissolved material is one source of cement.

3   Sandstones are commonly cemented by quartz overgrowths and intergranular calcite. Water percolating through quartz sandstones often becomes acidic, dissolving carbonates and inhibiting their precipitation. Clay mineral cements are less common, but they can have a drastic effect on permeability.

4   Red beds contain traces of hematitic dust that initially forms as yellow hydrated iron oxide (limonite) and very slowly converts to hematite. The process requires strongly oxidising conditions and is common in desert environments.

5   Limestones are much more susceptible to dissolution and cementation than siliciclastic sediments, creating characteristic karstic landscapes. Aragonite in particular is prone to dissolving in freshwater. Carbonate diagenesis begins early on the sea floor and continues during burial.

6   Contemporary cementation of marine sediment by aragonite or high-Mg calcite gives rise to submarine hardgrounds and beachrock. Low Mg-calcite is precipitated in freshwater and from groundwaters during burial as sparite; micrite is recrystallised. Dolomite may form as a primary cement or secondary replacement.

7   Mudstones undergo cementation and recrystallisation as well as appreciable compaction. The clay minerals kaolinite and smectite may be replaced by illitic and chloritic clays.

8   Nodules or concretions are hard structures that weather out from the surrounding beds. They often nucleate around a fossil fragment or burrow and are often concentrated along particular layers in the strata. Calcite nodules are common in calcareous sandstones or mudstones, siderite nodules are common in iron-rich mudstones, and chert is common in fine-grained limestone. Anhydrite nodules occur within supratidal evaporites.

9   Siliciclastic rocks are classified according to their grain size and mineral composition, which at least partly reflects their textural and compositional maturity. The terms mudstone, sandstone and conglomerate are purely grain size terms. Sandstones with less than 15% mud-grade matrix are further classified as quartz arenite, subarkose, sublitharenite, arkosic arenite (arkose or lithic arkose) or lithic arenite, based on the three main detrital constituents, quartz, feldspar and rock (lithic) fragments. Sandstones with more than 15% mud-grade clay matrix are termed wackes, specifically quartz wacke, feldspathic wacke or lithic wacke. Mudstones contain more than 75% mud.

10 Carbonate rocks are classified as limestones or dolomites (dolostones) and may be further subdivided according to the nature of the predominant grain types, such as bioclastic limestone, oolitic limestone or peloidal limestone. Further consideration of the cement and matrix gives rise to terms such as biosparite, biomicrite and oosparite. Rocks that have formed *in situ* by framework-building or sediment-binding organisms are termed biolithites.

9.8 Objectives for Chapter 9

Now you have completed this chapter, you should be able to:

9.1 Describe, with the aid of sketches some of the diagenetic processes associated with the lithification of sediments, and be able to recognise them in thin sections of sedimentary rocks.

9.2 Distinguish between depositional matrix and post-depositional cement.

9.3 Classify siliciclastic sedimentary rocks according to their grain size and mineralogical composition, using hand specimens or thin-section photographs.

9.4 Classify limestones according to their grain types and intergranular material, using hand specimens or thin-section photographs.

Now try the following questions to test your understanding of Chapter 9.

Question 9.4

What can you tell about the degree of compaction in Figure 9.18b and how might this provide information on the timing of cementation?

Question 9.5

How would you classify the following sedimentary rocks, based on their grain sizes and their grain composition?

(a) A rock containing around 80% clay minerals and 20% bioclasts.

(b) A rock containing 50% bioclasts and equal amounts of micrite and clay minerals.

(c) The rock shown in Figure 9.7.

(d) A rock containing 85% quartz and 15% feldspar, with a dominant grain size of 250–500 µm.

(e) A rock containing abundant peloids in a micritic matrix.

(f) The rock illustrated in Figure 6.4.

(g) The rock illustrated in Figure 6.5.

Chapter 10 Introduction to sedimentary environments

Chapters 10–16 of this book are about the subdivision and graphical representation of sedimentary successions and their fossil content, together with the different environments in which sedimentary deposits accumulate. The nine depositional environments that are discussed here represent the principal ones found around the world (Figure 10.1).

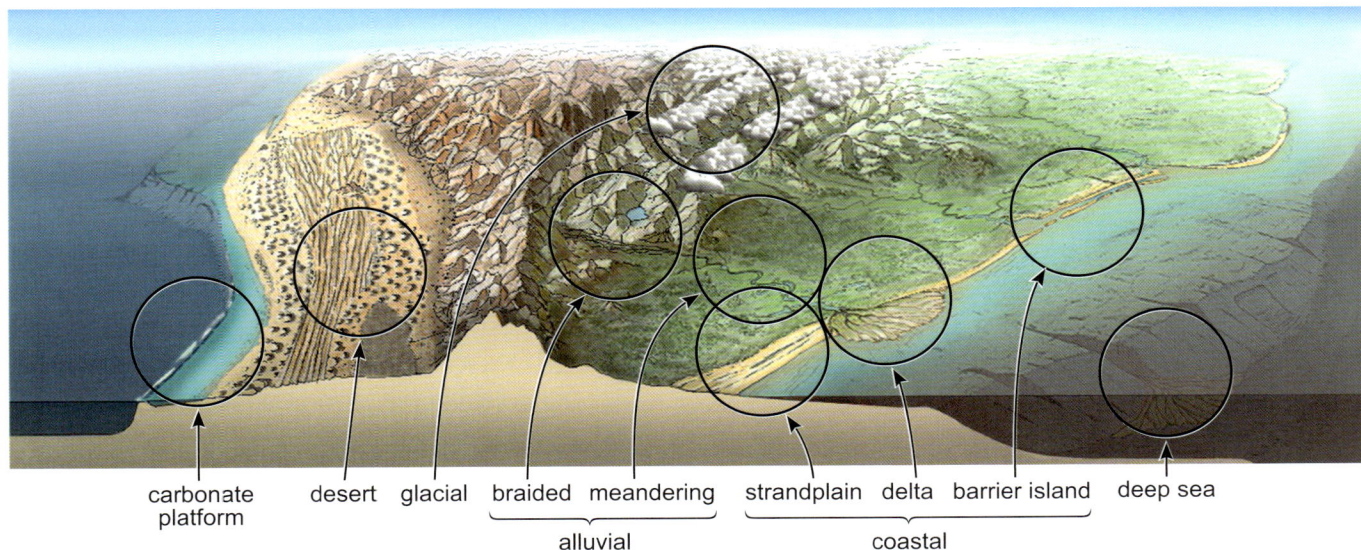

Figure 10.1   An imaginary panorama across a continent and adjacent oceans showing the nine principal sedimentary environments considered in this book. The climate on the left-hand side of the mountain range is arid and on the right-hand side is humid. The glacial environment could extend across much of the right-hand side of the panorama at high latitudes and/or high altitudes, especially during glacial episodes.

Chapter 11 examines rivers and their flood plains, otherwise referred to as **alluvial** environments. We shall consider two types of alluvial environment based on the most common river channel morphologies. These are: **meandering** systems where the river is confined to a single channel of sinuous form, and **braided** systems where the channel is split into a number of smaller channels that may join or divide again. Chapter 12 describes the main features of **deserts**. These are regions of land where the rainfall is less than the amount of evaporation. In Chapter 13, three types of coastal environment will be discussed. These are: **strandplains** where there is a beach attached to the coastline; **deltas** where sediment-laden rivers reach the sea and build out a fan of sediment; and **barrier islands** which comprise a sand bar situated just offshore with a protected lagoon or lake behind. Chapter 14 considers marine areas of mostly shallow-water depth where mainly carbonates are deposited. These are referred to as **carbonate platforms**. Chapter 15 describes the processes and features in marine environments where the water depth is greater than about 500 m; such settings are called **deep-sea** environments. Finally, Chapter 16 briefly looks at **glacial** environments, from land-based to marine settings.

These environments are all illustrated in more detail on the Sedimentary Environments (SE) Poster that you received in your mailing, and which is a larger version of Figure 10.1. The poster also shows 3D block models and idealised graphic logs of the nine principal environments that you will be introduced to in Chapters 11 to 16. (You will need to refer to this poster in several of the activities in this book.) In the text and on the poster the imaginary continent has been split up into these nine separate environments, but the boundaries are often gradational and a single sedimentary grain may visit many of the different environments during its life history. That is why we have shown you how the environments might connect together in the panorama. Imagine, for instance, chemical and physical weathering in the mountains leading to breakdown of rocks, which may then have been transported down the mountainside. Initially, the breakdown of the rocks may produce a grain of quartz that then gets carried away in a braided river system; from there it is carried into a meandering river system. Eventually, it is swept through a delta and then carried across a shallow sea by currents, finally to be incorporated into sedimentary deposits in the deep sea. At any point, the sand grain could become trapped for a while in a particular environment before moving on again, or, alternatively, it may stay there and become preserved in the geological record as part of the depositional succession of rocks representing the particular environment in which it became trapped. However, it is more than just chance that leads to the preservation of sediments; exactly where they are deposited in an environment depends on the nature of the processes acting at that time and also what changes occur in terms of tectonics and climate. These factors are examined as each sedimentary environment is described in the remainder of this book.

Chapters 11–16 consider the nine principal sedimentary environments and the associated features of each, but of course, in the field, a geologist has to work out the environment of deposition from observations on the rocks and the context in which they are set. When we discuss the environments, processes, and source of material, we will use the terms **proximal** to mean the area nearest to the source of sediment in a river system or desert, or nearest to land if it is a marine environment, and **distal** to mean the area furthest away from the source of sediment or furthest away from the land.

10.1 Factors that control sediment deposition

The supply, reworking and deposition of sediments are controlled by several factors. For sediments to accumulate there has to be:

- a supply of sediment either from the weathering of igneous and metamorphic rocks, or the weathering and reworking of older sedimentary rocks
- space in which to put the sediments.

The tectonic setting and climatic conditions are the most important factors controlling both the supply of sediment and the creation of space to accommodate the sediment because they, in turn, control three factors:

1. The rate and type of weathering and erosion and hence the *amount* and *type of sediment supply*.

   ■ Suppose a land mass were to be uplifted. Do you think there would be more sediment produced or less, and why?

   ☐ More sediment would be produced because increased elevation means that the rivers would have more potential energy and so cut down (erode) more rapidly to regain their equilibrium position.

- Climatic conditions can be either stable or changing. During which of these two conditions do you think the sediment supply will increase, and why?

  ☐ In general, when the climate changes, more sediment is produced because of the fluctuating conditions. The sediment supply will stabilise when climatic equilibrium is again reached.

2 The conditions in the seawater, including the temperature, amount of dissolved oxygen and salinity. These parameters are particularly important in controlling where carbonates are deposited because a large proportion of them are composed of the remains of biota, which are highly dependent on the local conditions.

- Suppose the oxygen content of the water is low. Do you think it will encourage or suppress the growth of carbonate-secreting organisms such as corals, and why?

  ☐ A low oxygen content would suppress the growth of carbonate-secreting organisms because they require oxygen to live; in fact, if there is not enough oxygen, they may not even exist in that environment.

- Do you think the majority of marine organisms live in normal salinity seawater or hypersaline (very salty) conditions?

  ☐ The majority of marine organisms live in water of normal salinity; hypersaline conditions require special adaptation. Algal and bacterial mats (Figure 8.6) are found in hypersaline water principally because the hypersaline conditions exclude other organisms that might otherwise graze on and destroy the mats, but also because the mats thrive in nutrient-poor conditions.

3 The position of sea level: this not only influences the marine environments, the type of marine sediments deposited and preserved, and how the whole sedimentary system moves around, but also the level of the water in the rivers and the water table beneath the land surface, and hence erosion levels and deposition in the non-marine, alluvial systems and deserts. Figure 10.2 shows the ideal stable longitudinal profile of a river in balance with sea level. If sea level falls (as shown by the blue line), then the river will respond by cutting down to a new equilibrium profile.

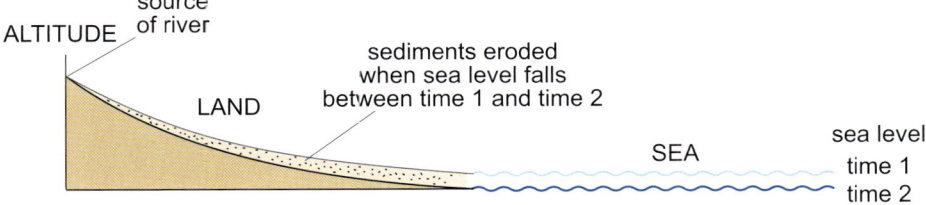

Figure 10.2 River profiles that are stable at two different positions of sea level.

- If sea level falls, will the beach move seaward (distally) or landward (proximally)?

  ☐ The beach will move in a seaward (distal) direction as sea level falls and a new coastline will be produced seaward of the previous one.

A further control that can influence some sediment deposition in both marine and non-marine environments is the type of biota, particularly carbonate-secreting organisms. For instance, during the late Mesozoic and Cenozoic, coccolithophores (a type of marine algae) became abundant in the oceans and their remains form the extensive chalk deposits we find today, whereas mid-Palaeozoic and mid-Mesozoic corals formed extensive reefs.

Organisms can also affect the deposition of siliciclastic and carbonate muds. Clay and silt-sized grains would normally remain in suspension under many conditions because they are so fine grained. However, many organisms eat clays and silt along with organic matter and create coarser faecal pellets from the indigestible residue that can be more easily deposited.

Another method of depositing fine-grained sediments relies on clay grains naturally sticking together to make larger grains; this process is called **flocculation** and is the result of charge-based attractive forces between the grains. In freshwater, the clay grains are negatively charged and repel each other, but in saline waters (e.g. in estuaries) these charges are neutralised by combining with other positively-charged grains (such as organic matter from bacteria and from dissolved ions) and so the clay grains flocculate into dense masses that are deposited at higher current speeds.

- ■ The cohesiveness of clay grains once they stick together is important in erosion of these sedimentary deposits. Explain why this is so.

- □ Because clay grains are cohesive, they are more difficult to erode, so a faster flow is required than you might expect for their grain size (Section 7.1.2).

10.2 Facies

So far, we have concentrated on the lithology of sedimentary rocks; in other words the texture and composition of a rock (e.g. mudstone, sandstone or quartz arenite). We have also considered the body and trace fossils that you may find in a sedimentary rock, and the sedimentary structures. The lithology, fossils and sedimentary structures can be grouped together to describe a particular **sedimentary facies**. The word 'facies' (pronounced 'fa-sees', 'fa' as in 'fact') literally means the aspect or appearance of something, in this case a sedimentary rock. You were introduced to metamorphic facies in Section 6.4 of Book 2. The concept of sedimentary facies is useful because it can be linked to a particular set of depositional processes and is a more holistic way to describe sedimentary rocks.

For instance, imagine two lithologically identical sandstones, both composed dominantly of quartz, being medium grained and well sorted with moderately well-rounded grains. In other words, they are sandstones that are lithologically very similar. However, one of these sandstones contains a few bivalves but shows no sedimentary structures and the other is planar-stratified but devoid of fossils. These differences define them as two distinct facies, deposited by slightly different processes. They may, of course, be closely related to each other and could, for instance, both have been deposited on a beach. The planar-stratified sandstone might represent the main part of the beach deposited by waves during fairweather conditions; the sandstone with bivalves might represent a small bank on the beach deposited by waves during a storm, as illustrated in Figure 10.3a. Facies may be identified by letters or numbers (e.g. Facies A), or else by brief descriptive names.

The latter is more informative, so the two facies described above could be called 'bivalve-rich sandstone facies' and 'planar-stratified sandstone facies'. Initially, facies characterisation is objective and descriptive, but ultimately, after interpreting the processes responsible for their deposition, facies may be grouped together to reflect the environment of deposition and changes in depositional conditions over time.

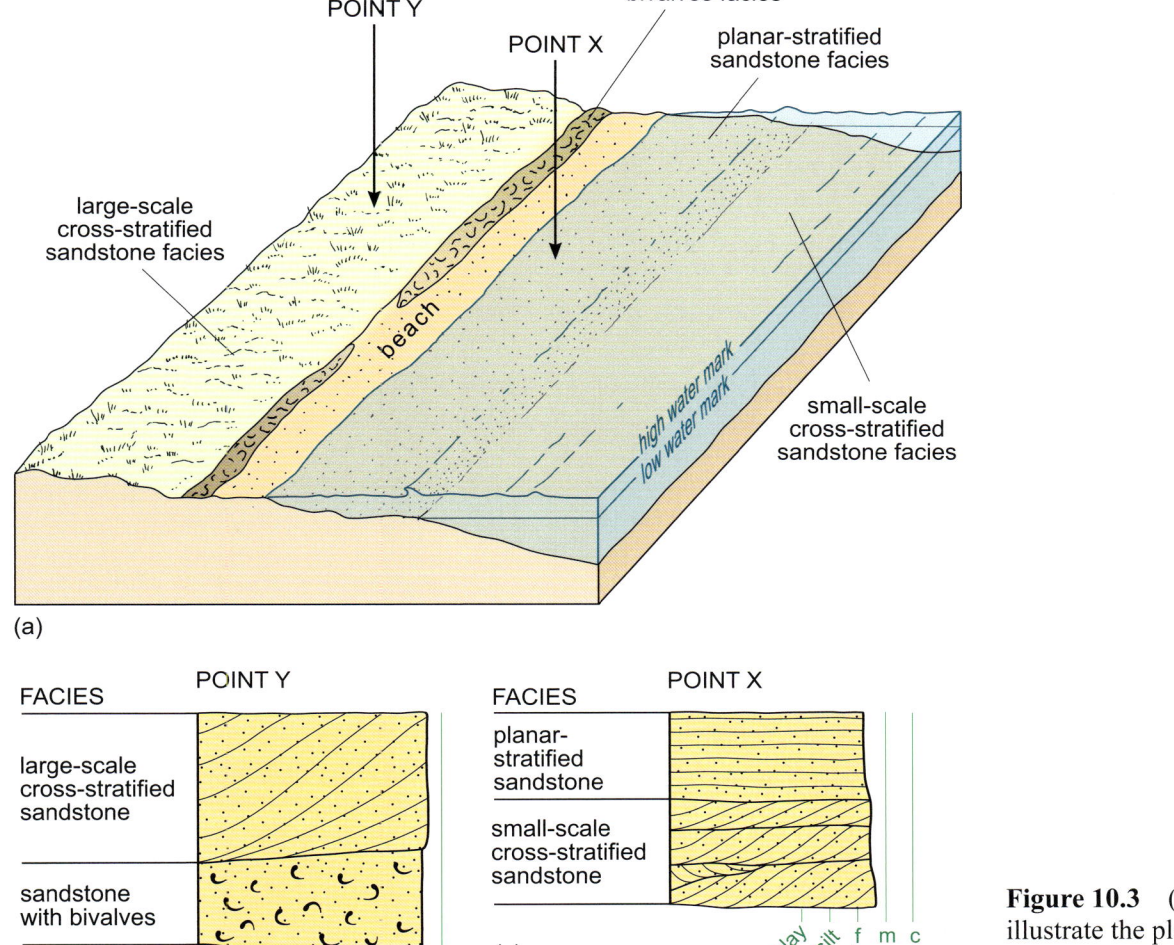

Figure 10.3 (a) Diagram to illustrate the plan view relationship of the various facies associated with a beach. (b) Vertical succession of facies at point Y in (a). (c) Vertical succession of facies at point X in (a). In both (b) and (c) it is assumed that the sedimentary deposits were building out in a distal direction.

Question 10.1

Examine RS 21 and RS 24 in the Home Kit and determine whether they represent different facies, giving reasons for your answer, and a brief description of each specimen. You may also find it useful to look at the labelled photographs available in the Digital Kit.

10.2.1 Vertical successions of facies

So far, we have considered only facies that are horizontally adjacent and pass laterally into each other in plan view. We also need to consider how one facies ends up on top of another in the geological record, i.e. how facies are related to each other in time as well as space.

First, two more facies in the beach environment need defining:
- sandstones deposited by wind on the landward side of the beach will have large-scale cross-stratification resulting from the migration of aeolian dunes. This facies is here called 'large-scale cross-stratified sandstone facies' (Figure 10.3).
- below low water mark, the sand is constantly being reworked by the breaking waves so it will also be cross-stratified – but on a smaller scale than that formed by wind at the landward side of the beach. This facies is here called the 'small-scale cross-stratified sandstone facies' (Figure 10.3).

Consider what would happen to these four facies, which have been deposited adjacent to each other along a coastline if factors controlling sedimentary deposition changed. Imagine that over geological time either the beach was gently uplifted tectonically, or more sand was supplied, or sea level was slowly lowered. In all of these scenarios, the facies would be deposited in a slightly more seaward (distal) position than it was before; this building of facies in a distal direction is called **progradation**.

■ If the beach has prograded and you were to dig a hole at point X shown on Figure 10.3a, what vertical succession of facies would you find?

☐ Planar-stratified sandstone facies overlying small-scale cross-stratified sandstone facies (Figure 10.3c).

■ Assuming that progradation has been occurring continuously, if you were to dig a hole at point Y, what vertical succession of facies would you find from the top downwards?

☐ At the top, large-scale cross-stratified sandstone overlying bivalve-rich sandstone, overlying planar-stratified sandstone, overlying small-scale cross-stratified sandstone (Figure 10.3b).

10.2.2 Walther's Law

The relationship between facies that are originally laid down next to each other and then end up in a vertical succession was first noted in 1894 by Johannes Walther (pronounced 'Val-ter') and so it is known as **Walther's Law**. Walther wrote:

> It is a basic statement of far-reaching significance that only those facies and facies areas can be superimposed primarily which can be observed beside each other at the present time.

In clearer terms, this means that facies which were originally deposited laterally adjacent to each other can, given the correct conditions, be preserved as a vertical succession.

■ Do you think that Walther's Law still applies if the contacts between the facies are sharp rather than gradational? Explain your reasoning.

☐ No, it does not always apply because a sharp contact is likely to indicate that some sediment is missing either through erosion or non-deposition.

So, the law must be applied carefully and if the contact between the facies is sharp, rather than gradational, then the facies cannot be assumed to have been laid down laterally adjacent to each other. Like all rules, there are some exceptions. Some facies have naturally sharp bases due to depositional processes, even though they are laid down next to each other, for example storm-deposited sandstone beds within an otherwise mudstone succession (such as in the video sequence *Coastal processes*, which you studied in Activity 1.1). This indicates again how important it is to understand the processes that have led to sediment deposition. Figure 10.4 illustrates what a cross-section through a beach profile might look like and shows how the facies relate to each other in time as well as in space.

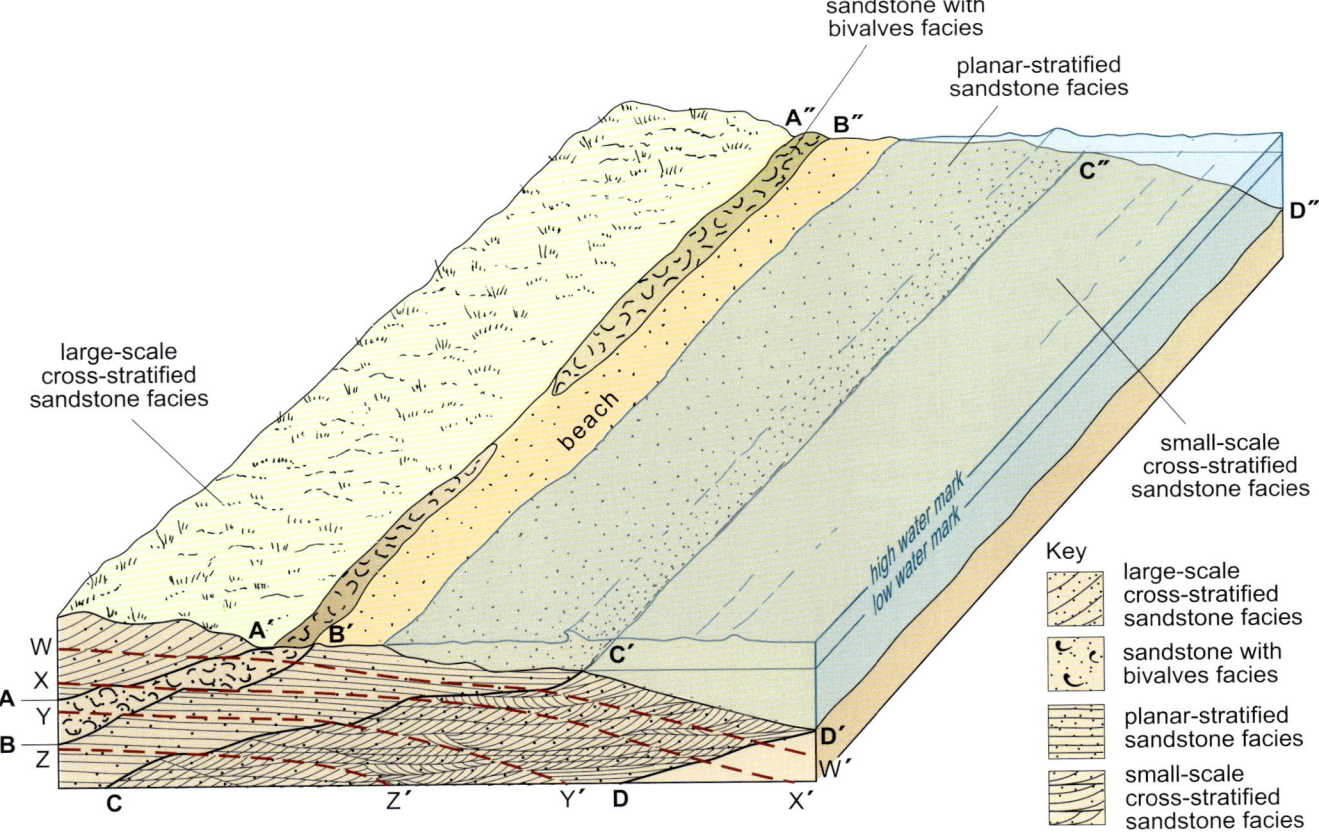

Figure 10.4 Block diagram showing the lateral and vertical relationship between adjacent facies typically associated with a beach. The boundaries between facies are gradational and the facies have prograded.

213

Question 10.2

(a) Examine the red dashed line labelled X–X′ in Figure 10.4. Explain whether it represents a time plane, i.e. a brief period of time, or a facies boundary.

(b) What do the lines A–A′–A″, B–B′–B″, C–C′–C″ and D–D′–D″ represent?

(c) What do lines W–W′, Y–Y′ and Z–Z′ represent?

10.3 Vertical successions of facies and graphic logs

A **graphic log** is simply a visual representation of the *important* and *common* features of a vertical succession of sedimentary facies. Graphic logs show the vertical thickness of the individual beds, and the nature of the contacts between each bed, as well as the lithology, grain size, sedimentary structures and fossils within the beds (i.e. the facies). Graphic logs are an excellent way of recording data because they display the main features of the sedimentary rocks much more readily than lots of written notes. They are the most common way of recording successions in the geological record. Diagrams similar to Figure 10.4 and to the graphic logs shown in Figures 10.3b and c, 10.5 and 10.6 are used throughout the

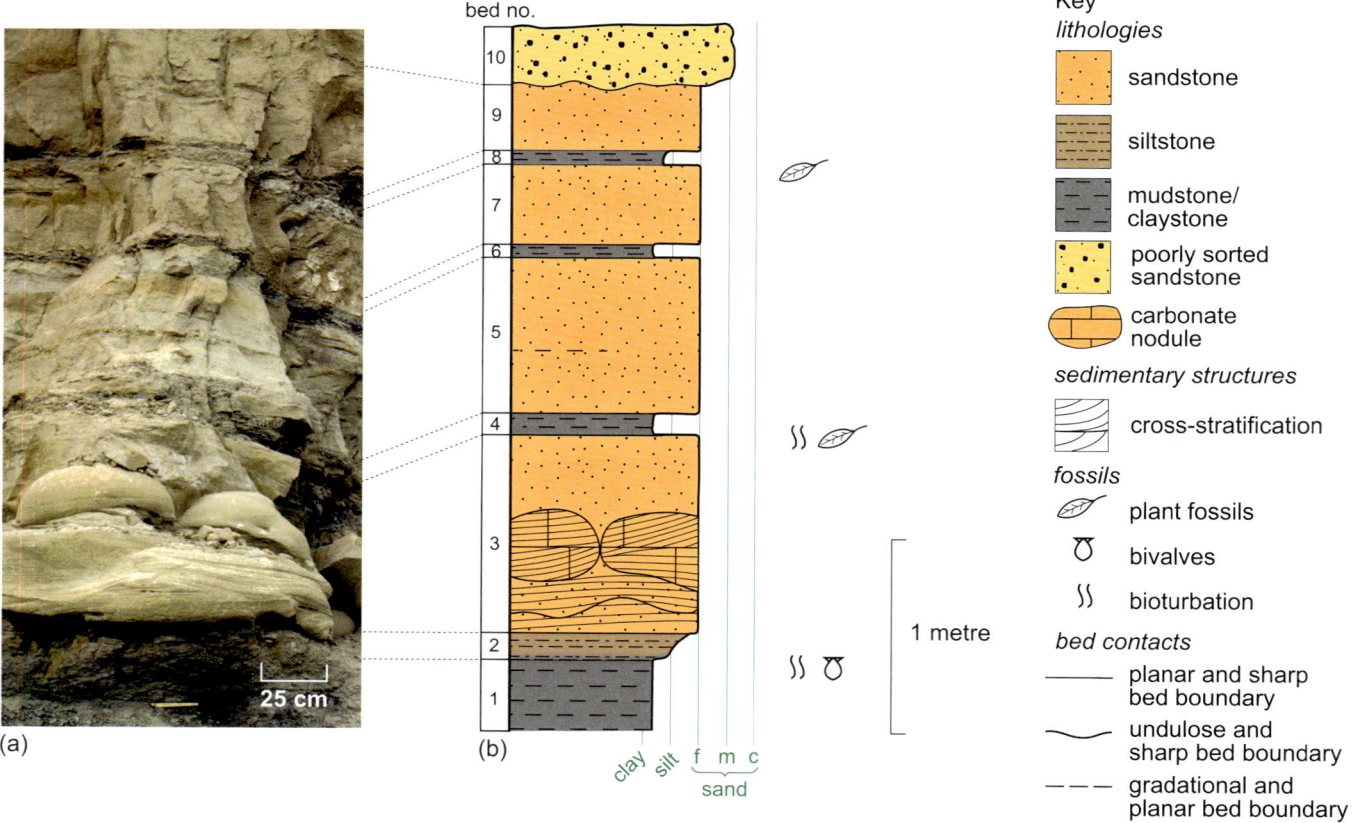

Figure 10.5 (a) A succession of siliciclastic rocks exposed in the sea cliffs at Osmington, Dorset, England. (b) A graphic log of the siliciclastic exposure in (a), incorporating some commonly used symbols.

rest of this book and on the SE Poster to characterise different environments and their associated facies. The facies in the graphic logs have been colour-coded to aid clarity, rather than to reflect standard practice, which is to produce them in black and white.

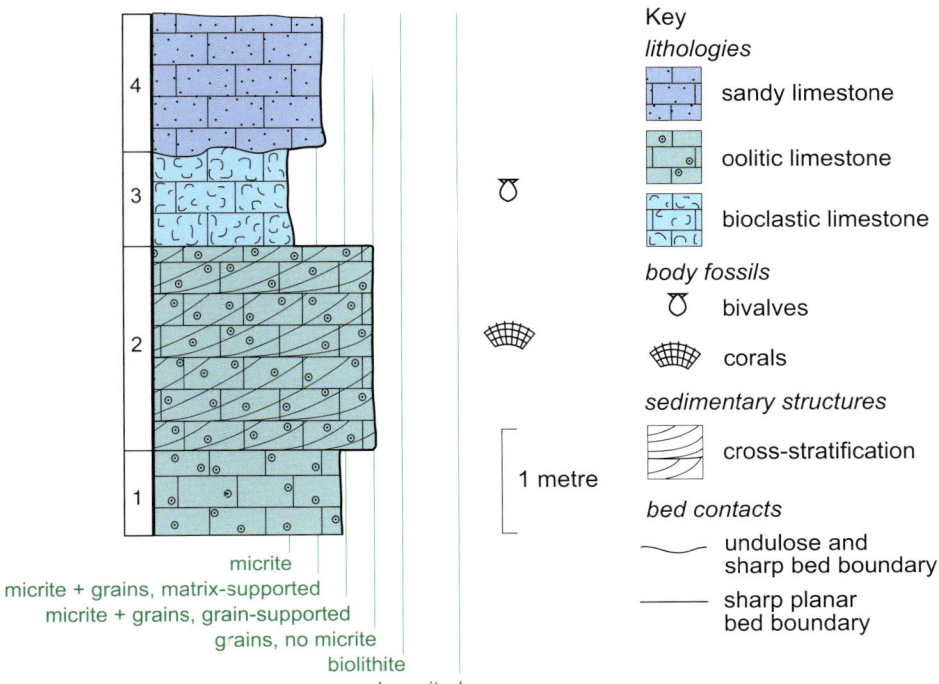

Figure 10.6 Example of a graphic log of a mainly carbonate succession and some commonly used symbols. Beds are numbered 1 to 4.

The different features of sedimentary successions are shown in graphic logs in the following ways:

Bed thickness

Graphic logs are usually drawn to scale vertically. The scale is most often given in metres (Figure 10.5) though it could be in 10 m or even 1 cm units depending on the succession. The vertical scale shows the cumulative thickness of the individual beds. If the bed is variable in thickness, an *average* is usually shown.

Bed contacts

These may be sharp or gradational and planar or undulose, and are shown by the nature of the line drawn between the beds. See key to Figures 10.5 and 10.6 for examples.

Lithology

The lithology of a bed is shown as a symbol within the bed, e.g. stippled for sandstones (as in bed 7, Figure 10.5), broken lines for claystones and mudstones (bed 4, Figure 10.5), alternating dashes and dots for siltstone (bed 2, Figure 10.5), and bricks for limestone (Figure 10.6). If it is a sandy limestone, then it would be shown as two overlapping symbols, in this case the brick ornament with some stipple (e.g. bed 4, Figure 10.6).

215

Grain size in siliciclastic rocks

This is an important feature, since it gives an indication of the energy involved in the depositional processes; it is shown as the horizontal scale on the graphic log (Figure 10.5). The scale used is basically a simplification of that shown in Book 1, Table 6.1 with the finest grain size, 'clay', on the left-hand side and the coarsest grain size on the right-hand side. The graphic log is constructed to show the beds as wide as the volumetrically most abundant grain size of which they are composed. So, if the lithology is a fine-grained sandstone, then the width of the bed on the log would extend as far to the right as the fine-grained sandstone vertical line shown for bed 7. If the rock is composed of a 50 : 50 mixture of fine-grained sand and coarse-grained sand, it would be shown as the volumetric average of these, i.e. a medium-grained sandstone (bed 10). A bed or group of beds that show a gradational coarsening upwards from, say, a siltstone at the base to a fine-grained sandstone at the top is shown by an inclined line that starts on the silt line and ends on the fine-grained sand line (e.g. bed 2).

- ■ How do you think variations in grain sorting, such as poorly-sorted and well-sorted sandstones, might be indicated on a graphic log?

- ☐ There are several ways that sorting could be indicated; the simplest is to show well-sorted sandstones with evenly sized stipple and poorly sorted sandstones with unevenly sized stipple (e.g. bed 10, Figure 10.5). Another way is to add an extra column to the graphic log to show this (i.e. to draw a line changing between 'well sorted' on the left-hand side and 'poorly sorted' on the right-hand side), or to add a few notes to the right-hand side of the graphic log.

Grain size in limestones

Generally speaking, this is not as important for determining the energy of processes involved because many of the grains are biologically or chemically formed *in situ*, so large bioclasts are more a function of which organisms were living at the time the sediment was deposited rather than the result of transport by strong currents. The size of ooids depends on the amount of carbonate available, the time they have been forming in the appropriate conditions and the amount of agitation in the water. However, the amount of carbonate mud (micrite) present in a succession of carbonates is useful, because under high energy conditions the carbonate mud tends to be kept in suspension, whereas under lower energy conditions it is deposited. This is similar to the behaviour of fine grains in siliciclastic systems. So the amount of carbonate mud is indicated on the horizontal scale (Figure 10.6) – the grain size lines from left to right then become:

- micrite (i.e. all carbonate mud)
- micrite with grains, matrix-supported; where the grains are supported by the matrix (Figure 7.13b)
- micrite with grains, grain-supported; where the grains are in contact with each other (Figure 7.13a)
- grains, no micrite; grains only, usually cemented together with sparite.

In this course, we recognise two other distinct types of limestone and, because they are both high energy, they are indicated by two further subdivisions to the right-hand side. These are:

- biolithite (i.e. organically bound carbonates preserved *in situ*, see Section 9.6.2)
- redeposited (e.g. breccias and conglomerates composed of redeposited limestone clasts).

Fossils

These are usually shown as symbols just to the right-hand side of the lithology column or, if they are particularly abundant and are forming many of the grains within the rock, e.g. RS 21, they may be shown as part of the lithology symbol itself. A few of the common fossil symbols used in this book are shown in Figures 10.5 and 10.6.

Sedimentary structures

A common way to show these on a graphic log in terms of both their scale and geometry is to sketch a *simplified* but *accurate* version of the structure onto the bed. If the sedimentary structure is too small to be shown at the scale that the log is drawn then, again, like the fossils, a small symbol can be added to the right-hand side of the log. This is also the place to add other features not recorded elsewhere, for example pyrite nodules, using an appropriate symbol that is added to the key.

When a graphic log and its symbols key are complete, the author, or anyone else with a trained eye, should be able to recognise easily any of the individual beds in the field. Graphic logs are used extensively throughout the rest of this book to illustrate the principal features of each environment that may be preserved in a vertical succession.

Question 10.3

Using Figure 10.5 and your grain-size scale, where on the horizontal scale would you plot beds of the same rock type as RS 10, RS 20 and RS 27?

Question 10.4

Using Figure 10.6 and a hand lens, where on the horizontal scale would you plot beds of the same rock type as RS 21 and RS 22?

Activity 10.1 Graphic logging

In this activity, you will watch the video sequence *Graphic logging*, which shows the process of graphic logging in the field using a Carboniferous succession exposed in northeast England.

Activity 10.2 Constructing a graphic log

In this activity, you will practise putting together a graphic log.

10.4 Facies and environmental models

Many facies may be grouped together into a model that summarises the features of a sedimentary environment. However, it must be remembered that models represent an idealised situation, and no real life environment, either today or in the past, will be exactly like these models. Nevertheless, models are exceptionally useful for four reasons:

- they act as the '*average*' for purposes of comparison
- each model provides a *framework* suitable for further observation on the real sedimentary successions
- they can help *predict* what is likely to occur in different geological situations
- they form a basis for an *integrated interpretation* of the environment they represent.

Figure 10.4 represents part of a model for a coastal strandplain environment. This model is developed further in Chapter 13, and facies models for the other environments are constructed in the remaining chapters of this book.

10.5 Summary of Chapter 10

1. Areas of the Earth's surface where sedimentary rocks are deposited can be subdivided into a number of different environments; these include alluvial (meandering and braided), deserts, coastal (e.g. strandplains, deltas and barrier islands), carbonate platforms, the deep sea and glacial environments.

2. For sediment deposition, there needs to be a supply of sediment and a space in which to accommodate the sediment. Sediment supply and accommodation space are controlled by tectonics and climate, which in turn affect weathering, erosion, seawater conditions and sea level. The type of biota present may also affect sediment deposition.

3. Sedimentary facies are used to describe all aspects of sedimentary rocks, including their lithology, fossil content and sedimentary structures.

4. Progradation of sediments occurs through continuous sediment supply, or tectonic uplift, or a lowering of sea level.

5. Walther's Law states that facies that were originally deposited side by side can, given suitable conditions, be preserved as a vertical succession. If the contacts between the facies are sharp, one or more facies may be absent and Walther's Law may not apply.

6. Vertical successions of sedimentary facies are conveniently and commonly represented by graphic logs. These are a shorthand pictorial way of representing sedimentary facies. The vertical axis represents the thickness of the beds and the horizontal axis represents the grain size. The lithology, fossils, sedimentary structures and nature of the bed contacts are all shown pictorially on graphic logs.

7. Environmental models are useful for four reasons: to act as the normal scenario for comparison, to provide a framework, to predict what is likely to occur, and as a basis for integrated interpretation.

10.6 Objectives for Chapter 10

Now you have completed this chapter, you should be able to:

10.1 Describe how climatic conditions, tectonic setting and the type of biota control the deposition of sediment.

10.2 Distinguish between the terms sedimentary facies and lithology, and give examples of each.

10.3 Explain with the aid of a sketch how facies found along a prograding coastline may be deposited on top of each other.

10.4 Summarise Walther's Law and explain when it is not applicable.

10.5 Explain the different components used to construct a graphic log.

10.6 Record your observations of sedimentary rocks (including their fossils and sedimentary structures) in the form of a graphic log and demonstrate how data can be assimilated from graphic logs.

10.7 State four reasons why models of environments are useful in the interpretation of sedimentary rocks.

Now try the following questions to test your understanding of Chapter 10.

Question 10.5

If a land area changes from being tectonically stable to subsiding, and sea level is rising, would more, or less, sediment be likely to reach the sea, assuming that all other factors, such as climate, remain stable?

Question 10.6

If the climate gradually becomes cooler and wetter, how might this affect an area of (a) evaporite deposition, and (b) a siliciclastic coastline adjacent to a poorly vegetated land mass?

Question 10.7

(a) Use the following notes to construct a graphic log summarising the main depositional features of the following succession. Considering the total thickness of the succession and the size of your paper, choose an appropriate scale (e.g. 1 m = 2 cm). Draw a vertical line of the appropriate total length and then mark the grain size divisions along the base. Do not forget to include a vertical scale and a key to the symbols that you use. The succession from the bottom (oldest bed) upwards is as follows:

Bed A: a grey mudstone, 2 m thick, with 30 cm spheroidal nodules of calcium carbonate with their bases 1.3 m above the base of the bed. The mudstone contains brachiopods and trilobites. There are no visible sedimentary structures. Bed A is gradationally overlain by Bed B over about 20 cm and with a horizontal contact.

Bed B: 80 cm of pale yellow siltstones with wave-formed ripples in the upper 20 cm. It also contains occasional brachiopods and is sharply overlain by Bed C. The contact between Bed B and Bed C is horizontal.

Bed C: 3 m of a red, coarse-grained, well-sorted sandstone. The surfaces of the sand grains are frosted and large-scale cross-stratification dips at 25°. The bed sets are 1.5 m thick and there are no fossils.

(b) Which beds in the succession described in (a) suggest a marine environment and which beds suggest non-marine? Give reasons for your answer.

(c) What geological term might be used to describe the contact between Bed B and Bed C?

Chapter 11 Alluvial environments

River systems are a familiar feature of the landscape and their importance in transporting sediment from areas of continental weathering and erosion to the sea cannot be underestimated. The term 'fluvial' is used to describe the river channel itself, whereas the term 'alluvial' includes the river channel, the adjacent, flat-lying **flood plain** and localised fans of sediment (alluvial fans), which accumulate at the junction between mountainous areas and flood plains.

The pattern of river development varies according to climate and topography but fluvial erosion and alluvial deposition operate to some extent in all types of continental environments, including deserts and glaciated regions.

11.1 River flow

The prime source of water for a river is precipitation in the area around the river. Rain or snow falling directly into a river channel makes only a minor contribution. Most of the water arrives in the river channel by various indirect routes through the drainage basin in which the river is situated, as illustrated in Figure 11.1. Another important factor in alluvial environments is the position of the water table. Permanent rivers form only where the water table intersects the ground surface. In order to maintain a high water table, precipitation needs to exceed evaporation. So, in regions with humid climates, it is the vast amounts of water stored beneath the water table that permit continuous river flow. However, temporary rivers can form when the rate of influx of water exceeds the rate of infiltration into the ground, as in arid or semi-arid regions. Here, rare, sporadic rainstorms create a lot of surface water and allow intermittent flow, and the results are often catastrophic.

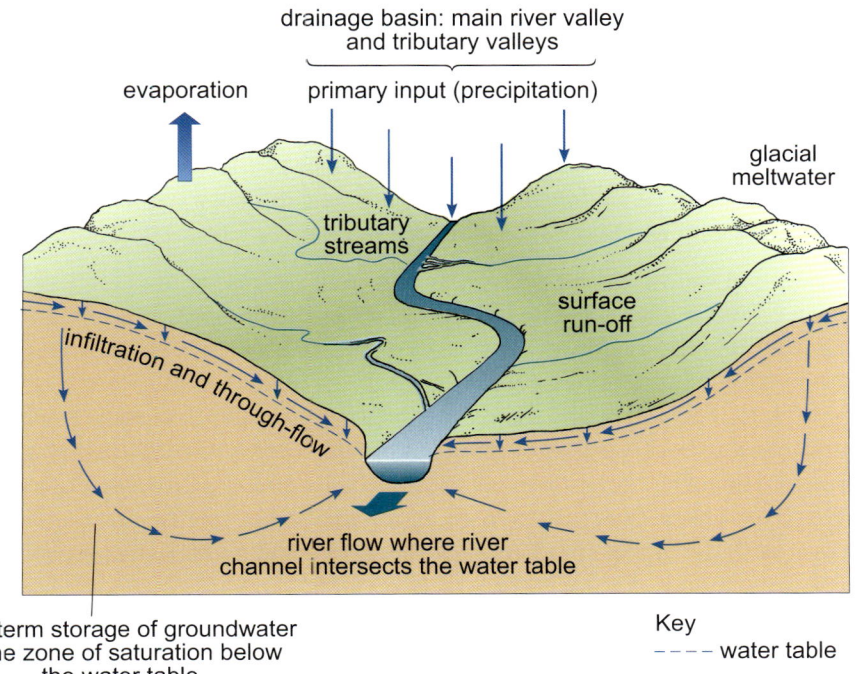

Figure 11.1 The different sources of water within a drainage basin that feed a river. A certain amount of water is lost by evaporation, but in humid climates, the primary input from precipitation always exceeds losses due to evaporation.

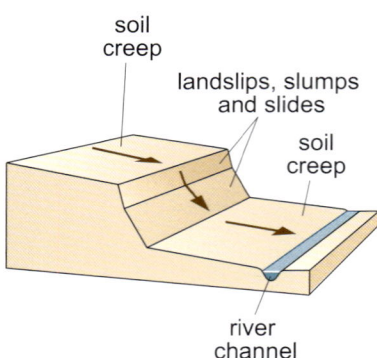

Figure 11.2 Cross-section to show the most important sediment gravity flow processes that lead to the development of river valley slopes and transport weathered rock debris into a river channel.

Question 11.1

Examine Figure 11.1 and explain whether you would expect the water table to remain at a constant depth below the ground surface (a) throughout the year, and (b) regardless of the underlying lithology.

The sediment derived from weathering is transported down the valley sides into a river channel by sediment gravity flow processes (Section 7.6) similar to those shown in Figure 11.2. This sediment is supplemented by material eroded from the river banks by gradual attrition and more catastrophic river bank collapse. During times of extreme weather, temporary tributary streams bring sediment directly into the main river channel.

The source of energy to move the water in a river and, indirectly, the sediment, is the gravitational potential energy that results from the difference in height between the upstream and downstream sections of a river channel. As the water flows downslope, much of this is converted to kinetic energy (the energy of the motion of the water), which transports sediment, erodes the river channel, and is capable of causing tremendous devastation at times of flood. The currents within the channel transport the sediment as dispersed grains.

- Recall the three ways in which sediments move in water according to their grain size.

- The largest grains (bedload) move by sliding or rolling, the medium-sized to smaller-sized grains (also part of the bedload) move by bouncing along the bed (a movement called saltation), and the smallest grains (suspension load) are carried in suspension (Section 7.1.1 and also Figure 7.3).

11.2 The river valley

In general, river valleys approximate to a 'V' shape in cross-section, particularly those in humid climates. They are wide at the top and the river occupies a channel at the narrow base. This is because mass movement of sediment, beginning with slumping of the channel sides, leads to the development of slopes on either side of a newly formed river channel. These slopes form new sites for water run-off and channel development, so that in time a drainage basin is developed (Figure 11.1) with its main river and many smaller feeders called tributaries.

If you were to follow a river downstream from the source to the mouth, you would see several changes. The volume of water flowing past in a given time, which is termed the **discharge**, increases exponentially (i.e. ever more rapidly), because water is being added continuously from tributaries and other sources (Figure 11.1). The valley and the river channel grow wider and deeper, which is not surprising because erosion is a continuous process related directly to the discharge. However, the gradient of the riverbed *decreases* downstream as the discharge *increases*. This results in a longitudinal valley profile that is generally concave, with the highest gradient near the source and a gradient approaching zero near the mouth (Figure 10.2).

At the source of a river, the discharge is small and, consequently, the channel is narrow and shallow. A lot of the energy is lost by friction at the base and sides of the channel, and by contact with the atmosphere, and so the discharge can only be maintained on a steep slope. Nearer the mouth of the river, the channel depth and width are greater, so the total area of water in contact with the base, sides and atmosphere in proportion to the volume of water in the channel is less. Energy losses due to friction are therefore less and so the discharge can be maintained on a shallow gradient. The fact that rivers flowing in narrow, shallow channels require a steeper gradient to maintain their discharge than rivers flowing in broad, deep channels is an important control on the pattern of river development, as you will see in Section 11.3.

Downcutting of a river channel should cease close to where it enters the sea because the gravitational potential energy available in the system approaches zero. However, as discussed in Section 10.1, sea level has not remained constant through geological time and changes in sea level affect the balance of erosion and deposition in rivers such that the ideal concave longitudinal profile (Figure 10.2) is rarely achieved. In fact, in some cases, what we see is a series of superimposed concave longitudinal profiles due to successive sea-level changes.

The river's flood plain is often very fertile due to moisture and nutrients from the river waters; it is therefore colonised by vegetation and soils develop. The type of soil formed depends on the climate and the composition of the bedrock and flood plain sediments. If the climate is humid, the area is often flooded, siliciclastic sediments are deposited, abundant vegetation grows, and peat may accumulate. During burial and compaction, the peat may form *coal*. The soil in which the coal-forming plants develop is often called a *seatearth*; these are particularly common in rocks of Carboniferous age in Britain (Figure 11.3). In hot, arid climates, evaporation of the water between the sediment grains in the flood plain area can lead to the deposition of carbonates and, more rarely, evaporites.

Figure 11.3 Thread-like, black carbonaceous strands that represent rootlets within a Carboniferous seatearth (fossil soil), Howick, Northumberland, England. The yellow colouration is hydrated iron oxide (limonite).

11.3 Fluvial channel styles and their typical sedimentary successions

A river does not flow in the same direction over its entire length. It will change course if it meets an obstacle, for example a resistant rock outcrop or a zone of more rapid weathering such as an active fault zone. Recall from Section 11.2 that the gradient of a river changes along its length and that broad, deep channels are found in areas of low gradient whereas narrow, shallow channels are found in areas of steep gradient. The amount of vegetation also governs the pattern of the channels: lush vegetation tends to bind each riverbank together and help to prevent erosion. The climate governs the amount and periodicity of the rainfall and therefore the amount of water in the drainage basin; this in turn will influence the shape of the river channel. In summary, the factors that govern the morphology of the river channel are: bedrock geology, tectonics, topography, vegetation and climate.

A wide range of different channel morphologies is possible based on their **sinuosity** or, at the other extreme, their **braiding**. These two end members provide a good model for comparing most ancient and modern river systems.

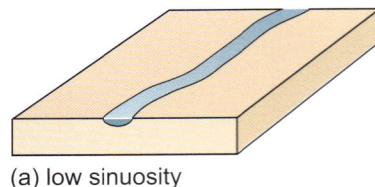
(a) low sinuosity

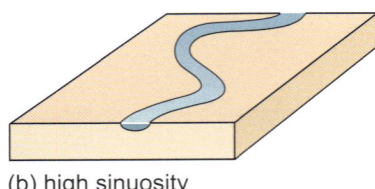
(b) high sinuosity

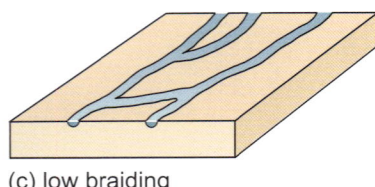
(c) low braiding

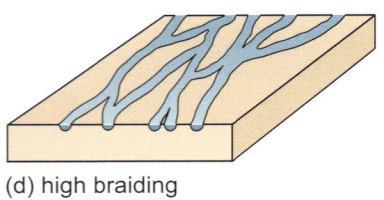
(d) high braiding

Figure 11.4 Sinuosity and braiding in river channels.

- The sinuosity of the channels depends on the number of twists and turns in the channel and the length of the loops. The river channel in Figure 11.4a has a low sinuosity, whereas the one in Figure 11.4b has a high sinuosity.
- Braiding reflects the tendency for channels to split and rejoin. Figure 11.4c shows a river with low braiding and Figure 11.4d shows a river with high braiding.

Rivers with high sinuosity but low braiding are called 'meandering' rivers. The pronounced bends in the river are **meanders**. Rivers with low sinuosity but high braiding are termed 'braided' rivers.

■ Which part of Figure 11.4 would you call a meandering river and which part a braided river?

□ Figure 11.4b is an example of a meandering river and Figure 11.4d is an example of a braided river.

11.3.1 Meandering river systems

Components of the meandering river system

Meandering river systems comprise a distinct sinuous channel and riverbanks (Figures 11.5 and 11.6). Meandering rivers are characterised by:

- relatively low overall topographic slopes both within the channel and its flood plain
- high suspension load to bedload ratio
- the area into which the channel has cut being made up of relatively cohesive deposits.

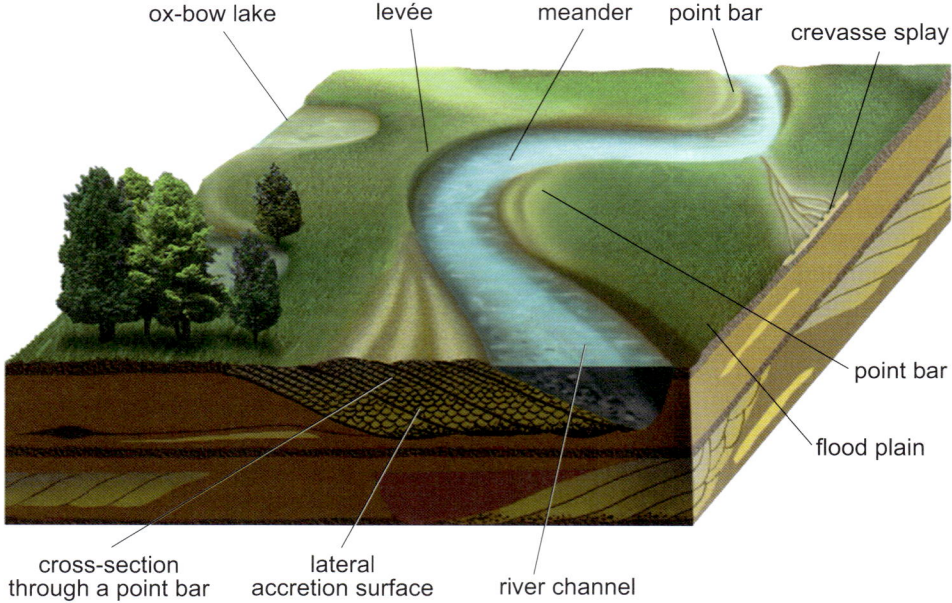

Figure 11.5 Block diagram of a meandering river showing typical features.

Chapter 11 Alluvial environments

Figure 11.6 Aerial photograph of a meandering river system in Alaska, USA.

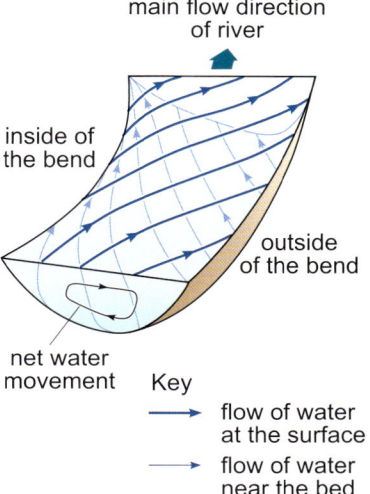

Figure 11.7 A plan view and lateral cross-section of a river meander to illustrate the helical flow of water. Water flows across the surface towards the outside of the bend and is returned by the flow across the riverbed towards the inside of the bend. Consequently, the main current swings from one side of the channel to the other.

As a meandering river develops through time, the number and size of the meanders will increase, in effect lessening its slope compared to a river that flows straight down a slope. If you find this hard to visualise, then remember that roads on very steep hillsides are built in a zigzag fashion with hairpin bends in order to lessen the gradient of the road. River meanders gradually increase in amplitude and migrate downstream, because a helical or corkscrew flow of water is superimposed on the overall downstream movement of water (Figure 11.7). This corkscrew flow is the movement of surface water across the meander towards the outside of the bend and its return across the base of the channel towards the inside of the bend. The result is that the main current does not flow straight down the river channel but impinges against each bank in turn causing erosion of the outside bend (Figure 11.8).

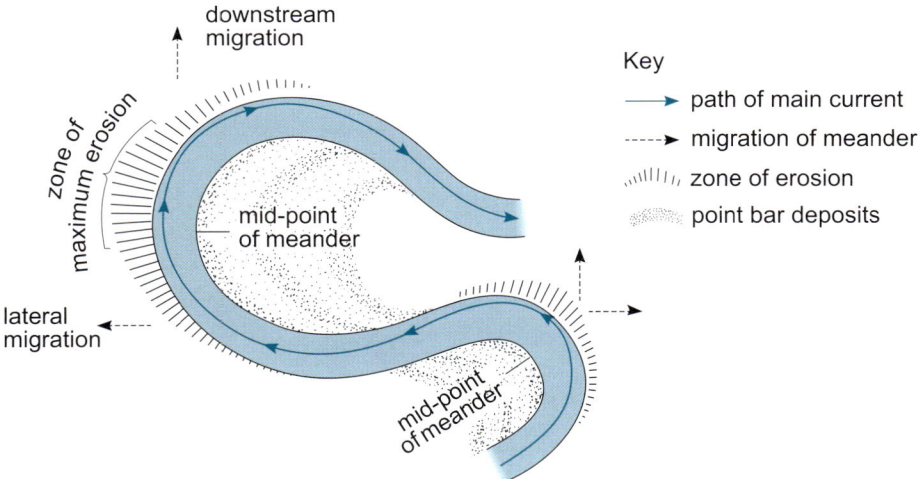

Figure 11.8 The lateral migration (left and right in the figure) of a meander caused by erosion of the outside bend and **point bar** accretion on the inside bend. Downstream migration (upwards in the figure) of the meander also occurs, because maximum erosion occurs downstream of the mid-point in the meander. The result is to increase the amplitude of the meander loop.

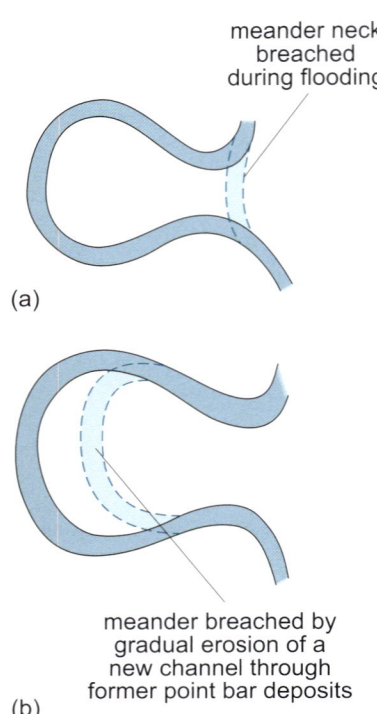

Figure 11.9 Two ways in which an ox-bow lake may form: (a) the river may breach the meander neck during flooding; (b) the river may gradually erode a new channel through former point bar deposits.

Figure 11.10 A meandering river system in Alaska, USA. On the right-hand side there are two distinct ox-bow lakes. The main river channel snakes across the middle of the photograph. Areas of active sediment deposition are pale coloured.

■ Will the maximum erosion occur at the mid-point of the meander or somewhere else?

☐ It will not occur at the mid-point, because the flow is helical. The point of maximum erosion actually occurs slightly downstream of the mid-point, thus extending the development of the meander further downstream (Figure 11.8).

So how does the amplitude of the meander increase? On the outside of a bend, water flows at a faster speed than on the inside. So, while the outside bank of a meander is actively eroded, material is deposited on the inside bend of the next meander downstream, where the flow is slower. The sediment is added slowly sideways on the inside of the bend and forms a bank of sediment called a point bar (Figure 11.5).

A meander is usually only a temporary landform. If the rate of downstream migration of one meander is greater than the one further downstream, then the upstream meander may get closer to the downstream meander leaving only a narrow neck of land between. This neck of land is easily breached at periods of high discharge so that the river cuts a new channel (Figure 11.9a). A meander may also be breached at periods of normal discharge by progressive erosion through the point bar deposits (Figure 11.9b). In either case, the entrance and exit to an old meander soon become plugged with deposited sediment so that it is abandoned and forms a lake termed an **ox-bow lake**, so-called because of its curved shape (Figures 11.5 and 11.10). The ox-bow lake will gradually fill with fine-grained sediments brought in during flooding and with decayed vegetation.

At times of increased water level in a river channel, water may flow over the top of the channel and any sediment carried in it will be deposited due to the sudden reduction in flow velocity outside of the channel. This sediment builds up to form **levées** (small ridges, often vegetated) on the edge of the river channel, analogous to the lava flow levées you met in Book 2, Section 3.2. If the water level rises further, the levées may be breached along much of their length, and widespread flooding occurs; this is called overbank flooding, and sediment and water will be carried out over the flood plain. Strong river currents during flooding may also breach the riverbank at one particular point and deposit sediment on the edge of the flood plain in a fan shape called a **crevasse splay** (Figure 11.5). In humid areas, meandering river banks and their flood plains are usually areas of lush vegetation and good soil development. The periodic flooding from the river carries fresh nutrients over the flood plain encouraging plant growth. Plant debris from natural decay, flooding and wildfires often becomes incorporated into the sediments being deposited. A variety of animal life, both invertebrate and vertebrate, will live in the confines of the river system.

■ How might evidence for animal life be preserved in alluvial sediments?

☐ Animals that live in the sediment on the point bars, such as bivalves, might easily be preserved in life position when they die. The point bar sediments may also preserve burrows made by various animals. The remains of vertebrates living on the river banks and flood plain will probably only be preserved during times of flooding when large volumes of sediment are deposited and they become buried.

Meandering river successions

Having considered the three-dimensional nature of alluvial systems within a meandering river system, the next step is to construct an idealised vertical succession of sedimentary deposits that might develop and to examine the types of succession present in the geological record.

The most volumetrically significant deposits are those associated with the migration of the channel. Re-examine Figure 11.8 and remember that, as the current is eroding the bank on the outer side of the bend, sediment is being deposited on the inner bank of the meander and on the next meander downstream, forming point bars. The net effect, as shown in Figure 11.11, is that as the outer eroding bank gradually migrates laterally, so too does the point bar on the inner bank, maintaining the same width and depth of the channel as it migrates.

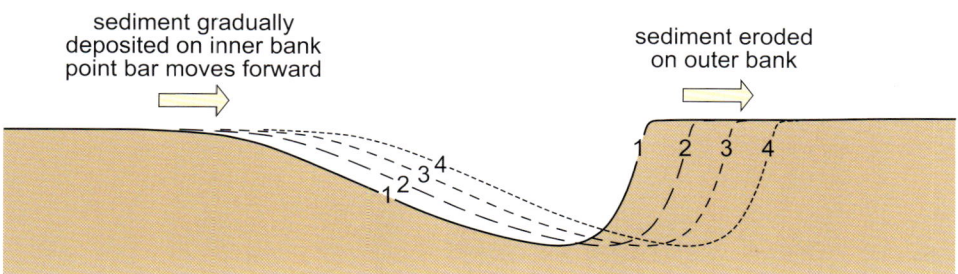

Figure 11.11 Cross-section of a meandering river channel showing the lateral movement of the channel through time caused by both erosion and deposition. 1 = oldest river channel surface; 4 = youngest river channel surface.

The movement of the channel is usually slightly greater in some periods than in others due to variability in discharge and sediment content, resulting in slight pauses in deposition followed by more continuous deposition. These pauses are recorded in the point bar as a number of **lateral accretion surfaces** that mark the former positions of the point bar surface (Figures 11.5, 11.12 and 11.13). These surfaces can be seen in plan view as a series of curved surfaces in meandering river systems preserved in the geological record (e.g. Figure 11.13c) and as large-scale cross-stratification in cross-section (Figure 11.13a and b).

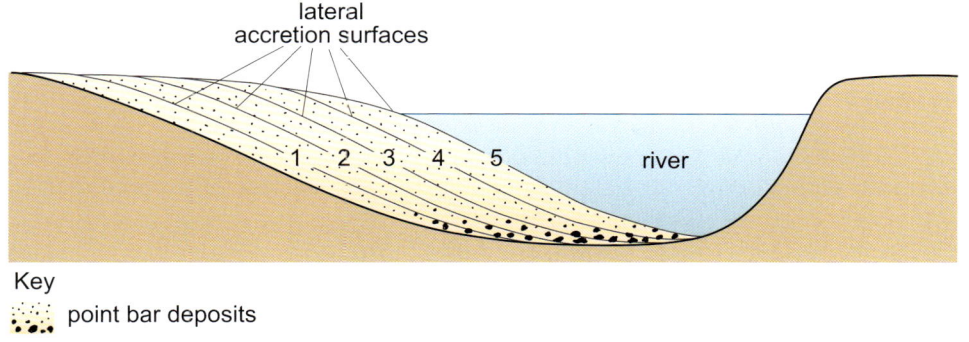

Figure 11.12 The formation of lateral accretion surfaces in point bars. 1 = oldest lateral accretion surface; 5 = youngest lateral accretion surface.

How do sediments accumulate in the point bar? As the outer bank erodes and collapses, fine-grained sediment is carried off in suspension, and the medium-grained sediment is moved by saltation. The coarsest-grained sediment will only move by sliding or rolling at times of high discharge so that most remains at the base of the channel, forming a lag deposit, which may be added to by large clasts that have rolled down the river channel.

Figure 11.13 Photographs from a fossilised meandering river system of Jurassic age preserved near Burniston, Yorkshire, England. (a) Lateral accretion deposits, overbank fines and crevasse-splay deposits in cross-section. (b) Line drawing of (a) showing the interpretation of the different features of the meandering river system. (c) Plan view of the curved beds representing a migrating point bar. The beach is about 500 m to the far point.

The currents in the channel may build up subaqueous dune structures composed of the medium-grained sediments, which migrate across the surface of the point bar.

■ What sedimentary structure will these dunes form?

☐ Cross-stratification, and in this case, because of the type of currents in the river channel and the 3D beds that they form, usually trough cross-stratification (Figure 7.23a).

The finest-grained sediments will be deposited on the upper part of the point bar where the currents are slower and weaker. They often form current-formed ripples (Figure 7.17), which give rise to small-scale cross-stratified sandstones and siltstones. Thus point bars are preserved as fining-upward successions with a lag of pebbles at the base overlain by cross-stratified sandstones, capped by small-scale cross-stratified sandstones and siltstones.

■ In what orientation will the cross-stratification surfaces in the point bar deposits be in relation to the point bar lateral accretion surfaces?

☐ The cross-stratification surfaces will dip downstream and be perpendicular to the slope of the lateral accretion surfaces. The dunes and ripples have migrated downstream as the point bar has built out sideways into the river.

If the meander neck breaks to form an ox-bow lake, the ox-bow lake will gradually infill with fine-grained sediment and plant material during flooding. The levées will be composed mainly of the suspension load fraction of the river deposited at the edge of the river channel when the water level is high. Overbank flooding will carry the finest-grained sediments over the flood plain and, as the flow slows down, a thin veneer of planar-stratified silts and clays is deposited. When the bank breaks at a confined point, sediments are deposited in a crevasse splay; these are usually composed of sediment with a mixture of grain sizes and are deposited quite rapidly so they may show no systematic change in grain size. After the flood subsides, the crevasse splay and levées may be colonised by vegetation. New channels cut through flood plain deposits and may erode old crevasse splays, resulting in a sharp contact between overbank and crevasse splay deposits and the overlying point bar deposits (Figure 11.5).

Overall, meandering river deposits typically comprise mudstones from the flood plain interbedded with thick sandstones that represent the point bars. An idealised graphic log summarising each of these features is shown in Figure 11.14 and a Jurassic example of a meandering river deposit is shown in Figure 11.13.

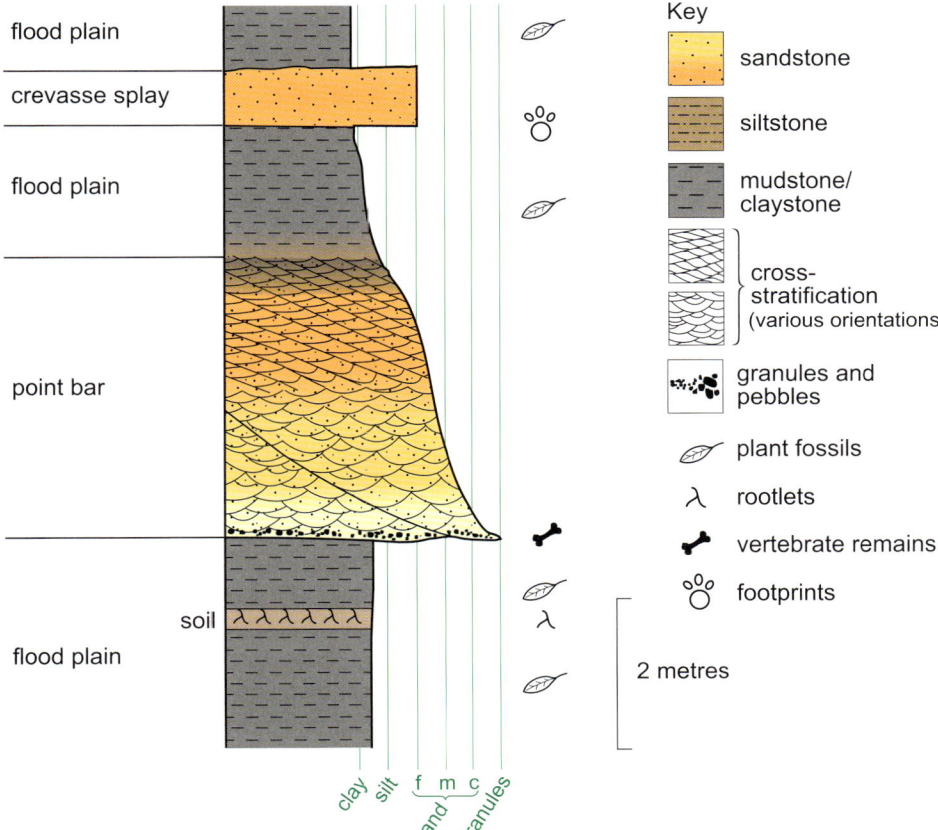

Figure 11.14 Idealised graphic log of a succession that might be produced by a meandering river.

11.3.2 Braided river systems

Components of the braided river system

In contrast to meandering rivers, braided rivers (Figure 11.15) develop when:

- the discharge fluctuates widely
- the volume of bedload is high and comprises coarse-grained, poorly sorted sediments
- the slope over which the river flows is relatively steep.

From what you know already about the erosion of sediment (Section 7.1.2) and river profiles (Section 11.2), you can begin to work out why these factors may be important.

■ Are medium- to coarse-grained sediments or very fine-grained sediments easier to erode?

□ Coarser-grained sediments are more easily eroded than very fine-grained sediments such as silts and clays because the latter are cohesive (Section 7.1.2).

At periods of high discharge, new channels are cut rapidly through the sediments that were deposited as the previous floodwaters abated.

Figure 11.15 Aerial photograph of a braided river system in Alaska, USA.

■ How does discharge relate to the shape of river channel cross-sections and channel bed gradients?

□ In a narrow, shallow channel, the discharge is maintained by a steep gradient. In a broad, deep channel, discharge is maintained by a gentle gradient.

Whereas a meandering river produces meanders to decrease the overall gradient and thereby accommodate its discharge, a braided river is able to disperse the discharge through a number of smaller channels and so the gradient of each channel bed is not reduced.

The sorts of conditions that favour braiding are often found where heavily debris-laden rivers flow from a steeply dissected mountain range onto a lowland plain and where there are seasonally high discharge periods such as in semi-arid and arid regions. The ancient braided river deposits in the sedimentary record are usually associated with such geographical features.

The braided river system thus comprises numerous braided channels, all of which are within a larger channel-like feature – usually with steep sides because of active erosion. Cones of sediment (Section 11.4) supplied via confined mountain valleys meet these steep valley sides and often the river channel dissects the end of the cone of sediment (Figure 11.16).

Chapter 11 Alluvial environments

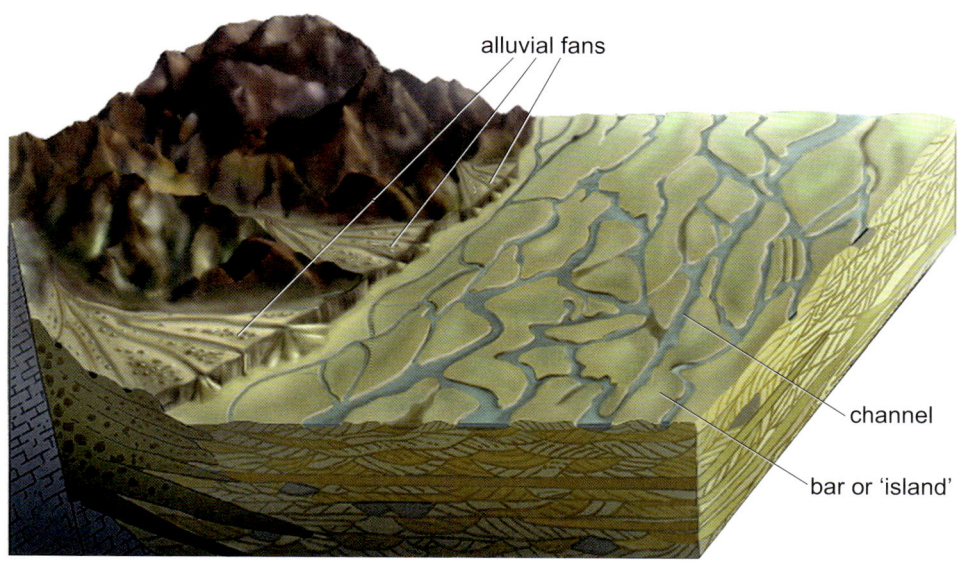

Figure 11.16 Block diagram of a braided river system showing the general features. This diagram only shows part of the larger channel-like feature in which all of the smaller braided channels lie.

Individual channels usually contain coarse-grained lags at the base that were deposited when the discharge was high. They are separated by 'islands' of sediment called **bars** (Figure 11.16), which develop in response to high sediment flux and variable discharge rates in three ways: by lateral accretion, by vertical accretion, or by downstream migration. The first is similar to point bars developing in a meandering river system.

- ■ Why do point bars in meandering river systems build up sideways, i.e. perpendicular to the flow of the river?

- ☐ Because of the helical flow of the water, the channel is constantly migrating sideways as it erodes on the outer bank and deposits sediment on the inner banks downstream to form the point bar.

High water levels allow the bars to build vertically upwards and subaqueous dunes develop on the bars and build forwards in a downstream direction. As the water level and, therefore, the current strength subside, current-formed ripples may develop and eventually the top surface of the bar may re-emerge. Flood plain sediments do not commonly accumulate in braided river systems because the continuous erosion and movement of the channels prevents this.

Braided river successions

Braided river successions are dominated by the sediments that comprise the bars and channel infill.

- ■ Braided river deposits tend to consist of coarse-grained sand and gravels. Why do you think this is so?

- ☐ Braided river systems tend to form where rocks are actively eroding, such as in mountain belts. The high discharge rate and steep gradient mean that the currents are strong and fast, and any fine-grained sediment tends to be carried downstream in suspension. However, the exact composition of the succession also depends on the nature of the rocks or sediments that are being eroded.

Figure 11.17 Channel fill of a Carboniferous braided river system in Northumberland, England. Erosive features and trough cross-stratification are particularly apparent in the upper part of the cliff face.

The migration of the bars will give rise to a wide variety of sedimentary structures including planar cross-stratification dipping downstream, and some trough cross-stratification from the migration of bedforms dipping both downstream and perpendicular to the main flow (Figure 11.17). The number of channels and the constant migration of channels and bars mean that there are many erosion surfaces and individual beds are often not laterally extensive. Variation from high discharge rates during flooding to lower discharge rates during 'normal' conditions will result in a broadly fining-upward succession in some of the channels.

So, to summarise, braided river successions comprise channel and bar sands and gravels with a variety of sedimentary structures and a high degree of lateral variability. Fine-grained sediments may be deposited on the top of the bars, forming thin bar tops. A typical braided river succession is shown in Figure 11.18.

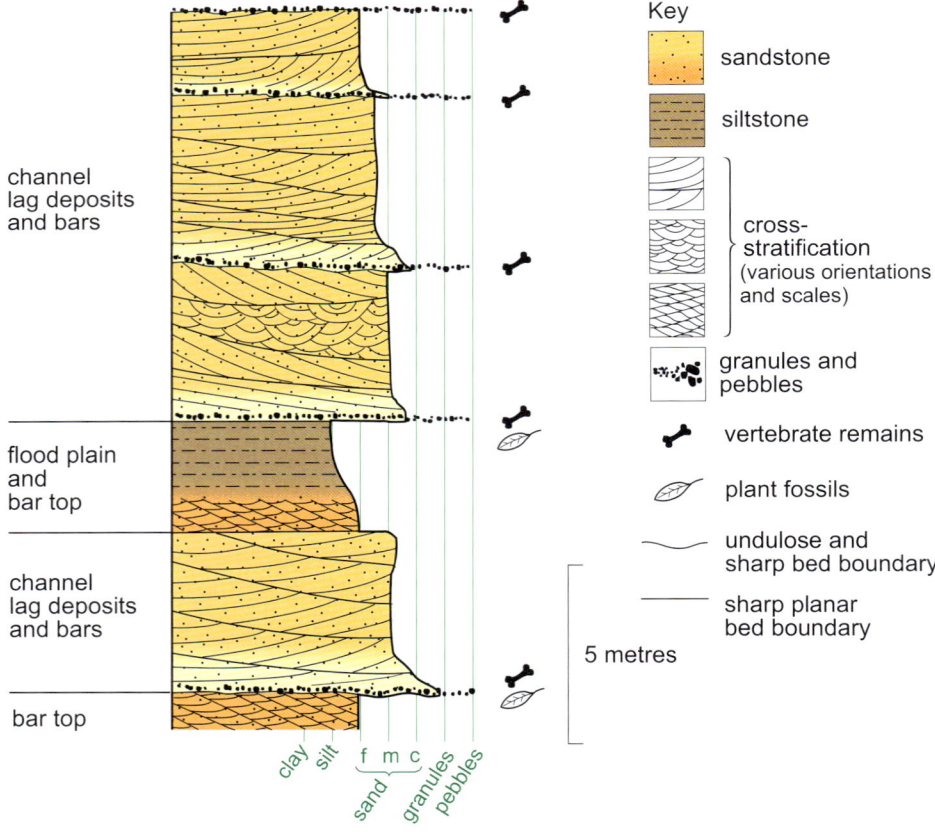

Figure 11.18 Idealised graphic log of a succession that might be deposited by a braided river system.

11.4 Alluvial fans

Alluvial fans are localised deposits that form where a river or stream loaded with sediment emerges from a confined mountain valley onto a topographically lower, flat lowland plain or major valley. Alluvial fans have a basic semi-conical shape. The apex of the cone points up the valley (Figures 11.16 and 11.19). As the water flows out from the steep valley and onto the flat plain, the sudden decrease in gradient is accompanied by extensive deposition from sediment gravity flows. Alluvial fans are most common in semi-arid and glacial climatic regimes where vegetation is sparse and run-off is seasonal, but alluvial fans can also form in humid climatic regimes.

Figure 11.19 A modern alluvial fan in the Himalaya feeding into a braided river system. Note that the apex of the cone points up the valley towards the source of the sediment.

Alluvial fan deposits are generally poorly sorted, matrix-supported and both texturally and compositionally immature (Figure 11.20), but they may show varying degrees of stratification or graded bedding, depending on how much water remains in the flow. The actual degree of grain roundness is a function of the sediment source, how far the sediment gravity flow travelled and whether it was fairly viscous so that grains were cushioned. The fan is traversed by a network of rapidly shifting braided streams and so the sediments in the channels comprise gravels, sandy gravels and sands, which show grain support and sedimentary structures characteristic of those already described for braided rivers. The sands may be arkosic if the nearby mountains contain feldspar-bearing rocks and chemical weathering and grain transport are limited. A feature common to all alluvial fans is that the sediments become finer-grained with increasing distance from the apex of the fan. The flow of streams across the fan predominates where the supply of water is more abundant and occurs throughout the year. Alluvial fans are a prominent feature of most braided river systems in mountainous areas and where mountains meet flat desert areas. You will see some examples when you do Activity 12.1 in the next chapter.

Figure 11.20 Quaternary alluvial fan deposits from the Sorbas Basin, Spain. Large boulders are fairly well rounded and there is some indistinct evidence of stratification across the centre of the photograph.

Activity 11.1 Alluvial environments

In this activity, you will consolidate your knowledge of alluvial systems.

11.5 Summary of Chapter 11

1. Sediment derived from weathering is transported into the river channel by sediment gravity flow processes. Material eroded from the banks of the river channel adds to the sediment load.

2. The water and sediment are moved downstream because of the gravitational potential energy caused by the river gradient, and the discharge increasing downstream. The river's bedload is moved by bouncing, rolling and saltation, while the finest grains are carried in suspension.

3. Meandering river systems are characterised by a distinct meandering channel cut into cohesive muds or channel sands, by a low overall topographic gradient and a high ratio of suspension load to bedload. The meanders increase in amplitude and migrate downstream by lateral accretion of the point bar on the inside of the bends and erosion on the outside of the bends. Ox-bow lakes form where the meander is breached. Levées may form on the edge of the river channel following increased water level and these may be breached during flooding to produce crevasse splays.

4. Meandering river successions comprise sharp-based, cross-stratified, fining-upward sandstones representing the point bar. These will be interbedded with fine-grained sedimentary rocks of the flood plain that may also contain coarser-grained crevasse splay deposits.

5. Braided river systems are characterised by high braiding and form where the discharge fluctuates widely, the bedload content is high, and where there is a relatively high gradient. The bars that divide the channels develop through lateral accretion, vertical buildup due to the high sediment flux, and deposition on the front of the bar.

6. Braided river successions mainly comprise stacked channel fills and bars. Occasional fine-grained bar top sediments are preserved. Braided river successions show a wide degree of vertical and lateral variability.

7. Alluvial fans are localised deposits that form where a mountain river or stream emerges into a flat, lowland plain or valley. They most commonly form under semi-arid and glacial climatic conditions. Deposition from sediment gravity flows is the most important process, reflected in deposits that are generally texturally and compositionally immature. The sediments become finer-grained away from the apex of the fan.

11.6 Objectives for Chapter 11

Now you have completed this chapter, you should be able to:

11.1 Describe the physical, chemical and biological processes that are important in alluvial environments.

11.2 Sketch graphic logs of braided and meandering river successions.

11.3 Describe the main features and the types of sedimentary deposits and structures formed in alluvial environments.

11.4 Distinguish between meandering and braided river systems.

11.5 Describe the main features of alluvial fans.

Now try the following questions to test your understanding of Chapter 11.

Question 11.2

Explain whether or not you would expect to find: (a) thick flood plain suspension deposits, and (b) organic debris, in a braided river succession.

Question 11.3

Figure 11.21 shows the progressive changes in the position of the channel of the River Mississippi during the periods 1765, 1820–30, 1881–1893 and 1930–32.

(a) Examine the pattern of the meanders and work out which of the channels A, B, C or D corresponds to each of these time periods. (*Hint*: notice the direction of flow of the river.)

(b) Name what feature X (indicated by the line) would have been in 1932.

(c) Describe the type of feature and sedimentary structures that would have been found at Y (indicated by the line) in 1932.

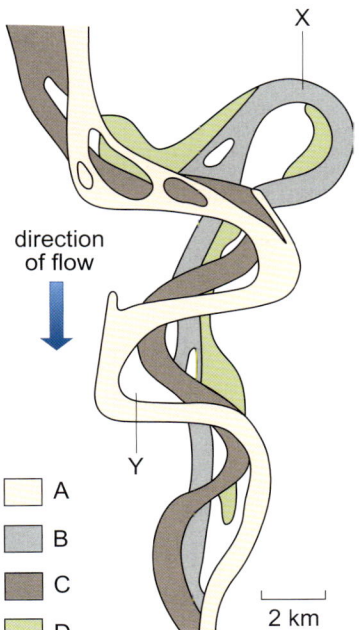

Figure 11.21 Shifts in part of the main channel of the River Mississippi during the periods 1765, 1820–30, 1881–1893 and 1930–32. (For use with Question 11.3.)

Chapter 12 Deserts

To many people, the word 'desert' conjures up an image of a vast area of sand dunes, but in fact this is only one of several sub-environments found in deserts. Deserts are defined as areas where the average rate of evaporation exceeds the average rate of precipitation (e.g. rainfall and snow). This lack of water is the reason why they contain very little flora and fauna. Lack of vegetation and water also means deserts are dominated by physical weathering and that when precipitation does come, erosion is rapid, and sediments can be easily moved around. Deserts occur in both hot and cold arid climates, with hot deserts covering a larger area of the Earth today than cold deserts – so we shall concentrate on these. Constant heating during the day and cooling at night in these deserts set up stresses within the rocks that lead to increased physical weathering (Figure 6.2).

The dominant agents of transport and deposition – wind and water – can be used to divide desert features and sedimentary deposits into two categories. Wind-dominated transport produces aeolian sand dunes, **ergs** (extensive sand seas typically greater than 125 km^2 in area), and large areas of bare rock and rock debris. Water-dominated transport, albeit intermittent, produces alluvial fans (Section 11.4), river valleys and temporary lakes.

Activity 12.1 Deserts

In this activity, you will watch the video sequence *Deserts* on DVD.

12.1 Wind in the desert

Aeolian transport processes and bedforms were described in Chapter 7, particularly Section 7.5. You may find it useful to re-visit this section now. Figure 12.1 shows a sketch cross-section through some ancient aeolian sand dunes and an associated sedimentary feature and deposit that formed in a desert environment.

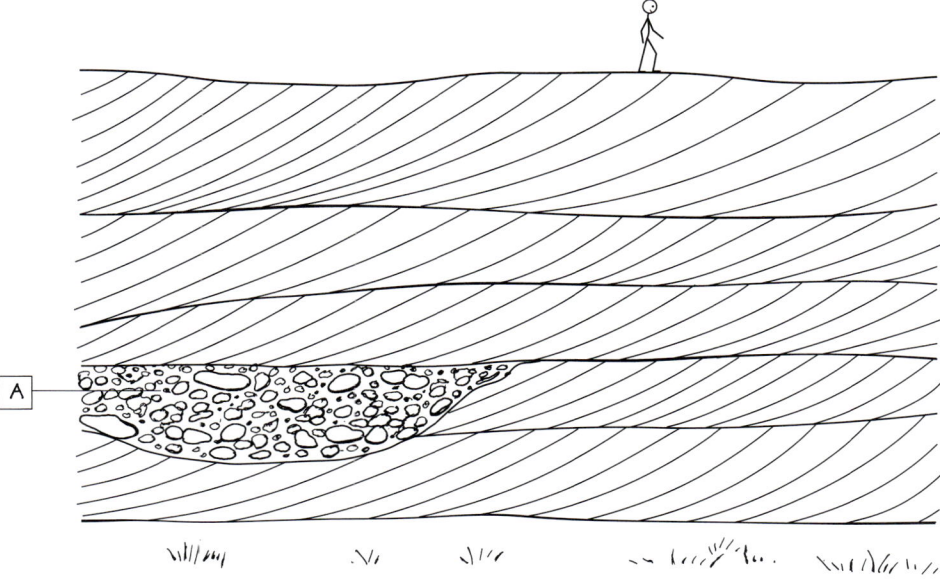

Figure 12.1 Sketch cross-section through some sedimentary rocks deposited in a desert. Label A is referred to in Section 12.2.

■ In Figure 12.1, what was the dominant wind direction? How do you know?

☐ From right to left, because the cross-stratification is dipping towards the left.

■ In what way is sand transported down the steep slope of dunes?

☐ Sand is transported down the steep slope of dunes by avalanching in a process called grain flow (Section 7.6.1), and also by grain fall as the grains fall out of suspension on reaching the sheltered (lee) side of the dune (Figure 7.36).

In order to preserve the dune sands so that eventually they become sandstones in the geological record (Figure 12.2), there has to be net accumulation of sand – so the sands have to be protected from the erosive and transportational power of winds. This can be achieved in several ways:

- through deposition at or near the water table (wet sands simply do not blow away), combined with subsidence, allowing more and more sand to accumulate
- by the growth of vegetation or the formation of surface cements that protect the sands and prevent them from blowing away. The red iron oxide coating around the grains of RS 10 could be a form of early cement that may well have helped to stabilise these Permian sands so that they could be preserved
- if the supply of sediment is exceptionally high, then the sands may accumulate because there is insufficient wind to remove them.

Figure 12.2 Large-scale cross-stratification in Jurassic aeolian sandstones, Nevada, USA. Varying dips in successive dunes point to changing wind directions. The exposure is about 40 m high.

■ Considering Figure 12.1 again, is it possible to determine the three-dimensional shape of these sand dunes? Give reasons for your answer.

☐ No, it is impossible to tell from this one sketch what their bedform was like. You would need to look at many faces cutting across the bedforms in different directions to determine the shape.

The three-dimensional shape of every sand dune depends on *two* variables – whether the wind pattern is variable or mainly blowing in one direction (unidirectional), and whether the supply of sediment is low, high or very high.

At first sight, the three-dimensional form of desert dunes may appear very complex, but the two variables described above result in four basic dune morphologies. **Barchan dunes** are crescent-shaped dunes when viewed from above. The ends of the crescent shape point downwind. They form where the wind pattern is unidirectional and the supply of sediment is low (Figures 12.3, and 12.4a and b). If the wind remains unidirectional but the supply of sand increases, the crests of adjacent barchan dunes will join up and grade into long, linear, asymmetric ridges called **transverse dunes** (Figure 12.4c). They are termed transverse dunes because the crests of the ridges are orientated perpendicular to the dominant wind direction. Where the wind direction is variable, very complex dune forms are found; one morphological type is the linear dune (Figure 12.4d), which forms where the sediment supply is low (similar to that for barchan dunes), and the other type is the star dune (Figure 12.4e), which forms only where the sediment supply is very high. In order to distinguish these dune morphologies in aeolian sandstones, it is necessary to collect many measurements of the direction of dip of the cross-stratification over a wide area.

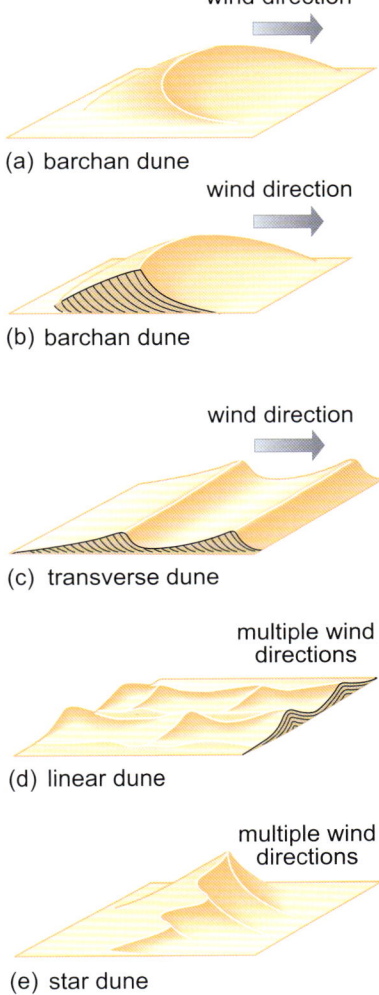

(a) barchan dune

(b) barchan dune

(c) transverse dune

(d) linear dune

(e) star dune

Figure 12.3 A barchan dune in Oman. The steep slope is facing towards the right and the dune is encroaching onto an interdune area characterised by finer-grained sediment and stabilising vegetation.

Figure 12.4 The morphology of the four basic types of desert dune in relation to the wind pattern and amount of sediment supply: (a) and (b) barchan dunes; (c) transverse dunes; (d) linear dunes; (e) a star dune. The typical cross-stratification pattern for barchan, transverse and linear dunes is shown. The cross-stratification pattern in star dunes would be complex and multi-directional.

Interdune areas consist of depressions or flat areas between the dunes and may be dry, damp or wet areas, depending on the position of the water table and the presence or absence of stabilising vegetation (Figure 12.3). The interdune sediment is very fine grained, often contains evaporite minerals and the interdune morphology is similar to playa lakes described below, but on a much smaller scale. When sand input to the system is minimal, wind erosion and transport will produce large expanses of bare, loose rocks. Such areas are simply referred to as stony deserts. They yield characteristic three-sided clasts called **dreikanters**, which are clasts sandblasted by the wind (Figure 12.5).

Figure 12.5 Two dreikanters from the stony desert near Luxor, Egypt. Field of view is 9 cm across.

So, in summary, wind processes are important in shaping, transporting and depositing sand grains, by moving all loose sediment away from an area and leaving behind bare rock and clasts that are too large to be moved. The wind is also important in a desert in helping any water that may be present to evaporate. The next section considers the features that might form when water does enter the desert.

12.2 Water in the desert

Water causes some of the most spectacular and dramatic features of desert erosion because most of the sediments are not bound together by vegetation, fine-grained sediment or cement. Rainfall is sporadic, but when it occurs over a short seasonal period (e.g. over the Australian desert) or every few years (e.g. over the Sahara) it may be torrential. In desert regions within or close to mountain ranges, a rainstorm can be catastrophic. The sudden rainfall carries large volumes of accumulated sediment down from the mountains and into the desert area, and may erode a steep-sided, narrow valley called a **wadi** (*wadi* is Arabic for watercourse) (Figure 12.6a). As the floodwaters subside by infiltration and evaporation, sediment is deposited. It is usually poorly sorted and may show only crudely defined bedding, although, if conditions are favourable, planar stratification and cross-stratification similar to that in other fluvial deposits may be present (Figure 12.6b). For much of the time, most wadis are dry valleys that are only periodically occupied by water.

Figure 12.6 (a) A steep-sided desert valley or wadi formed during heavy rainfall in Oman. (b) Wadi deposits in Oman showing crude stratification and grading.

- Consider Figure 12.1 again. What do the rocks labelled A most likely represent and why?

- ☐ Wadi deposits. The clasts are too large to have been moved by wind processes and the cross-cutting nature of the contact between the cross-stratified sandstones and the wadi deposits suggests erosion of the sand dunes.

- On the video sequence *Deserts*, two other morphological features formed by the presence of water in a desert are described. What are they?

☐ Alluvial fans and temporary, playa lakes.

Alluvial fans were described in Section 11.4. Temporary lakes, often termed **playa lakes**, can form in any depressions in the desert surface as it rains and are common at the foot of alluvial fans, as shown on the *Deserts* video sequence. The playa lakes become infilled with fine-grained sediments that have been transported in suspension either by wind or water. If sediment has been carried by water down the alluvial fans, the finest-grained sediments will be carried the furthest and will be deposited at the distal part of the fan in the flat area where the playa lake forms. The resulting sedimentary rocks are typically homogeneous or bedded mudstones and siltstones (Figure 12.7). Desiccation cracks are widespread in playa lake deposits and form as the sediments periodically dry out (see Figure 7.45c). The water may support organisms and both trace and body fossils may be found. If the water does not drain away, but is allowed to evaporate slowly, evaporite minerals such as gypsum and halite will be deposited and a **saltpan** is formed. Similar processes operate on a smaller scale in interdune areas.

Figure 12.7 Triassic playa lake deposits, St Audries Bay, Somerset, England.

One rare, but rather special feature that is preserved in some desert sediments is the product of lightning strikes that weld the sand grains together to produce characteristic tubular structures called **fulgurites** (Figure 12.8).

Activity 12.2 Desert environments

In this activity, you will complete a graphic log of a succession typical of a desert environment.

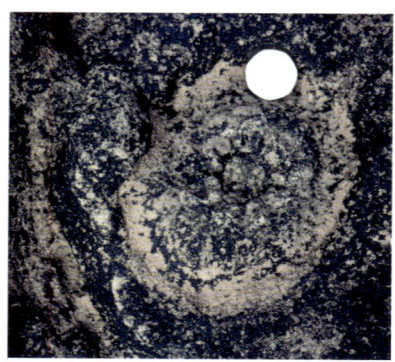

Figure 12.8 Fulgurite with a welded texture typical of lightning strikes, Permian aeolian sandstone, Arran, Scotland.

12.3 Summary of Chapter 12

1. Deserts form where the average rate of evaporation exceeds the average rate of precipitation.

2. Deserts occur in both hot and cold climatic areas of the world.

3. Frosted sand grains, good sorting and large-scale cross-stratification are diagnostic of aeolian processes.

4. Sand dunes usually start to form by grains accumulating around an obstacle and producing a drift. When sufficient sand has accumulated, the dune begins to migrate as saltating sand grains move up the shallow slope and then avalanche down the steep slope in a process called grain flow.

5. The wind pattern and size of the sediment supply determines the morphology of the dunes that form.

6. Barchan dunes are crescent-shaped with the ends of the crescent pointing downwind. Transverse dunes are long linear asymmetric ridges with the steep side pointing downwind. In both cases, the cross-stratification is dominantly inclined in one direction.

7. Sand dunes may be preserved in the desert due to deposition near the water table combined with subsidence, through the growth of vegetation, precipitation of cement or due to high sediment supply.

8. Interdune areas of much finer sediment form in the troughs between the dunes and are stabilised by moisture or vegetation.

9. Sporadic rainfall in the desert leads to the formation of wadis in which crudely bedded and often coarse-grained sediments are deposited. The presence of water also means that playa lakes can form in depressions; these become infilled with fine-grained sediments and evaporites may precipitate.

10. Where mountains fringe the edge of the desert, alluvial fans and playa lakes are commonly found.

12.4 Objectives for Chapter 12

Now you have completed this chapter, you should be able to:

12.1 Describe very broadly what governs the formation of deserts.

12.2 List the features that together are diagnostic of sediments deposited by aeolian processes.

12.3 Describe the processes by which sand dunes form.

12.4 State two factors that are important in determining the morphology of the sand dunes that form.

12.5 Describe the shape and cross-stratification pattern produced by the migration of barchan and transverse dunes.

12.6 Describe how sand dunes might be preserved in the desert.

12.7 Name and describe the features that may be formed by water in the desert.

Now try the following questions to test your understanding of Chapter 12.

Question 12.1

In the form of a table, summarise the main morphological features of deserts and the common sedimentary facies they produce.

Question 12.2

Describe the features you could look for in large-scale cross-stratified sandstones to help determine that they were deposited by aeolian, rather than tidal, processes.

Chapter 13 Siliciclastic coastal and continental shelf environments

The coastline and continental shelf both today and throughout geological time represent a complex natural continuum of different environments (Figure 13.1) – in other words, the type and shape of an environment changes to another as the relative dominance of fluvial, tidal, wave, wind and biological processes changes. Recall from the video sequence *Coastal processes* (Activity 1.1) and Chapters 7 and 8 that these processes also govern the characteristics of the sedimentary rocks that are deposited in coastal environments. In this chapter we will consider coasts dominated by siliciclastic deposition and, in Chapter 14, those dominated by carbonate deposition.

Figure 13.1 A rocky coastline with bays and beaches, east Dorset, England.

Siliciclastic coastlines include deltas, strandplains and barrier islands that are illustrated on the right-hand side of the continent on the SE Poster (estuaries are not covered in any detail). These coastal morphologies are intimately associated with the adjacent **continental shelf** which itself forms the edge of the continent and is covered by relatively shallow water (usually between a few tens of metres and 200 m deep).

13.1 Subdivision of the coastal and continental shelf environment

Coastal environments can be subdivided into a number of zones, defined by the position of mean high and low tide, and the depth to which wave processes affect the sediments. Figure 13.2 shows an idealised cross-section through the coast and continental shelf, leading down the adjacent continental slope and out into the deep ocean. These two deeper water sub-environments are discussed in Chapter 15.

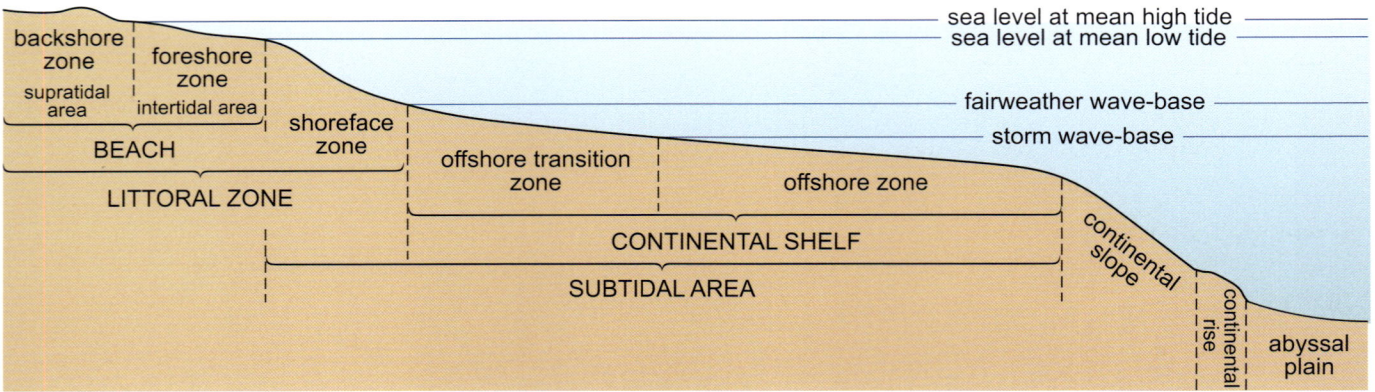

Figure 13.2 Vertical cross-section to show the morphological zones in a coastal area and adjacent continental shelf. The terms are explained in the text. Note that the shoreface is usually a concave-up shape and that at the base of the shoreface there is a decrease in gradient. The intertidal area is between mean high and low tide and the subtidal area is below mean low tide.

Figure 13.3 Backshore dunes stabilised by vegetation and foreshore sands at Tràigh Iar, Isle of Harris, Scotland.

The mean high tide is the average of the high spring and neap tides and, similarly, the mean low tide is the average of the low spring and neap tides. (The concept of spring and neap tides was introduced in Section 7.4.2.) The mean high and low tidal marks are important because they define the **intertidal** area of the shore, which is the area influenced directly by breaking waves (see video sequence *Coastal processes*, Activity 1.1). Above the mean high tide is the **supratidal** area, i.e. the area affected only by the highest spring tides or occasional storm waves. At the other extreme, the **subtidal** area is the area below mean low tide, in which the upper part is totally exposed only during the lowest spring tides. However, you should note that the subtidal area is still influenced by tidal processes because tides produce strong currents (Section 7.4.2).

We shall consider each zone in turn from the land to the deep sea. The **backshore** zone (supratidal) is the area above mean high tide and is only covered by seawater during very large storms (Figure 13.3). The **foreshore** zone (intertidal) is seaward of the backshore and is the area between mean high tide and mean low tide. This is the area of the coastline that is usually covered by seawater twice a day at high tide and uncovered twice a day at low tide. Consequently, it is a very dynamic area with the sediments and organisms constantly moving around to adjust to the changing conditions. The foreshore zone and backshore zone together comprise the **beach** and they are often separated by a ridge or **berm**.

The **shoreface** is the area below mean low tide but above fairweather wave-base, and is frequently very rich in organisms (Figure 13.4).

■ Explain what fairweather wave-base represents.

□ Fairweather wave-base is the maximum depth at which there is water movement caused by waves during fairweather conditions (Section 7.4.3).

The seabed above fairweather wave-base is affected by waves on an almost daily basis. With increasing depth down to storm wave-base, wave-related processes operate with decreasing frequency. The area below fairweather wave-base but above storm wave-base is known as the **offshore transition zone**.

■ Explain what storm wave-base represents.

□ Storm wave-base is the maximum depth at which there is water movement caused by waves during storm conditions (Section 7.4.3).

■ Bearing in mind that the amplitude of storm waves will be considerably greater than that of fairweather waves because the winds are stronger during storms, why is the storm wave-base deeper than the fairweather wave-base?

□ The greater the amplitude of a wave, the larger the diameter of the orbital motion, and so the deeper the orbital motion will penetrate. The depth that waves penetrate is equivalent to about half their wavelength (Section 7.4.3).

Figure 13.4 Richly fossiliferous Pleistocene sandstones accumulated in submarine sand waves in the shoreface zone, Walton-on-the-Naze, Essex, England. The coin is 3 cm in diameter.

The area below storm wave-base to the distal edge of the continental shelf is called the **offshore zone**. The offshore and offshore transition zones both form part of the continental shelf. Beyond this is the continental slope, leading down to the deep ocean basin.

The tidal range, i.e. the average height by which the tide rises and falls on a coastline, can also be described as being small (<2 m), medium (2–4 m), or large (>4 m).

13.2 Interaction of fluvial, tidal and wave processes

Section 7.4 and the video sequence *Coastal processes* (Activity 1.1) introduced you to the different types of water and sediment movement that result from fluvial, tidal and wave processes. The interaction of these processes results in a variety of bedforms and sedimentary structures in coastal and shelf areas and it is the dominance of particular processes that produces the spectrum of coastline morphologies observed today. Both tides and waves produce strong currents. To recap on these processes, try the following questions.

Question 13.1

Complete Table 13.1 by indicating which water movements are applicable to each of the processes.

Table 13.1 Type of water movements applicable to fluvial, tidal and wave processes. For use with Question 13.1.

	Fluvial	Tidal	Wave
oscillatory flow			
unidirectional flow			
bi-directional flow over ~12 hours			

Question 13.2

Complete Table 13.2 by indicating which process(es) may lead to the formation of the listed sedimentary structures and bedforms.

Table 13.2 Processes that may lead to the formation of particular sedimentary structures and bedforms. For use with Question 13.2.

	Fluvial	Tidal	Wave
herringbone cross-stratification			
current-formed ripples			
wave-formed ripples			
cross-stratification			
tidal bundles			
planar stratification			
trough cross-stratification			
planar cross-stratification			
mud-draped ripples			
climbing ripples			
hummocky cross-stratification			

Consider the effect on the coastline of these fluvial, tidal and wave processes by examining the coastline shown on Figure 10.1 (and reproduced at a larger scale on the SE Poster).

- Which of the three environments – delta, strandplain or barrier island – will form if the dominant process at the coastline is fluvial, there is a constant high sediment supply, and strong current flow from a river into the sea?

- ☐ A delta will develop – provided the wave and tidal processes are not strong enough to immediately transport the sediment away.

For fluvial processes to be dominant, wave and tidal processes need to be relatively weak and there also needs to be a well-developed alluvial system on land, providing a moderate to high sediment supply. This supply may result either from a mountain chain that is being uplifted, or because there is a huge drainage basin with a high rainfall, such as the Mississippi delta forming off the southern coast of North America (Figure 13.5). The Mississippi River drains much of the continent and the delta is largely protected from waves because it lies within the Gulf of Mexico.

Chapter 13 Siliciclastic coastal and continental shelf environments

Figure 13.5 The Mississippi delta from space in 2001. Part of the delta was destroyed by hurricanes Katrina and Rita in 2005. The pale blue areas between the river distributary channels represent suspended sediment discharged by the rivers that flow above denser marine waters.

Along all coastlines, no matter what their morphology, there is always some influence from tides and waves and, to a lesser extent, from fluvial processes.

■ Carefully examine Figure 13.6 and identify the process that is much more important for the formation of strandplains than for the formation of barrier islands.

☐ Waves are much more important for the formation of strandplains than barrier islands.

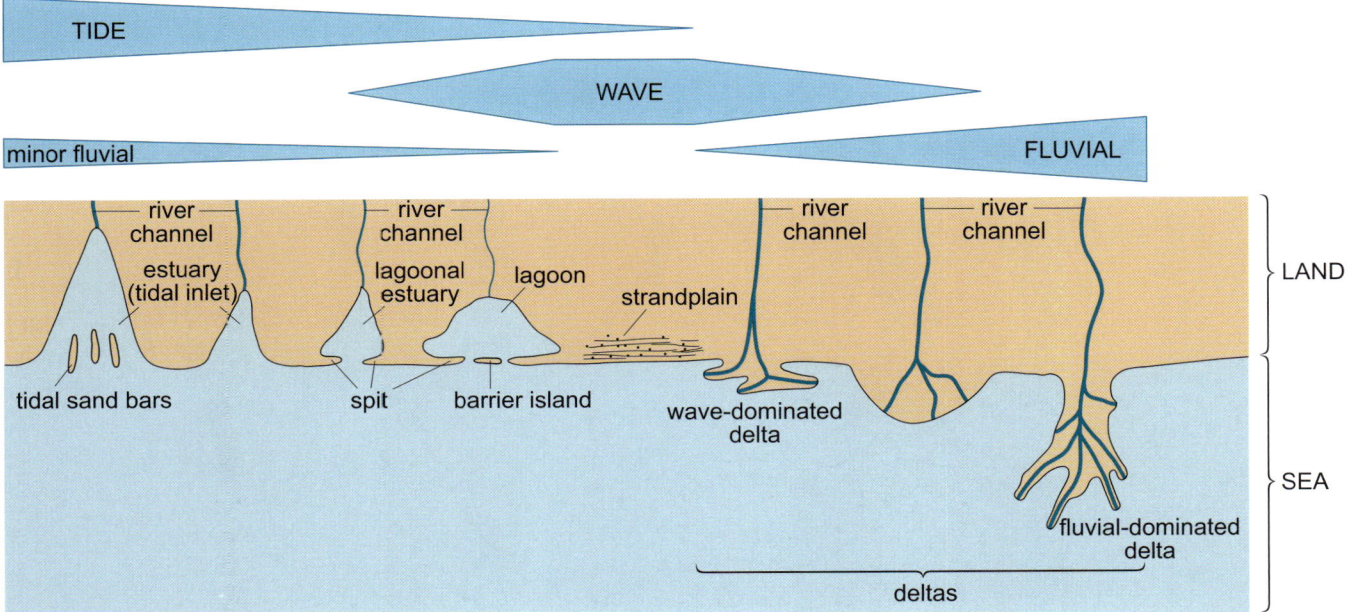

Figure 13.6 Spectrum of coastal morphologies. The relative importance of waves, tides and fluvial input for each morphology is shown schematically by the boxes at the top of the figure; tapering of the box indicates less influence.

■ Why do you think this might be?

☐ As waves break, they move the sediment landward, thus the sediment becomes attached to the coastline rather than being separated from it.

In addition, the formation of strandplains is favoured by a high sediment supply and a steep coastal gradient, which will cause the waves to break much more abruptly.

The dominance of either wave or tidal processes along a particular piece of coastline is generally complicated since it depends on several factors. These include:

- the size and shape of the ocean adjacent to the coastline
- the width and gradient of the continental shelf
- the shape of the coastline.

■ How will the size of the ocean affect the relative strength of the waves?

☐ If the ocean is large and the coastline is exposed to open ocean wave processes, wave strength will be generally greater because the wind has had thousands of kilometres over which to blow (the *fetch*) and build up waves.

The coastlines of northwest and southwest Britain, Brazil and west Africa are examples of predominantly wave-dominated coastlines because they face the open Atlantic Ocean. The width and gradient of the continental shelf also affects the dominance of wave and tidal processes since narrow continental shelves favour wave processes, whereas wide continental shelves dampen wave effects and amplify the action of tidal currents. In a similar manner, the tidal range and strength of the tidal currents vary according to the shape of the ocean basin and the effect of the spin of the Earth. Thus, tidal ranges around the world differ, as shown in Figure 13.7.

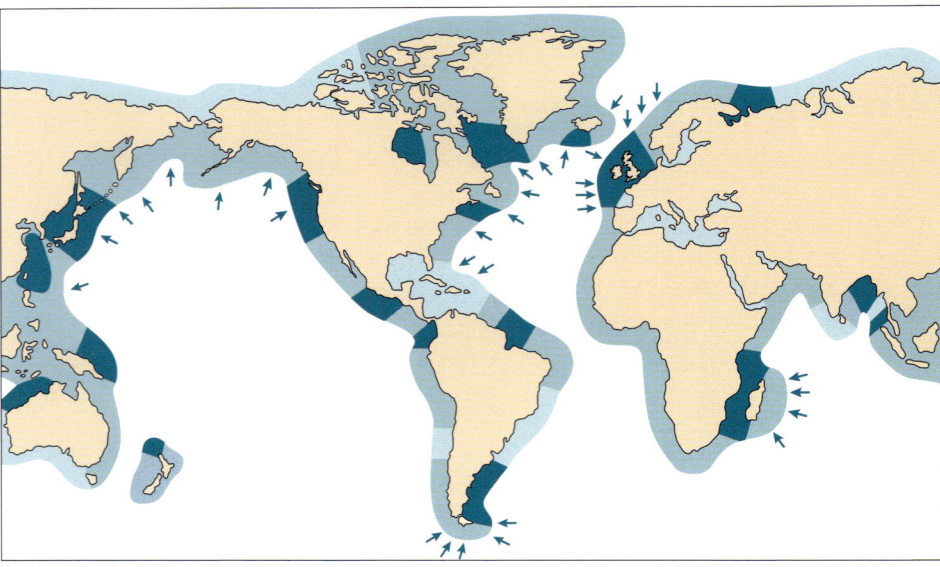

Figure 13.7 A classification of present-day coastlines around the world, based on tidal ranges.

Key
■ large tidal range > 4 m
→ coastline subject to storm waves
■ medium tidal range 2–4 m
■ small tidal range < 2 m

The shape of the coastline determines whether the tidal processes are enhanced or reduced. For instance, natural ranges can be enhanced where flow is funnelled into a narrow passage. For this reason, estuaries tend to be dominated by tidal processes, though of course there is also some fluvial influence (Figure 13.8). If tidal processes are dominant along a linear strip of coastline without rivers, then extensive areas of the foreshore (i.e. the tidal flats) will be covered in interbedded thin layers of sand and clay deposited during different parts of the tidal cycle (video sequence *Coastal processes*, Activity 1.1 and Section 7.4.2) and offshore, tidal sandbodies will form.

Waves and tides produce strong currents within the coastal and continental shelf areas above storm wave-base. As waves strike the coastline and dissipate their energy, water flowing onto the foreshore must be balanced by the amount that returns. Waves striking the shore at right angles produce **rip currents** within a few hundred metres of the shoreline, whereas oblique waves generate **longshore currents** parallel to the coastline (Figure 13.9). Both types of current are also created by variations in the height of the breaking waves, which are governed by the weather and morphology of the coastline.

Figure 13.8 Tidal creek at low tide, Blakeney, Norfolk, England.

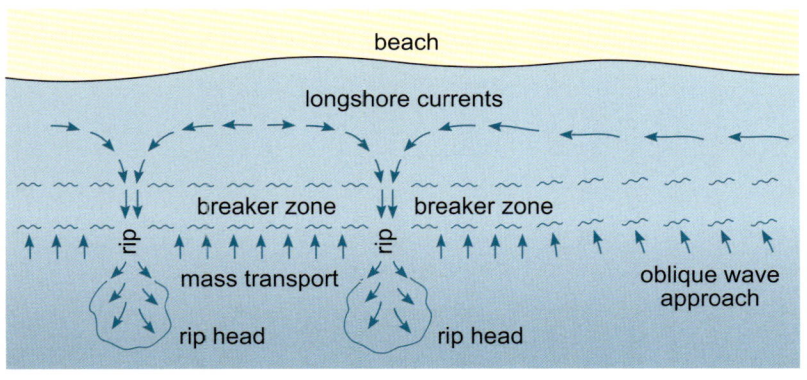

Figure 13.9 The shoreface is a zone of onshore mass transport of water by waves. This is balanced either by longshore currents (if the wave hits the shoreline obliquely), or the formation of rip currents (if it hits the shoreline at right angles).

The effect of longshore currents can be seen along many stretches of coastline where groynes (often called breakwaters) have been erected to trap sand on the beach. As the longshore current moves the sand along the beach, the sand becomes trapped against the upcurrent side of each groyne (Figure 13.10). Evidence for rip currents can sometimes be seen in the geological record in the form of elongate scours called gutter casts (Section 7.2.2), but longshore currents are more difficult to recognise.

- In Figure 13.10, what is the principal direction of the longshore currents along this sandy beach?
- ☐ From bottom left to top right of the photograph, parallel to the shoreline. This is evidenced by the sand trapped on the upcurrent side of each groyne.

Figure 13.10 A series of groynes built perpendicular to the shore to reduce the amount of shore-parallel sediment transport due to longshore drift, Walton-on-the-Naze, Essex, England.

Tides also create currents when they act in confined spaces; this can occur for example where the sea-floor topography changes, resulting in inequalities between the ebb and flood tides over the tidal cycle (Section 7.4.2). In such cases, sediment will be moved differentially in one direction and net transport will occur. Such tidal currents acting in subtidal areas can deposit large dunes or sand waves. Figure 13.11 shows some large-scale cross-stratification of Cretaceous age produced by migrating sand dunes. They have been interpreted as tidal sand waves or sandbanks deposited in a tidal estuary (river-dominated sediment) and adjacent marine bay area (marine-dominated sediment), similar to the Wash today.

Figure 13.11 Cretaceous large-scale cross-stratification produced by the migration of tidal sand dunes exposed in a quarry in Leighton Buzzard, Bedfordshire, England. Note that in the upper bed, the cross-stratification indicates a strong, dominant flow direction to the right because the cross-stratification is unidirectional rather than bidirectional. The long axis of the dune would have been orientated perpendicular to the tidal current.

- Name another process that forms large-scale dunes.
- ☐ Aeolian processes also form large-scale dunes (Section 7.5).

13.3 Strandplains

Strandplains are essentially linear accumulations of either sand or gravel running parallel to the coastline and attached to the land (Figure 13.6). They account for most of the beaches around Britain and include the modern beach that featured in the video sequence *Coastal processes* in Activity 1.1.

Question 13.3

What are the dominant processes and conditions necessary for the formation of a strandplain?

Figure 13.12 Pebbles on a foreshore, Budleigh Salterton, Devon, England.

Whether the strandplain is made up of sand or gravel depends on the sediment supply available and the wave energy. Coastlines that frequently experience high energy storm waves (more common in middle and high latitudes) tend to be made of pebbles (Figure 13.12).

Along the foreshore, the strandplain comprises one or more coast-parallel ridges that are separated by slight depressions or runnels (Figures 13.13 and 13.14). The ridges and runnels are built up by wave action. As waves approach the shoreline, the orbital motion of the parcels of water reaches as far as the seabed (Figure 7.28b). The frictional resistance of the seabed will cause the waves to increase in height and decrease in wavelength. As they break, they expend their energy by driving a thin, turbulent sheet of water (called the **swash**) that carries sediments up the foreshore. As the less energetic water moves back down the beach as the **backwash**, some sediment is deposited whilst the remainder is carried further seaward again. The ridges build up most when the waves are particularly large, for example during storm conditions. If the waves are smaller and break gradually as they approach the shoreline, they have very little energy left by the time they reach the foreshore, so sediment is gently pushed up the beach by the swash but is not scoured and returned again during the backwash. The position of the ridges relates to the variations between storm and fairweather conditions, together with spring and neap tides (Section 7.4.2).

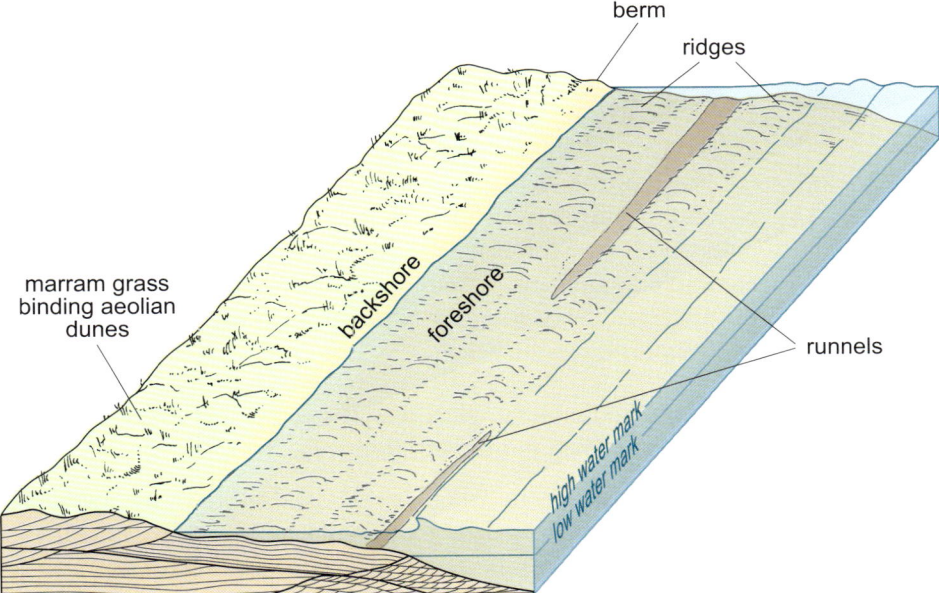

Figure 13.13 Block diagram of the foreshore showing ridges and runnels. The berm is a larger, often flat-topped ridge that marks the boundary between the foreshore and backshore zones.

Several factors affect the slope of the foreshore. The most important of these is sediment grain size: the coarser the sediment, the steeper the foreshore profile. This is because when a wave breaks on the foreshore, the swash is more intense than the backwash. Water percolates down through sediments on the swash so there is less water, and therefore less energy to flow back over the surface of the beach as backwash. As a result, less material is moved offshore and there is net movement of sediment onshore.

Figure 13.14 Well-developed ridges and runnels on Chesil Bank where it joins the Isle of Portland, Dorset, England.

■ Would you expect water to percolate more readily into coarser-grained sediment or finer-grained sediment?

☐ Into coarser-grained sediment because the pore spaces between the grains are larger.

This means that the coarser the grains, the less water is available to move sediment offshore from the surface of the beach and so the foreshore profile is steeper.

Figure 13.15 Gently seaward-dipping profile of a modern-day foreshore in South Carolina, USA, showing planar-stratified sands in a trench in the foreground. The sea is towards the right.

Wave energy also affects the foreshore profile. Higher energy waves tend to flatten the profile and remove finer-grained sediment from the foreshore. Sandy beaches tend to have a flatter profile than pebble beaches. In regions such as Britain, both types of beach are likely to have flatter profiles in winter than summer because of the greater incidence of high energy waves associated with winter storms.

The constant, high-speed movement of the swash and backwash means that the foreshore sediments are planar stratified, with current lineations indicative of the upper flow regime (Section 7.4.1). Beds are usually inclined gently seawards due to the natural slope of the foreshore (Activity 1.1 and Figure 13.15). On sandy beaches, slight changes in the angle of the beach profile between summer and winter result in subtle planar erosion surfaces between groups of planar beds (see block diagram on the SE Poster). Wave- and current-formed ripples are also common and antidunes may be present; antidunes form in small channels during the backwash and where shallow rivers and streams cross the beach.

■ Explain which textural and mineralogical features you would expect in sandstones that were deposited in the foreshore area.

☐ The foreshore area is constantly reworked, so the sandstones of the foreshore area are typically well-sorted, rounded, glassy and usually compositionally mature. Just such mature textures are found in some Carboniferous sandstones with low-angle planar stratification, which are interpreted as ancient foreshore deposits (Figure 13.16).

Figure 13.16 Planar-stratified sandstones in Carboniferous foreshore deposits, Northumberland, England. Vertical thickness of section is approximately 1 m.

The foreshore area is often colonised by a range of biota, depending on whether the surface they are living on is hard or soft. Epifaunal or infaunal bivalves and gastropods are common, together with various annelids (e.g. lugworms and ragworms) and arthropods (e.g. lobsters and crabs). In ancient foreshore deposits, you might expect to find the hard parts of similar animals associated with their trace fossils. The high energy conditions of the foreshore result in many of the body fossils being broken. The common trace fossils found along ancient shorelines are illustrated in Figure 3.68. The trace fossils are not exclusive to each zone and local physical, chemical and biological factors will, ultimately, determine what is found at each site.

Question 13.4

The holes in the intertidal zone illustrated in Figure 13.17 are the top of vertical feeding tubes made by worms; their faecal casts are also shown. Explain whether the vertical feeding tube or the faecal cast made by the worm is more likely to be preserved in the geological record.

The backshore area of the beach (i.e. the area above mean high tide level), is only occasionally swept by waves – at high spring tides and during storms. Since this area is subaerial for most of the time, the sand dries out and is then reworked by aeolian processes into aeolian coastal dunes (Figure 13.13). The area is also stabilised by salt-tolerant grasses (Figures 13.3, 13.18 and the video sequence *Coastal processes*).

Figure 13.17 Worm tubes and casts on an intertidal zone, for use with Question 13.4.

The shoreface zone lies below the effects of breaking waves, but above fairweather wave-base, so sediment is moved by the oscillatory motion of the water as each waveform passes. Wave-formed ripples (Section 7.4.3) are common in the shoreface, resulting in small-scale cross-stratification, and storm waves sometimes create swaley cross-stratification (Figure 7.31) just above fairweather wave-base. In addition, currents created by the waves will form both straight-crested and sinuous or curved-crested ripples and dunes.

■ Describe what type of cross-stratification results from (a) straight-crested dunes, and (b) sinuous or curved-crested dunes.

□ (a) Straight-crested dunes result in planar cross-stratification; (b) sinuous or curved-crested dunes result in trough cross-stratification (Section 7.4.1).

Figure 13.18 Aeolian sand dunes stabilised by salt-tolerant grasses, Northumberland, England. The foreshore is to the left.

A great abundance of biota lives in the shoreface zone. This means that the primary sedimentary structures may often be destroyed by bioturbation as a consequence of the organisms burrowing and moving through the sediment. However, the frequency of reworking by wave currents means that much of this bioturbation may not be preserved. Figure 3.68 shows some of the trace fossils typical of the shoreface. Biota from many phyla are found in both modern and ancient shoreface areas. They include benthic communities of arthropods (e.g. crabs, lobsters and trilobites), molluscs (bivalves and gastropods), brachiopods, cnidaria (e.g. sea anemones and corals), as well as pelagic organisms such as molluscs (e.g. cephalopods, such as ammonites) and vertebrates (e.g. fish).

The area immediately below the shoreface and therefore below fairweather wave-base is the offshore transition zone. The fact that this zone is above storm wave-base means that sediment is moved around, but only during storm conditions. The offshore transition zone is generally characterised by finer-grained sediments than those that make up the adjoining shoreface because the offshore transition zone is a lower energy environment. This zone is also heavily colonised by many organisms and the sediment is usually intensely bioturbated because of the infrequent disruption by waves, though wave-formed ripples are sometimes preserved.

■ What is the name of the almost entirely unique sedimentary structure formed by storm waves in the offshore transition zone?

☐ Hummocky cross-stratification (HCS) (Section 7.4.3; Figures 7.31 and 7.32). This is further illustrated in Figure 13.19.

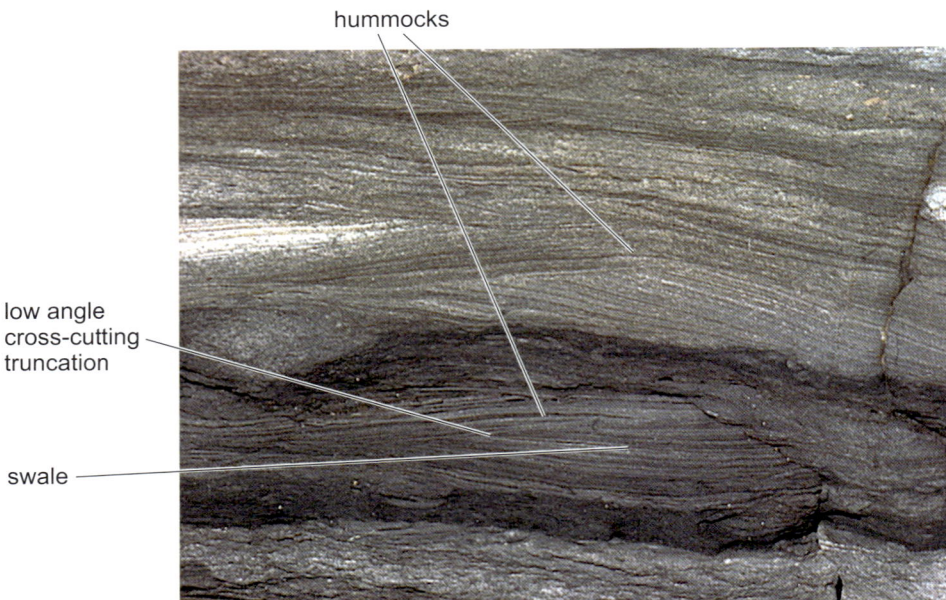

Figure 13.19 Hummocky cross-stratification in Carboniferous rocks from Northumberland, England. The exposure shown is about 1 m high.

Beyond the offshore transition zone is the offshore zone; this comprises the rest of the continental shelf (Figure 13.2). The bulk of the sediments deposited on the outer part of the continental shelf are silts and clays that are carried there in suspension. Final deposition is aided by flocculation and by filter-feeding organisms ingesting the mud and redepositing it as faecal pellets.

RS 27 represents a typical shelf mudstone that was deposited in an extensive continental shelf sea during the Jurassic Period. Calcite-cemented mudstone nodules are common in continental shelf deposits and are thought to form during the early stages of burial of the mudstones in response to changing chemistry of the water in the sediment pores.

Marine mudstones are very important in the geological record because they preserve the delicate details of marine biota. Their fine-grained nature means that conditions within the sediment are often anoxic and thus organic matter is more likely to be preserved. The organic matter may be the source of oil and gas. Mudstones that are rich in organic matter are dark brown to black in colour and they often smell bituminous. Organic-rich mudstones (Figure 13.20) often contain pyrite (FeS_2), which is produced in response to the high amounts of sulfur preserved in these organic-rich sediments during anoxic conditions at the time of deposition. The lateral equivalents of the beds shown in Figure 13.20 occur in the North Sea and are the source of much of Britain's oil reserves.

Figure 13.20 Jurassic organic-rich mudstones from the Kimmeridge Clay Formation, Dorset, England. The brown layers that stand proud of the cliff face contain the most organic matter.

Siliciclastic sandbodies can be found on the outer parts of the continental shelf and they are formed by a number of different processes, which include offshore transport of the sands during very intense storms and strong tidal currents.

The main facies of each of the coastal zones in the strandplain environment can be summarised in the form of an idealised graphic log, as shown in Figure 13.21.

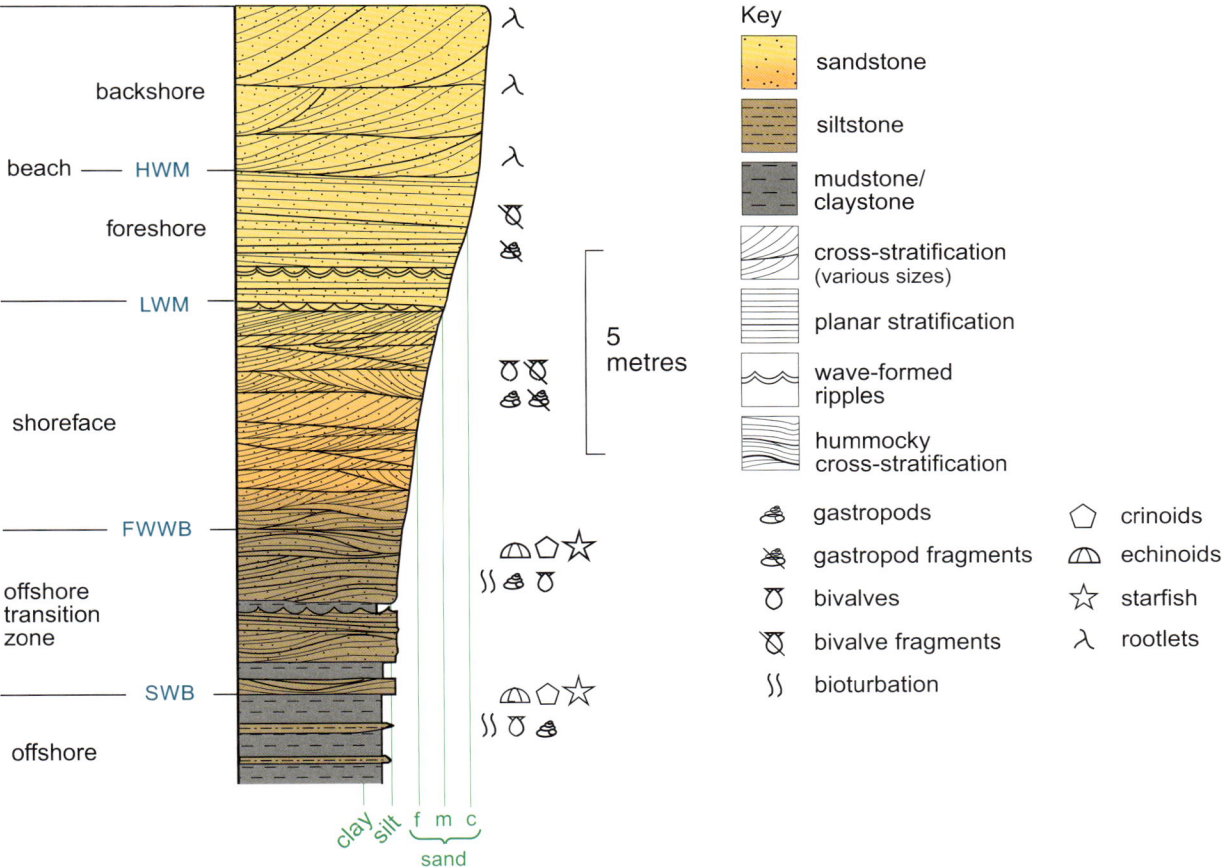

Figure 13.21 An idealised graphic log of a prograding strandplain environment. A block diagram of a prograding strandplain is shown on the SE Poster.

13.4 Barrier islands

Barrier islands are similar to strandplains in many of their features. What is different is that there is a long, linear barrier offshore that is separated from the land mass by a **lagoon** (Figure 13.6). This consists of a shallow-water area protected from major coastal currents and high amplitude waves, except during storm conditions when the barrier may be breached. The barrier can be thought of as a long, linear island situated anywhere between tens and thousands of metres offshore. Breaks in the barrier occur where rivers enter the open sea, where currents change, tidal currents flow in and out of the lagoon, or where storms have broken through the barrier (Figure 13.22). Flow between the lagoon and the open sea is restricted so that conditions in the lagoon are brackish rather than fully marine.

Figure 13.22 A barrier island system called Kiawah Island, South Carolina, USA. The photograph was taken at low tide. The barrier runs across the middle of the image, the lagoon is to the left and the open sea to the right, where waves are breaking. Complex sandbodies are being deposited and channels are being cut where there is a break in the barrier, about three-quarters of the way up the photograph.

■ What features and processes favour the formation of barrier islands compared with strandplains?

☐ A lower coastal gradient, some fluvial input, longshore drift, a slightly higher tidal range, slightly lower wave power and a moderate, rather than an abundant, supply of sand (see Section 13.2, especially Figure 13.6).

Seawards from the barrier, the facies are very similar to those of the strandplain system (compare Figure 13.21 with Figure 13.23). However, there is usually less wave energy than in strandplain environments, so HCS may be entirely absent or only poorly developed in the offshore transition zone. This is because barrier islands do not tend to form along storm-dominated coastlines, where the energy is too great to favour preservation of the barriers. The facies that are characteristic of barrier islands are those deposited within and on the shores of the sheltered lagoon, notably fine-grained sediments settling out of suspension and abundant trace and body fossils (including plants).

■ Explain whether or not the body fossils will be marine forms like those found on the barrier shoreface.

☐ Because the lagoon is not usually fully marine, species that are adapted to live in brackish water conditions will be present. They will include bivalves, such as oysters, and brackish water arthropods and gastropods.

The brackish but fairly quiet water conditions mean that usually there are a lot of individuals but with low species diversity. The topographic slope around the shoreline of the lagoon is gentle and the wave action weak, and so if there is also a medium to high tidal range, tidal flats will develop. These comprise flat, laterally extensive areas of coastline dominated by tidal processes where mainly silts and clays are deposited. The bedforms and sedimentary structures produced here will include planar stratification, small-scale cross-stratification, mud-draped ripples and clasts made from mudstone layers, which have been subsequently broken up by current action.

If the barrier is breached during storms, then sand from the barrier will spill out into the lagoon in a fan shape, a feature usually referred to as a **washover fan** (see block model on the SE Poster). These washover fans will then be preserved as discrete lenses of cross-stratified and planar-stratified sandstone within the lagoonal mudstone (Figure 13.23). At the back of the lagoon, the sheltered conditions and water supply are favourable to colonisation by marsh plants. Accumulation of large amounts of organic material may lead to the development of peat that, with subsequent burial, may become coal. An idealised graphic log showing the succession of facies typical of a barrier island environment is shown in Figure 13.23.

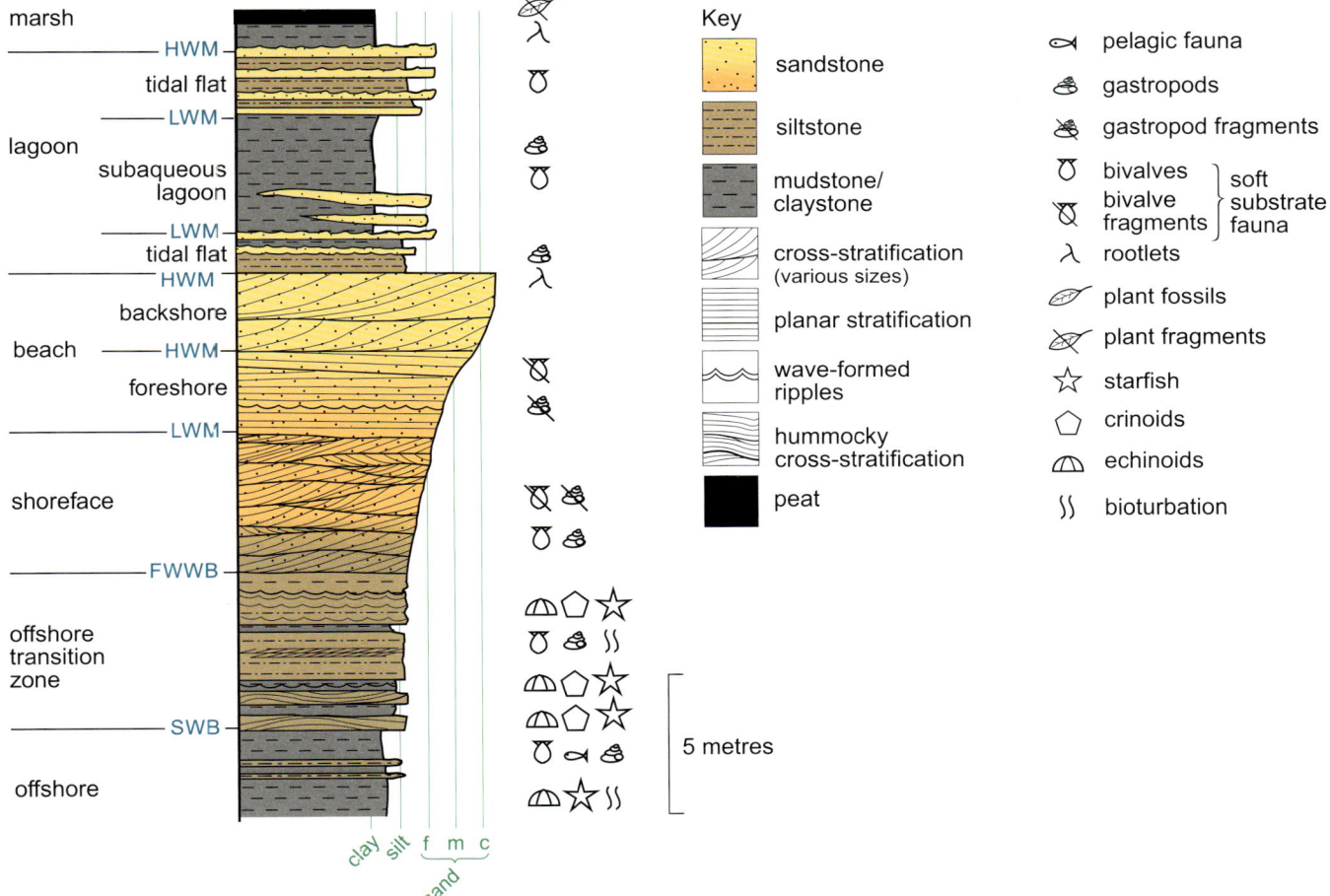

Figure 13.23 An idealised graphic log of a barrier island environment. A block diagram of a barrier island is shown on the SE Poster.

Activity 13.1 Strandplain and barrier island environments

In this activity, you will consolidate your understanding of strandplains and barrier islands.

13.5 Deltas

Deltas form along coastlines where alluvial processes are dominant. The sediment supply being discharged by a river into a lake or the sea is so plentiful that it cannot be totally dispersed by wave, current or tidal action. It was the historian Herodotus who, in about 454 BC, first used the Greek capital letter delta (Δ) to describe the broadly triangular area of sediments where the River Nile discharges into the Mediterranean Sea (Figure 13.24).

Figure 13.24 Landsat false-colour image of the Nile Delta, Egypt, showing the characteristic triangular deltaic shape.

Ancient deltaic successions are of great economic importance because they contain abundant coal, oil and gas reserves. These include the Niger delta in Africa, the Mississippi delta in the USA and some deposits in the North Sea. The Mississippi delta is covered in more detail in Activity 13.2 later in this section. During the Carboniferous Period, Britain was the site of extensive deltaic sedimentation and these sedimentary deposits form the coal-bearing rocks that dominate parts of the Pennines, north-east England, the Central Valley of Scotland and South Wales today (units C5 to C8 on the Bedrock Geology UK Maps) and are the source of almost all of the British coal deposits.

■ Why are no deltas forming around the coastline of Britain today?

☐ Because the conditions in Britain today must be different from those in the Carboniferous. Today, most of the rivers entering the sea around Britain drain land areas of low to moderate relief and the climate is stable. The rivers therefore carry relatively small sediment loads, and the sediment that does reach the sea is easily redistributed by waves and tides.

13.5.1 Controls on delta building

■ Why do rivers deposit their sediment load as they enter the sea (or lake)?

☐ As the river moves from its restricted channel to the sea (or lake), the flow expands laterally, energy is dissipated and therefore the flow decelerates so that it is no longer fast enough to move the bedload. In addition, the mixing of river water and seawater increases the frictional resistance between these two water masses and also aids flocculation of the suspended clay sediment (Section 10.1), which is deposited seawards of the coarse sediment.

On entering the sea, the river bedload is deposited rapidly and the suspended load is carried along by the lower-density river water across the top of the higher-density seawater, transporting a plume of suspended sediment far out to sea (Figures 13.5 and 13.25).

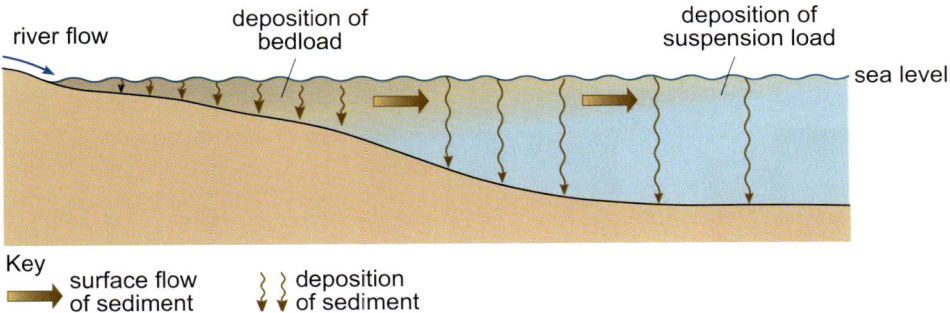

Figure 13.25 Cartoon to show what happens as sediment-laden river water enters the sea.

■ What other factors besides density differences in the water might govern the size of a delta?

☐ The amount of sediment supply, the power of the waves and tides, and the water depth will govern the areal extent of the delta. If the water is shallow, then the delta will build out over a greater area than if the water is deep.

13.5.2 Delta architecture and deltaic successions

Deltas can be divided into three topographic regions, which are each characterised by a particular type of deposition (Figure 13.26). The top of the delta behind the shoreline is called the **delta plain**. The upper part of the delta plain (i.e. the landward portion) is characterised by alluvial depositional processes and sediments similar to those described in Chapter 11. There is a main river channel that splits into smaller channels called **distributaries**. The positions of the channels change as levées and bars build up, and meanders get cut off. Crevasse splays are formed when the levées are breached during periods of high discharge. The lower part of the delta plain may experience

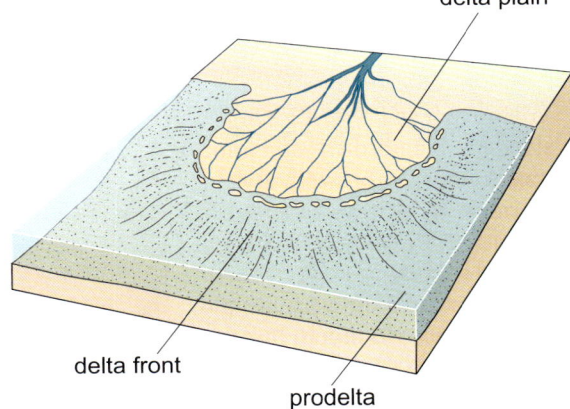

Figure 13.26 Three-dimensional block diagram to show the three main regions of a delta: delta plain, delta front and prodelta.

some marine influence, especially tides. The distributaries divide and become more numerous on the lower part of the delta plain, and interdistributary bays develop between the distributary channels (Figure 13.27). Migration of river channels on the upper delta plain sometimes results in meander cut-off, which in turn leads to the formation of ox-bow lakes in the abandoned river channels. The lakes progressively fill with mud during flooding of the river channels and thus become swamps or marshes where, provided the climate is humid, favourable conditions of water and nutrients from a constant replenishment of sediment will support an extensive flora and fauna. The prolific growth of new vegetation means that a vast amount of dead vegetation accumulates beneath the new growth on the marsh or swamp and the lack of oxygen will lead to the formation of peat. Sediment will no longer be supplied to the swamp or marsh area because the river channel has changed its course. As the sediments and the plant matter become compacted, sedimentary rocks rich in organic matter are formed, and on burial the peat may become coal. Isostatic subsidence (Book 2, Section 2.3.1) due to the weight of accumulated sediments leads to lowering of the area and inundation by the sea. Eventually, the delta may build back out over this area and hence the whole process repeats itself.

Figure 13.27 (a) A succession of Carboniferous sedimentary rocks showing a distributary channel that has eroded into sandstones and mudstones deposited under tidal conditions in an adjacent interdistributary bay. Point bar deposits within the meandering channel have formed lateral accretion surfaces as the point bar migrated across the channel (dipping down to the left above flat-lying interdistributary bay deposits at the base of the section). Note rucksack for scale. (b) The tidal deposits of the interdistributary bay have also been intensely bioturbated, Howick Bay, Northumberland, England.

The region where the river channels meet the sea is called the **delta front**. This is a slightly inclined area rimming the outer edge of the delta top (Figure 13.26).

■ Recalling Section 13.5.1, describe what happens to the bedload as a river channel enters the sea.

☐ The bedload will be deposited rapidly because of the slowing down of the currents caused by lateral expansion of the flow and frictional resistance between the freshwater and marine water masses.

Bedload sediments are thus deposited at the mouths of the river channels, to produce accumulations of sediment known as **mouth bars** that form on the delta front. The flow of the river current gives rise to dunes and ripples on the mouth bar, which will be seen as different scales of cross-stratification in cross-section. Waves and tides may also rework the mouth bar sediments, giving rise to other types of sedimentary structures (Figure 13.28). Benthic marine organisms are sparse because of the high sedimentation rates and constantly shifting sand bars.

Figure 13.28 Carboniferous sandstones interpreted as forming within a delta front mouth bar that has been reworked by tides, Howick Bay, Northumberland, England. Note that the cross-stratification dips in both directions and tidal bundles are visible near the coin.

Slightly further offshore, on the seaward edge of the mouth bar, finer-grained sediment will be deposited (Figure 13.29). The sedimentary rocks of the distal area of the bar are usually planar-stratified siltstones and silty claystones (mudstones).

Plant debris carried by the river will also be deposited in these fine-grained sediments. The abundant organic material, together with the fact that the distal part of the mouth bar remains subaqueous, encourages and supports a variety of infauna such as bivalves, gastropods, worms and crustaceans, which leads to extensive bioturbation.

The third, outermost region of the delta is the lower part of the slope, and is called the **prodelta**. The prodelta is the site of deposition of the finest-grained sediments, chiefly clays and silts. Plant debris may also be carried here in suspension. As on the distal bar, numerous marine organisms will also inhabit the prodelta muds.

Through time, the delta top, delta front and prodelta will gradually build out seawards as more and more sediment is deposited. This process of building out more and more sediment in a distal direction is another example of progradation (Section 10.2.1; Figure 13.30).

The presence of a slope on the front of a delta means that the sediments are not always stable. Submarine landslides and turbidity currents (Section 7.6) may be generated and sediment will be carried across the continental shelf and down the continental slope. The fate of these sediments is discussed in Chapter 15.

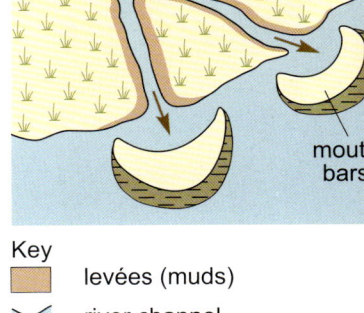

Figure 13.29 The delta plain comprises distributary channels with levées and interdistributary marsh areas. The distributaries carry sediment to the coast where it is dumped (due to the significant drop in current energy) to form mouth bar sands and muds on the delta front.

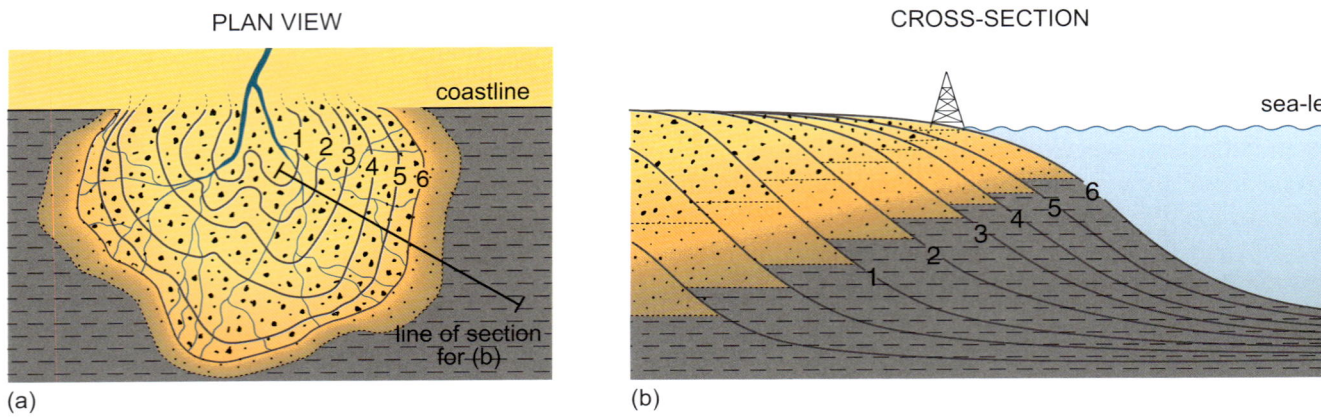

Figure 13.30 Diagram to show how a delta builds out seawards (progrades): (a) plan view; (b) cross-section.

Figure 13.31 Carboniferous example of peat (now preserved as coal that forms the dark brown layer above the ledge near the top of the photograph) overlying a yellowish-brown soil horizon (seatearth), with a characteristic mottled appearance (forming the whole area below the coal), Northumberland, England. The compass-clinometer is 10 cm long.

Lastly in this section, consider what an idealised vertical succession of deltaic sediments might look like. You can do this by imagining what would be found if a hole was drilled through the delta at the point shown by the drilling rig in Figure 13.30b and assuming that Walther's Law applies (see Section 10.2.2). The first layer encountered would be peat deposited in the marsh area. The peat would be sitting on top of a soil horizon in which the plants grew (Figure 13.31).

■ In Figure 13.30, where did the marsh form?

□ In an abandoned distributary channel or interdistributary bay.

Underneath the soil, the deposits of the old distributary channel should be present. There would usually be a fining-upward succession of cross-stratified sandstones, representing point bar deposits. The base of the channel may be marked by a lag of coarser-grained material similar to that found at the base of the channel fill in alluvial deposits (Section 11.3.1). The base of the channel is erosive and will overlie the delta front deposits of the mouth bars. This is because through time, as the delta progrades seawards (Figure 13.30b), the alluvial deposits of the delta plain will build out further and further seawards on top of the previously deposited delta front which, of course, is at a slightly lower topographical level (Figure 13.26). Continuing down through the borehole, the next deposits would be the mudstones and siltstones deposited on the prodelta.

Thus, overall, deltaic successions usually coarsen upwards from mudstones and siltstones of the prodelta to slightly coarser-grained sediment of the mouth bar. These will be capped erosively by the coarsest-grained sediments of all – the alluvial sediments of the delta plain. An idealised deltaic succession is shown in Figure 13.32 and a block diagram of a delta is illustrated on the SE Poster.

Activity 13.2 Deltaic environments

In this activity, you will explore further some modern and ancient examples of deltas.

Chapter 13 Siliciclastic coastal and continental shelf environments

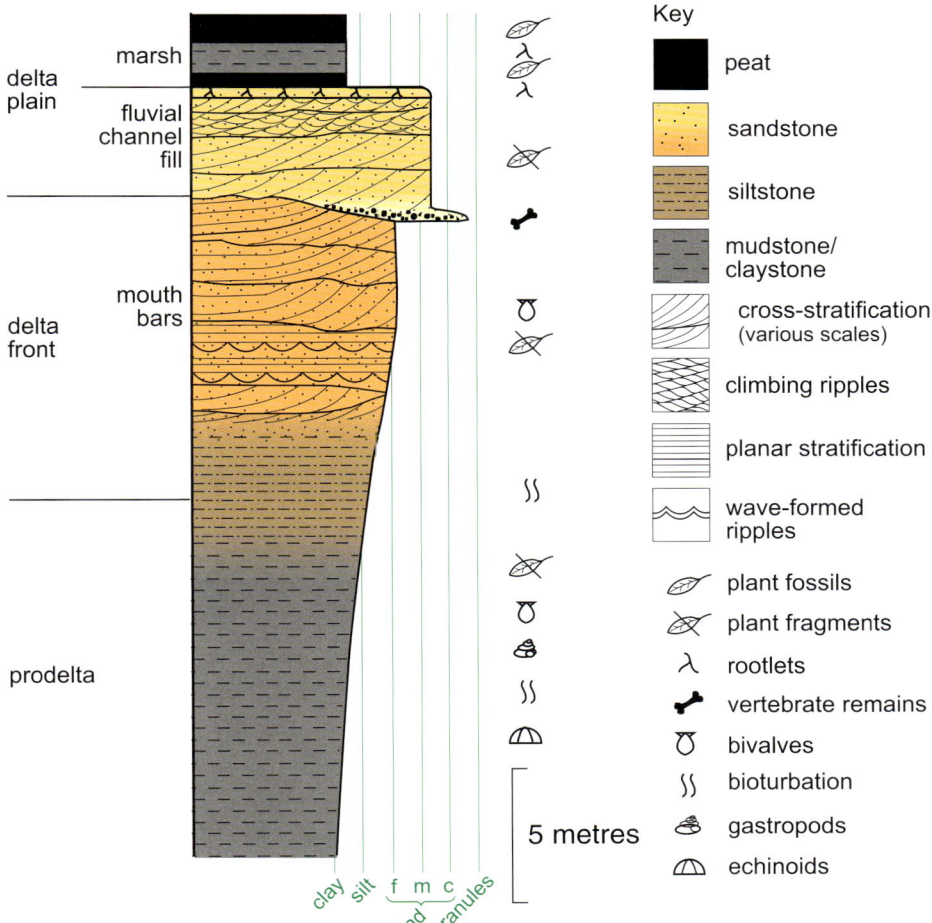

Figure 13.32 Graphic log of an idealised deltaic succession.

13.6 Summary of Chapter 13

1. Physical processes along coastlines include waves, tides and currents. Currents may be derived from fluvial processes, from waves (e.g. longshore and rip currents) or from tides. Chemical processes along the coastline include flocculation and the formation of soils. Biological processes include the growth of plants, which stabilise sediment, and the actions of animals.

2. The coastline may be split into the backshore and foreshore (constituting the beach), shoreface, offshore transition zone, and offshore zone. These areas are determined by the position of mean high tide and low tide, fairweather wave-base and storm wave-base.

3. Strong alluvial processes, combined with tide and wave processes that are insufficient to transport all the sediment away, favour the formation of deltas. Strandplains tend to form in areas of strong wave action. Barrier islands form where there is moderate wave action and some tidal and fluvial action.

4. Strandplains comprise linear accumulations of sediment attached to the land. On sandy strandplains, aeolian dunes often develop in the backshore area. The foreshore is characterised by planar-stratified sands, the shoreface by trough and planar cross-stratified sands with some swaley cross-stratification

near the base, the offshore transition zone by hummocky cross-stratified sands and muds, and the offshore zone by muds. Faunas are found throughout but will be most common in the offshore transition zone and offshore zone where they are not being constantly reworked by wave, tide or storm processes. The foreshore may show ridges and runnels, and the profile is governed by the grain size of the sediment and by storm or fairweather conditions.

5   Barrier islands comprise an island separated from the mainland by a lagoon. The sediments and structures of barrier island successions are similar to those for strandplains. However, barrier islands display less evidence of storms and the backshore barrier sands will be capped by muds deposited in the lagoon and on tidal flats. During storms, the barrier may be breached, resulting in washover fans forming in the lagoon.

6   Deltas can be subdivided into three distinct areas. The *delta plain* consists of a flat area dominated by alluvial deposition and the resulting vertical deposits include alluvial channel fills, overbank muds and fine-grained lake sediments. The *delta front* forms at the distal edge of the delta plain where the rivers enter the sea; sediments are deposited in mouth bars. The *prodelta* forms the most distal part of the delta, where the finest-grained sediments are deposited.

13.7   Objectives for Chapter 13

Now you have completed this chapter, you should be able to:

13.1 Describe the physical, chemical and biological processes that are important in siliciclastic coastal environments.

13.2 Explain how the interaction and relative dominance of fluvial, wave and tidal processes influence the morphology of the coastline.

13.3 Recognise and describe the main facies and features of deltas, strandplains and barrier islands.

13.4 Describe the different areas present along the coast (backshore, foreshore, shoreface, offshore transition zone and offshore zone) and the processes and types of sediment typical of each.

13.5 Recognise and sketch simple graphic logs summarising the main biological and sedimentological features of an idealised succession that ranges from the offshore to backshore for (i) a strandplain, (ii) barrier island and (iii) a deltaic coastal morphology.

Now try the following questions to test your understanding of Chapter 13.

Question 13.5

Why is the identification of hummocky cross-stratification in a sedimentary succession important when you are trying to work out the positions of storm wave-base and fairweather wave-base in the field?

Question 13.6

What type of sedimentary structures in mouth bar deposits would suggest that they have been reworked by (a) waves; (b) tides?

Chapter 14   Shallow-marine carbonate environments

The sedimentary environments discussed so far are dominated by siliciclastic sediments. However, shallow-marine carbonate sediments are an important feature both of the sedimentary record and of modern environments. In the past, they thrived during warmer climatic conditions and periods of high sea level when shallow seas covered extensive areas of the continent, for example during the Carboniferous and Cretaceous periods. Modern shallow-marine carbonates are common in the tropics and subtropics between 30° north and 30° south of the Equator, particularly in the Bahamas, South Florida shelf and the Arabian Gulf, and around volcanic islands in the Pacific (Figure 14.1). Temperate, cool-water carbonates are particularly well developed along the southern coast of Africa, Australia and New Zealand (Figure 14.1).

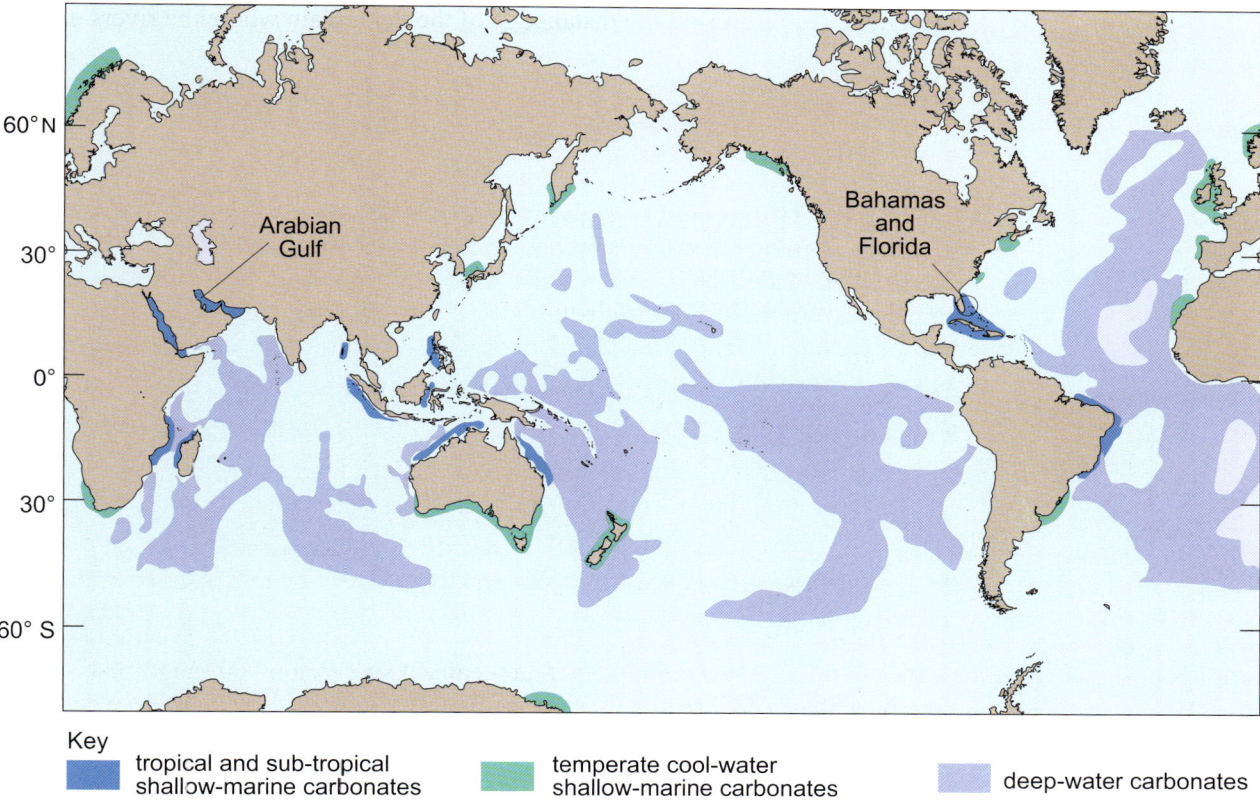

Figure 14.1   Important areas of deposition of marine carbonates around the world today.

Carbonate and siliciclastic sediments are not mutually exclusive, but one is usually dominant because the presence of siliciclastic sediments inhibits carbonate production. Mixed siliciclastic and carbonate successions are found in the geological record, where they are thought to represent changing conditions, or where sediments are derived from two sources (Figure 14.2).

Figure 14.2 A mixed siliciclastic and carbonate succession of Jurassic age, Dorset, England. The white bed in the middle of the cliff section is an oolitic limestone and the pale grey beds overlying this and near the base of the cliff are sandy limestones. These are all interbedded with darker grey mudstones. The beds exposed at the very base of the cliff on the left-hand side of the photograph are sandstones. These sandstones are the same as those shown in Figure 10.5. The cliff is about 30 m high.

14.1 Controls and processes affecting carbonate deposition

In Section 10.1 the controls on sediment deposition were discussed. Tectonic and climate controls profoundly affect both siliciclastic and carbonate deposition, but for carbonates, the biology of the carbonate-producing organisms is equally important. Whereas siliciclastics are transported over large distances, most carbonate sediment is produced more or less *in situ* by various organisms. The main products and processes involving biota include:

- bioclasts – the calcareous parts of the organisms preserved as clasts
- peloids – most peloids consist of faecal pellets. Organisms ingest the carbonate sediment, consume the organic matter and then excrete the waste products as ovoid-shaped carbonate peloids
- precipitates – clasts such as ooids and micrite may be produced chemically
- framebuilders – some organisms, e.g. corals, build strong skeletal frameworks of carbonate as they grow. This framework then supports other organisms
- bafflers – some organisms, such as seagrasses, act as sediment baffles, trapping sediment around them
- binders – algal and bacterial mats are examples of binders that hold successive layers of micrite together (and also promote carbonate precipitation). Thread-like algal filaments within the mats grow through the fine-grained sediment to form irregular to columnar structures (Figures 8.6 and 14.3)
- bioerosion – erosion by organisms probably accounts for most of the micrite produced. Both vertebrates (e.g. parrot fish) and invertebrates (e.g. gastropods and echinoids) graze on a carbonate host (e.g. a coral reef) and in the process they break down the carbonate into fine-grained carbonate sediment, micrite. Other animals drill holes into carbonate rocks, also producing micrite.

Figure 14.3 Permian algal stromatolites display small columnar structures and have trapped and/or precipitated successive layers of micrite, Durham, England. Note top of a pencil (approximately 5 cm) for scale.

■ What is the other chief way of producing micrite?

☐ Direct chemical precipitation of calcium carbonate from seawater (Section 8.1.3).

For maximum carbonate production and accumulation, the controls have to be just right: not too deep, not too shallow, not too warm or too cold, the correct salinity and nutrient supply, not too much siliciclastic supply and, probably most importantly, the right types of organisms must be present. Table 14.1 summarises the different controls that influence carbonate deposition, and why each control is important.

Table 14.1 Controls and subcontrols on carbonate deposition.

Control	Subcontrol	Influence and production
biota	form and function	The type and amount of carbonate clasts produced depends partly on the dominance of bafflers, binders, framebuilders and the amount of bioerosion.
	evolution	Different organisms have evolved and dominated through geological time; these control the type of carbonate produced.
	cool-water biota	Less sediment produced than by warm-water faunas; cool-water biota include coralline algae and bryozoans (Figure 14.4).
	warm-water biota	Major carbonate production, e.g. corals (Figure 14.5), green (calcareous) and blue–green (non-calcareous) algae.
climate, and to a lesser extent tectonics, which control the shape and depth of the ocean basin	light penetration	Light penetrates approximately the upper 70 m of water, but light penetration is greatest in the upper 10–20 m. The amount of light affects organism productivity.
	water temperature	Affects organisms. Warm waters occur generally down to 50–100 m water depth. Temperature is also controlled by latitude and the presence of warm- and cold-water currents. There are cold- and warm-water faunas, therefore water temperature controls the type of carbonate produced (see biota above). Ooids and micrite are precipitated only in warm waters.
	water circulation	Influences the type of sediment produced. Good water circulation is required for some organisms to survive.
	oxygenation	Well-oxygenated waters are essential for the growth of all skeletal invertebrates.
	salinity	Both high and low salinity conditions reduce the biotic diversity because only some organisms are adapted to these conditions. At very high levels of salinity, most invertebrates disappear.
tectonics and climate	siliciclastic sediment supply	A high siliciclastic sediment supply from the land inhibits carbonate production because filter-feeding organisms become swamped with sediment and cannot feed.
tectonic setting		Controls the area available for carbonates to accumulate. Also controls the latitude of the area of carbonate production because tectonic plates move. Although some shallow-water carbonates form at higher latitudes, the majority do so in the tropical–subtropical belt between 30° N and 30° S of the Equator.

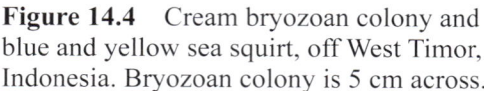

Figure 14.4 Cream bryozoan colony and blue and yellow sea squirt, off West Timor, Indonesia. Bryozoan colony is 5 cm across.

Figure 14.5 Framework-building corals, Great Barrier Reef, Australia. Field of view approximately 4 m.

The dominance of biological and chemical processes in the production and accumulation of carbonates means that most of the sediment is produced *in situ*. However, like siliciclastic sediments, the carbonates may be affected by both unidirectional and oscillatory currents within the depositional environment.

- ■ Would you expect some carbonate sediments to show cross-stratification, wave ripples and other current-formed sedimentary structures, and, if so, why?

- □ Carbonates will show all the same types of sedimentary structures as siliciclastics; some carbonates such as ooid shoals behave in almost the same way as well-sorted quartz sandstones because they are composed of small grains that can be easily moved around (Section 8.1.2).

Figure 14.6 shows an example of a cross-stratified oosparite of Jurassic age from the Dorset coast.

Figure 14.6 Cross-stratified oosparite of Jurassic age, Dorset coast, England. Scale bar = 20 cm.

Any transport process that affects siliciclastics can also affect carbonate sediments, so carbonates, like siliciclastics, may also be affected by sediment gravity flow processes and become re-deposited as turbidites and debris flows. These types of sedimentary deposit are discussed in Chapter 15.

14.2 Similarities and differences between shallow-marine carbonates and siliciclastics

■ In this chapter, one similarity and one difference between siliciclastics and carbonates have already been introduced. What are they?

☐ The similarity is that they are both moved around by currents and sediment gravity flows and thus contain similar sedimentary structures. The difference is that the majority of carbonates are produced *in situ* whereas siliciclastics are transported *over significant distances* prior to deposition.

There are, however, several other differences which are summarised in Table 14.2. Study this table carefully and then answer the questions that follow.

Table 14.2 The differences between shallow-marine siliciclastic and carbonate sediments.

Feature	Siliciclastics	Carbonates
sediment production	transported from the nearby land mass or along the coast	Mainly produced *in situ*. Some organisms may build up structures that modify the sea-floor topography.
grain size	reflects the size of the particles in the source rock and the duration and type of processes that have acted on the particles during transportation	Reflects the size of the original carbonate skeletons and chemically precipitated grains, and sometimes the type and duration of the processes of transportation and deposition.
cementation	sediments remain unconsolidated on the sea floor, and are usually cemented only once they are buried	Sediments are commonly cemented on or immediately below the sea floor because of the amount of carbonate in solution.
periodic subaerial exposure of sediments	minor or no alteration of the siliciclastic grains	Due to the change from seawater to freshwater in the sediments, dissolution of the carbonate and widespread cementation may result.
mud	indicates settling from suspension of particles derived from the chemical decomposition of rocks	Final deposition is from suspension but a large amount of carbonate mud (micrite) is produced from bioerosion. The amount of micrite also relates to the organisms that are around to produce it, either from bioerosion or from the growth of micro-organisms whose skeletons are made up of mud-sized carbonate particles.
geological age	erosion and deposition processes are the same at all times.	Relative tendency for calcite or aragonite to precipitate or dissolve on the sea floor in response to different saturation levels at various times in geological history.
latitude and climate	sediments occur worldwide.	Most shallow-marine carbonates are deposited in warm shallow-water environments between 30° N and 30° S of the Equator.

Question 14.1

Compare the grain size of RS 22 and RS 10. What does this tell you about the strength of the transporting current in each case?

Question 14.2

RS 18 and RS 21 were both deposited relatively close to their source area. What features allow you to deduce this?

14.3 Carbonate platforms

The term carbonate platform is a general one used to describe a thick succession of marine, mostly shallow-water carbonates. There are five different morphological types of carbonate platform defined on the basis of their extent, general geometry (shallow or steep edges) and any connection with the continent (Figure 14.7).

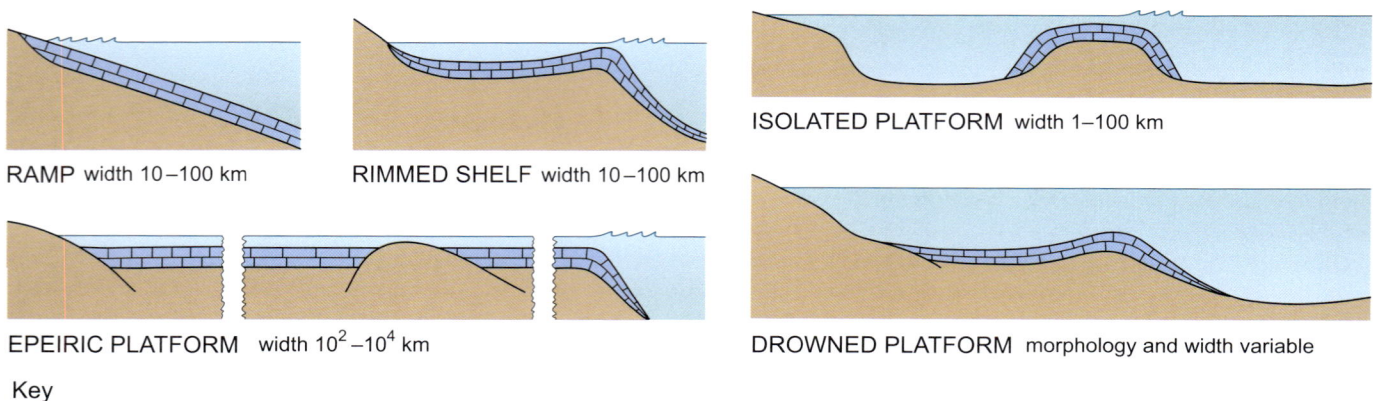

Figure 14.7 Cross-sections through the different morphological types of carbonate platform, each of which shows substantial vertical exaggeration.

The morphological type of carbonate platform that develops depends on the pre-existing topography, the organisms present, and the rise and fall of sea level. The majority of the facies will be similar on different types of carbonate platform. However, the thickness and lateral extent of the facies will change. For instance, rimmed shelves favour the growth of reefs at the edge, whilst epeiric carbonate platforms, formed in laterally extensive shallow seas, have widespread lagoonal deposits.

At the seaward edge of rimmed shelves, epeiric carbonate platforms and isolated carbonate platforms, where the depositional slope is steep, the carbonates may be redeposited by sediment gravity flows (see Section 7.6 and Section 15.3). Such redeposition is much less common or absent on ramps because the depositional slope is not as steep.

14.4 Peritidal facies

Peritidal is the term used to mean 'near the influence of tides'; it includes the area above mean high water and just below mean low water; Figure 14.8), but excludes organic buildups (Section 14.5) present in this zone. There are three principal areas of extensive peritidal carbonate deposition, namely tidal flats, lagoons, and carbonate sandbodies. The presence, dominance or absence of each of these facies depends on the latitude, climate, tidal influence, wave influence and the morphology of the carbonate platform.

Figure 14.8 From left to right: the supratidal (brown), intertidal (white) and shallow subtidal (submerged) areas of the Florida Keys, USA.

Figure 14.9 The desert rose form of gypsum. The specimen is about 8 cm across.

14.4.1 Tidal flats

The shoreline, or most proximal part of the carbonate platform, usually includes a tidal flat area that is periodically inundated by weak tidal currents and wave action. Micritic limestones predominate because of the low energy conditions, but some pelmicrites, and occasionally oomicrites and biomicrites (Section 9.6.2) may be preserved due to storms.

If the climatic conditions are arid, evaporite minerals – usually gypsum and anhydrite – will precipitate out of water held in the pore spaces of the sediment in the high intertidal and supratidal areas. Initially, gypsum ($CaSO_4.2H_2O$) may precipitate as well-developed crystals, typically 0.1–25 cm across and similar to the desert rose variety shown in Figure 14.9. Continued high evaporation rates lead to increased salinity that converts gypsum to its anhydrous equivalent, anhydrite ($CaSO_4$), with its characteristic nodular form called chicken-wire anhydrite (Figure 14.10). Extensive areas of evaporites and shallow-marine carbonates are particularly well developed along the Arabian Gulf coast today (Figure 14.1).

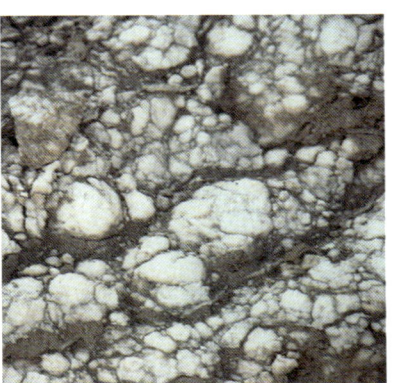

■ Where do you think all the mineral salts come from to form the evaporites?

☐ The continued re-supply of seawater during the rising and falling tide on the tidal flats.

Figure 14.10 Chicken-wire texture, showing grey micrite between evaporite nodules (originally anhydrite, now gypsum because of exposure to modern porewaters). These Jurassic deposits are from Mount Carmel, Utah, USA.

Figure 14.11 Algal mats, Oman. Note how the mats curl up at the edges of the polygonal desiccation cracks. Person in the distance shows scale.

Figure 14.12 Laminated planar stromatolites (dark colour) interbedded with micrite (pale colour) and subvertical, tapering desiccation cracks. Glacier–Waterton National Park, USA.

If climatic conditions are humid, then vegetation grows on the tidal flats and supratidal marshes; this is dominated by algal growth and, in the Recent past, mangrove trees.

The periodic exposure of tidal flat sediments gives rise to a number of other distinct sedimentary structures. These include desiccation cracks, which form due to repeated subaerial exposure, and a dried crust of sediment at the surface, which may curl up into thin flakes on drying (similar flakes form in siliciclastic muds as in Figure 7.45c). The cracked sediment and mud flakes may then be incorporated into intertidal sediments during later flooding to form flat-pebble conglomerates, or eventually to form a soil in the supratidal zone.

The fauna are an important element of tidal flat facies. In general, the range of species is fairly restricted due to the rapid fluctuation in salinity and water depth. Typically, gastropods, bivalves, worms and crustaceans adapted to the conditions may be found together with some microfossils.

- ■ Will there be any trace fossils in the tidal flats, and, if so, why?
- □ Yes – because of the fluctuating conditions, the worms and crustaceans will leave feeding and dwelling trace fossils.

Reworking of gastropod and bivalve shells during storms can give rise to biomicrites and biosparites. Other important elements of the biota of tidal flats are bacterial and algal mats, which when fossilised form stromatolites (Figure 8.6 and Section 3.5.3). These thrive in the exceptionally saline conditions, and, importantly, they bind together the fine-grained carbonate sediment. The mats with a simple planar morphology are most common, although small domes (30–50 cm in diameter) may also form. The mats and intervening sediments are often cut by desiccation cracks. Modern-day mats with desiccation cracks are shown in plan view in Figure 14.11, and a cross-section through a fossilised example is shown in Figure 14.12.

All the features of the tidal flat facies described above can be put together into an idealised 3D model and vertical facies succession; these are shown in Figure 14.13. In a similar fashion to siliciclastic tidal flats, as the tide rises over the carbonate tidal flats, first it fills the shallow tidal channels (Figure 14.13a) in the intertidal area, and then it floods across the entire intertidal zone.

14.4.2 Lagoons

Lagoons are protected subtidal areas that lie on the landward side of a barrier. A barrier can be either some kind of organic growth, such as a reef, or a stabilised bank of oolitic and/or bioclastic sediments (Figure 14.14).

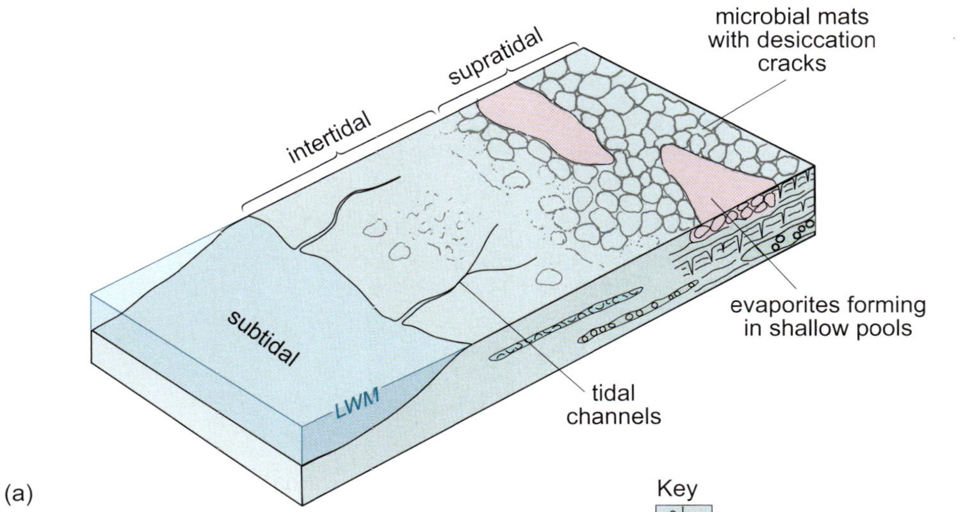

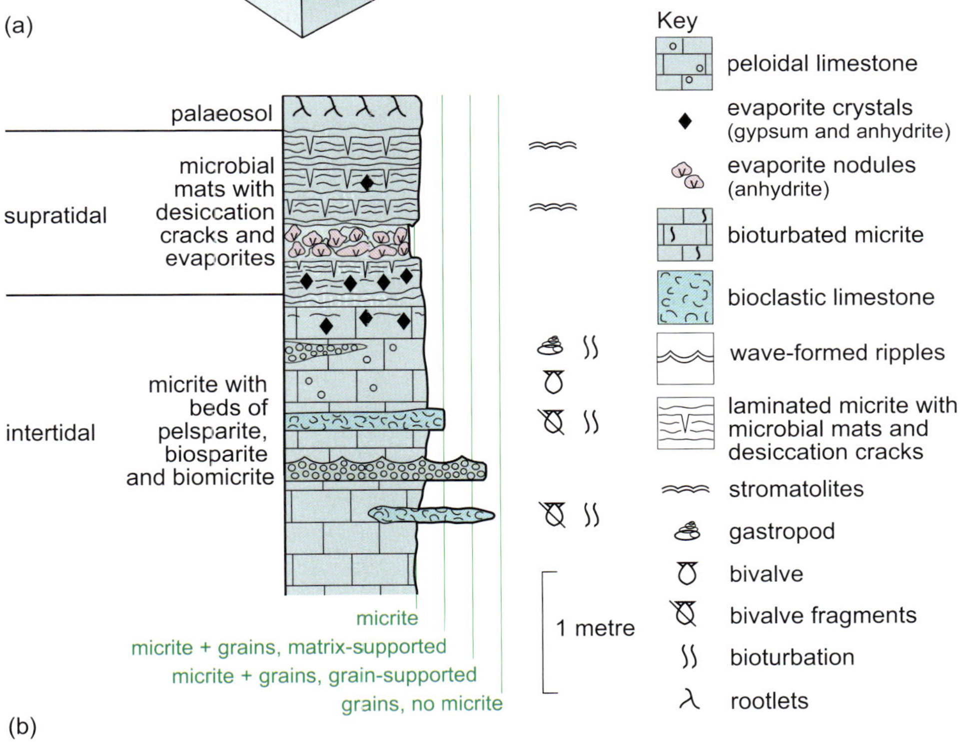

Chapter 14 Shallow-marine carbonate environments

Figure 14.13 (a) Idealised block model of the facies and (b) graphic log of facies succession for an arid tidal flat. *Note*: palaeosol is a fossilised soil.

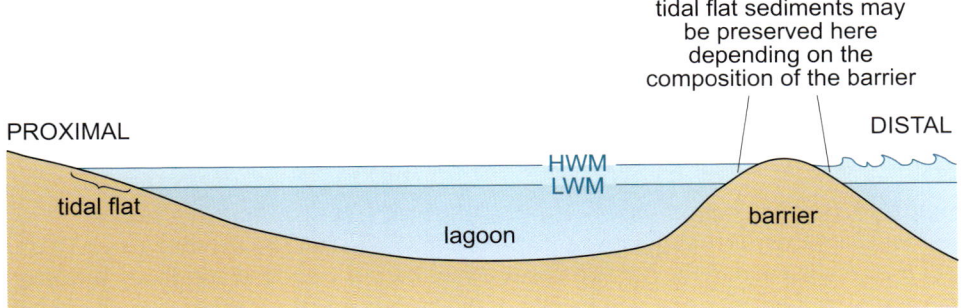

Figure 14.14 Cross-section through a lagoon showing its relationship to a barrier and tidal flat sediments.

Rimmed shelf carbonate platforms always have lagoons, whereas ramps do not necessarily develop these features. Lagoons can extend to many hundreds of kilometres in length and occasionally up to 100 km in width. They are connected to the open sea through gaps or channels across the barrier (Figure 14.15), but are largely protected from coastal currents and wave action (Figure 14.16). Low energy micrite is the typical lithology, often stabilised by seagrass, and the rich biota may be preserved in life position. Bivalves, gastropods, crustaceans and algae are common and faecal pellets abound because of the lack of current activity. During storm conditions, however, waves may break through the barrier and surge through the channels, reworking the sediment into storm beds that are typically preserved as thinly bedded biomicrites.

Figure 14.15 An isolated carbonate platform, Majuro, Marshall Islands, Pacific Ocean. A barrier of coral reefs protects the lagoon area (lower part of photograph) from the breaking waves, which can be seen on the open sea side of the barrier. The barrier is broken by channels through which water and sediment may pass. Reworked carbonate sand derived from the reef supports vegetation in the subaerial part of the barrier. The barrier shown is several kilometres in length.

Figure 14.16 The barrier reef of Majuro, Marshall Islands, Pacific Ocean is running diagonally across the lower part of the photograph. The exposed part of the reef is vegetated and to the right of this is a subaqueous carbonate sandbody forming in the lagoon. The open sea is to the left of the barrier, and the field of view is several kilometres in length.

14.4.3 Carbonate sandbodies

Carbonate sandbodies develop in both the subtidal and intertidal areas of carbonate platforms subject to high tidal currents and/or wave activity. They are composed of ooids and/or rounded and sorted marine bioclasts (e.g. fragments of corals, bivalves, foraminiferans, algae, echinoderms and brachiopods).

■ What carbonate rock types are likely to form if a carbonate sandbody is preserved, and why?

☐ The likely rock types are oosparites and biosparites. This is because the constant scouring and reworking by currents is unlikely to allow the settling out of much carbonate mud (micrite) and the grains are usually either ooids or bioclasts, which will become cemented by sparite during burial.

Carbonate sandbodies are very similar to siliciclastic submarine sandbodies (as discussed in Sections 13.3 and 13.4) in terms of their architecture and sedimentary structures. For instance, sandbodies may form part of the foreshore or shoreface or build up to form a barrier at the edge of the carbonate platform (Figure 14.14). Sedimentary structures are very common, mostly as cross-stratification on a variety of scales (Figure 14.6). Herringbone cross-stratification may develop where there are tidal currents and scours, and channels may form through the sandbodies; if they are deposited in the subtidal offshore transition zone, they may also show hummocky cross-stratification. The carbonate sandbodies may also be bioturbated.

Carbonate sandbodies may be orientated either parallel or perpendicular to the coastline. Those that form barriers or are developed on the continental shelf tend to lie parallel to the edge of the continental shelf and shoreline and are common on the protected landward side of shelf margins where bioclastic debris may be derived from nearby reefs (Figure 14.16). Where tidal currents are strong, the sandbodies may be orientated perpendicular to the coastline and edge of the continental shelf. In the area between the sandbody ridges, muddy (micrite) carbonate sands often accumulate.

14.5 Carbonate buildups (including reefs) and biostromes

In tropical latitudes, modern carbonate platforms with a steeply dipping seaward margin are often characterised by coral-rich reefs. They thrive in isolated platforms such as the Bahamas, or rimmed shelves like the Great Barrier Reef. The reefs form rigid structures made from framework-building organisms such as corals and coralline algae, and support a myriad of sessile (fixed) and vagrant (mobile) species in and around the reef, including such organisms as brachiopods, sea anemones, fish and octopus. Modern corals rely on a symbiotic (mutually beneficial) relationship with single-celled algae in their tissues, and since the algae can only photosynthesise in clear, warm waters, then the reefs are also confined to shallow water in the tropics. The reefs build upwards sufficiently to protect the relatively quiet lagoonal waters in the back-reef (Figure 14.17). Reef debris is transported by gravity-driven processes to the fore-reef slope in front of the reef.

Book 3 Fossils and Sedimentary Rocks

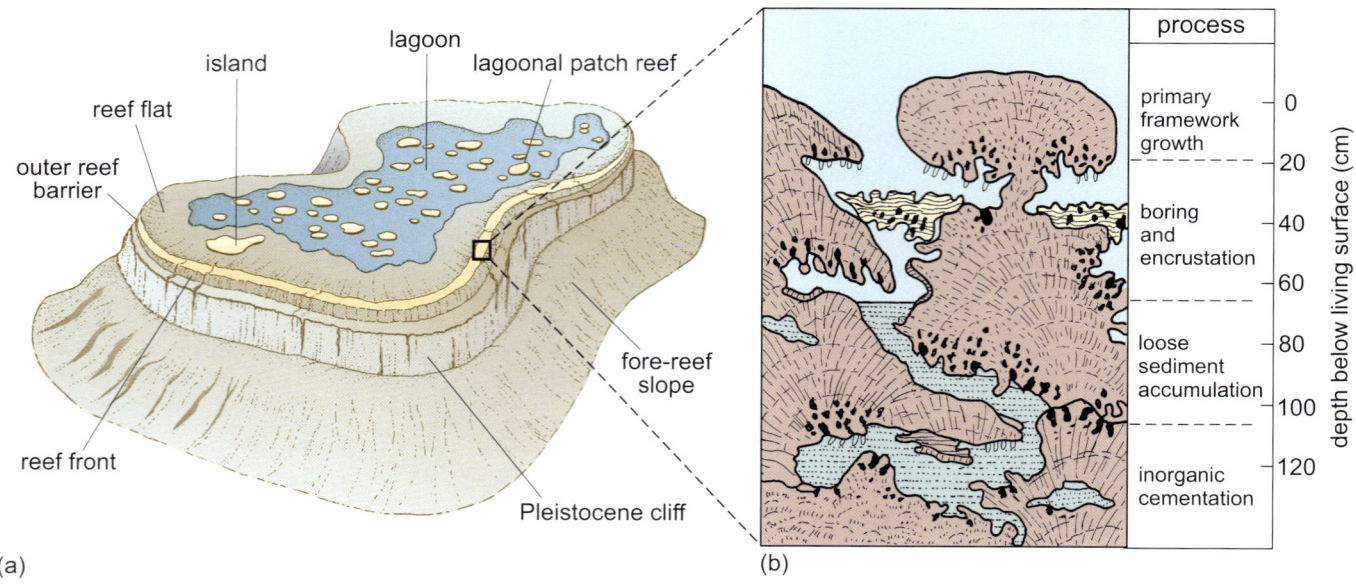

Figure 14.17 Diagram showing the typical form of a modern tropical coral reef along a platform margin. (a) Overall structure (all below sea level, but ocean water omitted for clarity). (b) Reef framework and the processes involved in creating it. Seawards of the outer reef barrier, the water depth increases markedly. Pleistocene is a stratigraphic term encompassing all but the most recent strata of the Quaternary Period.

Figure 14.18 Parrot fish on a coral reef. A single fish may be responsible for removing and grinding up as much as 5 tonnes of coral in a single year. The coral is pulverised with grinding teeth in the fish's throat in order to get at the algae-filled polyps inside. This fish is about 30 cm long.

The reef rocks, though characterised by corals and other framework builders, are buffeted by ocean waves and grazed and bored by various organisms. Much of the original structure is broken up and reduced to bioclastic debris that eventually becomes cemented together to form a rigid structure. A single parrot fish (Figure 14.18) can remove 5 tonnes of coral in one year, ingesting it and eventually depositing a stream of fine carbonate sand and mud (micrite) onto the reef. A borehole through an existing reef would therefore encounter mostly broken up coral debris and interstitial bioclastic sand and micrite, cemented by aragonite or calcite, instead of entire coral heads in life position. In addition, a borehole through a rimmed-shelf reef would also encounter the rocky substrate on which the relatively thin veneer of reef material was built – a substrate sculpted by subaerial erosion during the last glaciation, when sea level was significantly lower (Figure 14.19).

Much less well known are the carbonate **mudmounds** up to 300 m high (Figure 14.20) that thrive in water depths in excess of 500 m off the continental margins of Europe, such as in the Rockall Trough off Ireland and Northwest Scotland. The mounds are the product of rapid growth and breakdown of cold-water corals, and coral thickets are commonly associated with the mounds. Mounds that are stabilised by algae are common in the geological record, such as during the Carboniferous.

Both coral reefs and mudmounds develop a significant relief on the sea floor, and they are often referred to as **buildups**. Not all of them are rich in corals and, in the geological past, other colonial organisms have been involved, including

Chapter 14 Shallow-marine carbonate environments

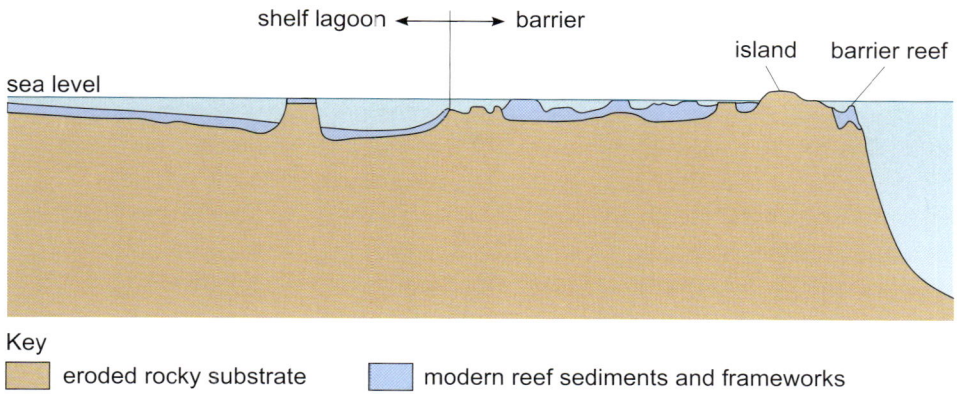

Figure 14.19 Position of modern-day reef and associated sediments built upon the irregular, previously eroded, antecedent, rocky substrate.

bryozoans, calcareous algae, and the now-extinct stromatoporoids (which are very different from algal stromatolites). Other organisms form rich concentrations on the sea floor, but lack a rigid framework – these much flatter structures are called **biostromes**. They include oyster beds (Figure 14.21) and the now-extinct group of molluscs called rudists that flourished in tropical, shallow-marine 'meadows' during the Cretaceous. Part of a Pleistocene coral reef from the Bahamas is illustrated in Figure 14.22a, with its modern counterpart in Figure 14.22b.

Figure 14.20 Cross-section through carbonate mudmound. These can reach up to 300 m high.

Figure 14.21 A Pleistocene oyster bed, Jamaica. Exposure shown is about 1 m across.

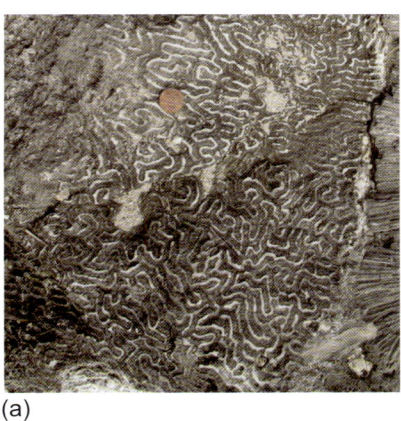
(a)

(b)

Figure 14.22 (a) A Pleistocene coral reef, San Salvador Island, Bahamas, illustrating a distinctive brain coral. The coin is approximately 2 cm in diameter. (b) Part of a modern coral patch reef, San Salvador Island, Bahamas, sited two kilometres from the fossil reef; note the brain coral in the centre. Field of view approximately 50 cm.

279

14.6 Carbonate platform facies models

In terms of the different morphological types of carbonate platform introduced in Section 14.3, the peritidal facies, carbonate sandbodies and carbonate buildups form the basic building blocks of carbonate facies models. However, it is important to note that the development and proportion of the different facies making up the platforms have varied over geological time due to changes in climate, tectonic setting and biology. In this section, you will consider two carbonate platform examples: a ramp from the Carboniferous Period and a modern-day rimmed shelf from the South Florida shelf.

14.6.1 Carboniferous carbonate ramp model

The Carboniferous carbonate ramp can be divided into four facies groups or zones, as illustrated in Figure 14.23. They comprise:

- distal micrites and biomicrites deposited mainly below fairweather wave-base
- mudmounds (Section 14.5)
- oolitic sandbodies of the shoreface and foreshore
- backbarrier peritidal sediments.

Question 14.3

(a) Explain whether the succession shown in the graphic log in Figure 14.23b represents a shallowing-upward or a deepening-upward succession.

(b) Based on your answer to (a), decide whether or not the ramp was prograding seawards.

Question 14.4

What is the true dip on the ramp shown in Figure 14.23?

Chapter 14 Shallow-marine carbonate environments

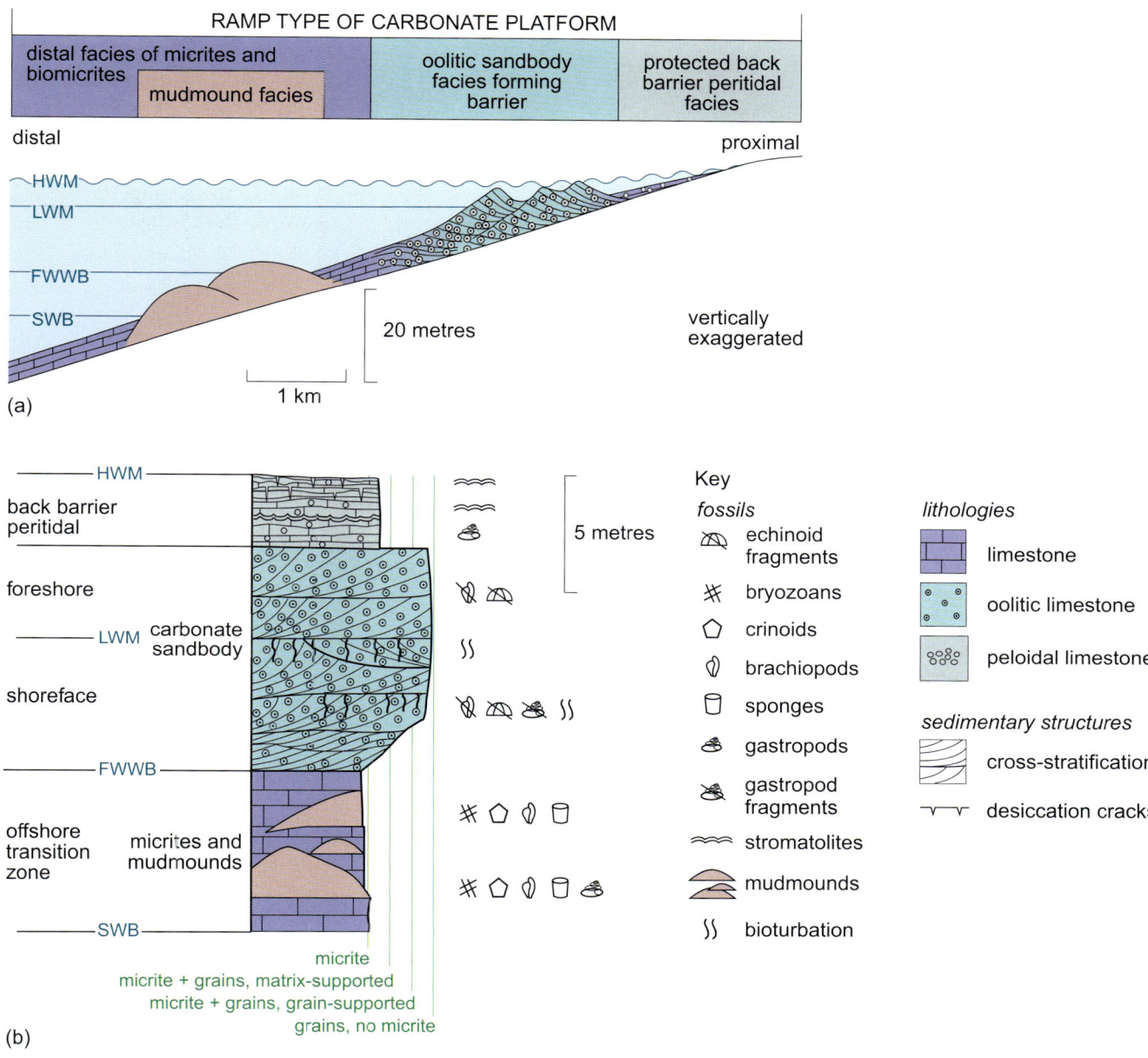

Figure 14.23 Idealised Carboniferous ramp-type carbonate platform showing: (a) 2D model, and (b) graphic log of a typical succession of facies that conforms to Walther's Law.

14.6.2 South Florida carbonate rimmed-shelf model

Rimmed shelves comprise a shallow lagoon area on the shelf that is protected from the open ocean by a barrier. The barrier develops at the edge of the continental shelf and slopes away into the ocean. Figure 14.24a shows an idealised cross-section of a modern-day carbonate-rimmed shelf based on South Florida. This example comprises a 5–30 km wide shelf, the distal edge of the shelf occurs in 8–18 m of water depth and this descends at a gradient of between 1° and 18° into water that is 200–400 m deep. The shelf is about 300 km long and extends south and southwest from Miami; it is covered by a discontinuous string of islands (the Florida Keys) marking the inner margin of the shelf (Figure 14.25).

281

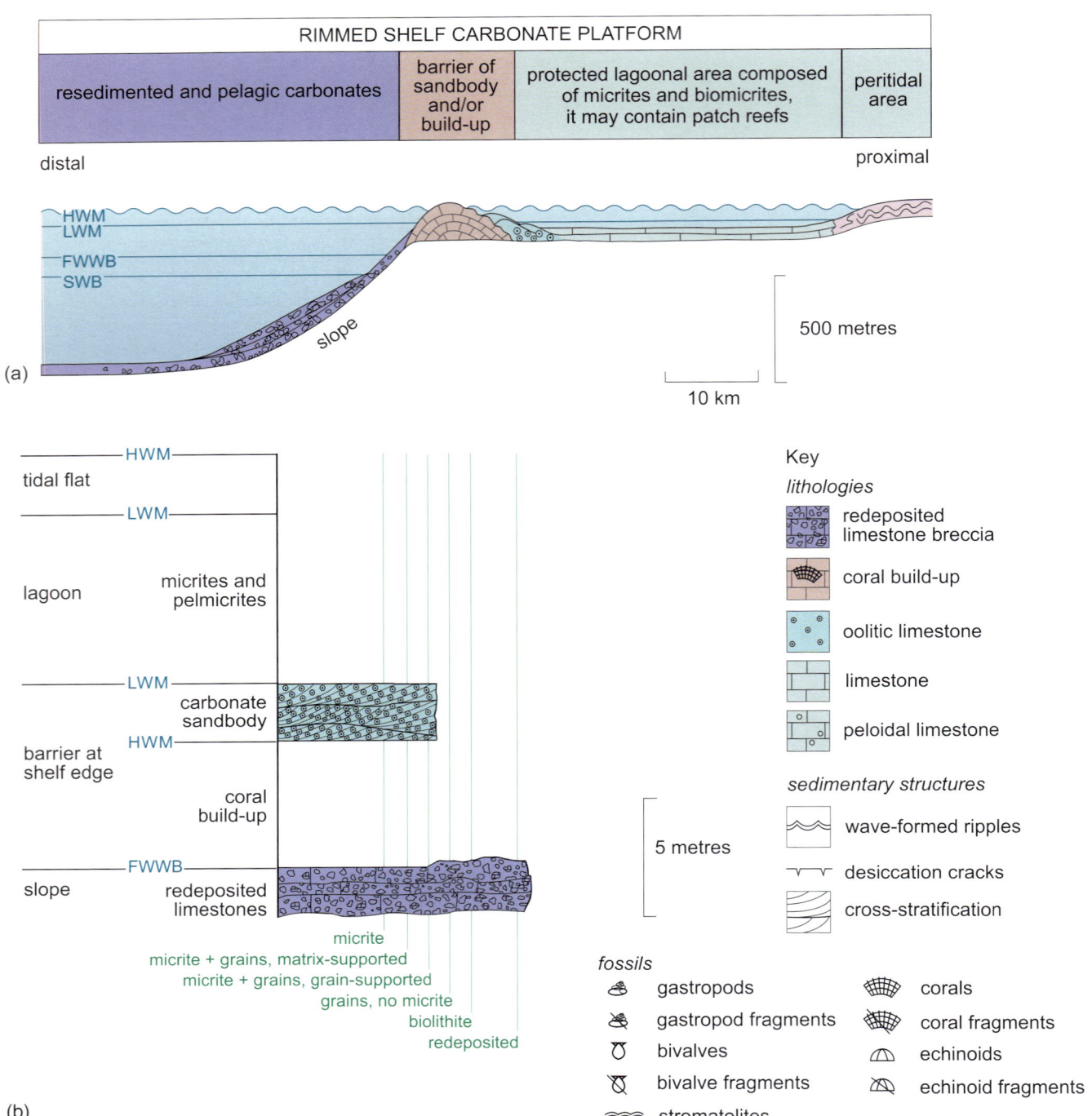

Figure 14.24 (a) An idealised cross-section through a rimmed shelf carbonate platform. (b) Partially completed graphic log for use with Question 14.5.

Chapter 14 Shallow-marine carbonate environments

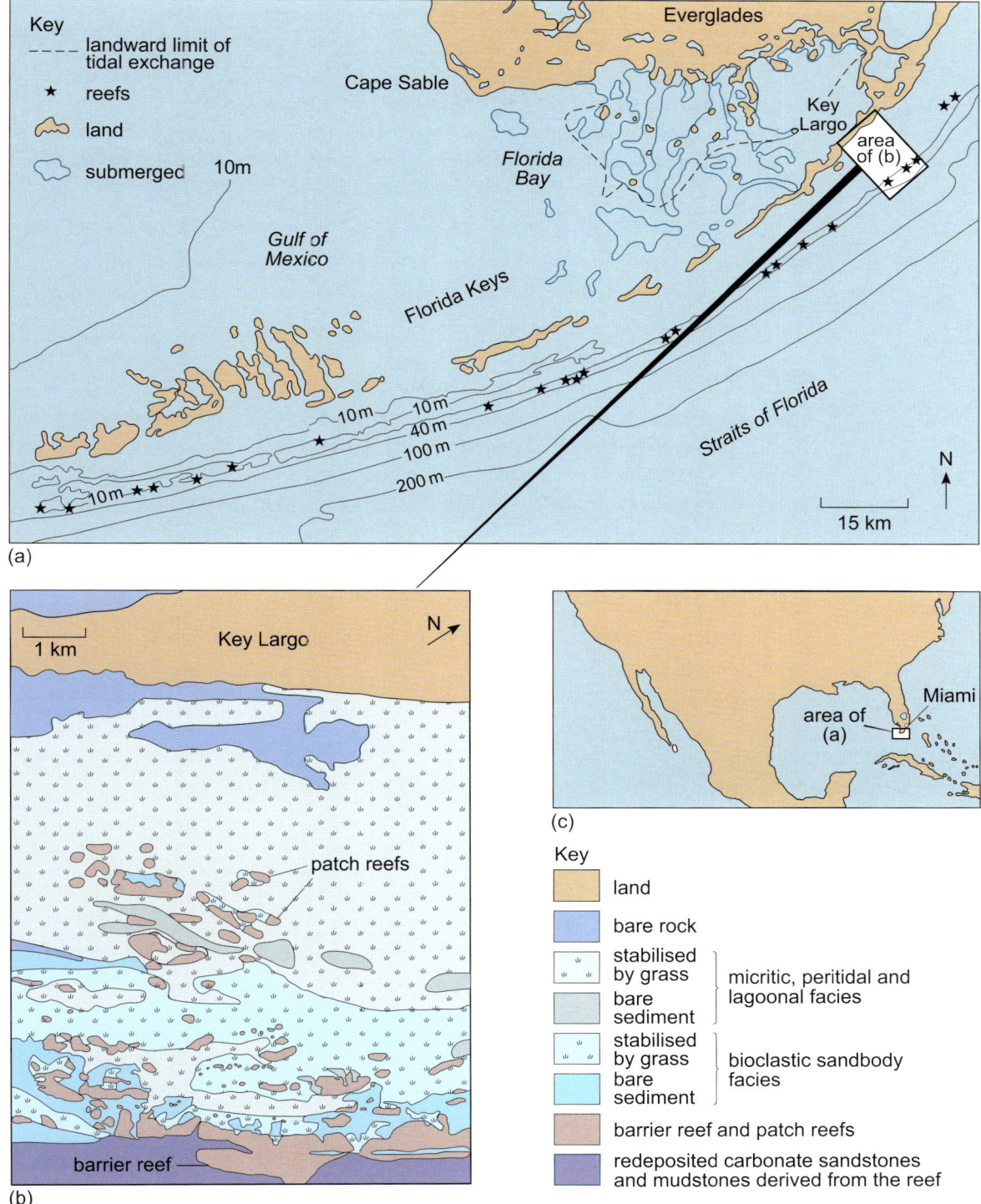

Figure 14.25 (a) Map of the Florida Keys area, USA, showing the principal topographic features. The landward limit of tidal exchange shown in the key refers to the proximal limit of the area under tidal influence. (b) Detailed map of the Key Largo area showing the distribution of facies with key. (c) Location map of the Florida Keys, USA.

283

Facies typical of carbonate-rimmed shelves include:

- peritidal and lagoonal facies, which are dominated by micrites and biomicrites. These have been stabilised by grasses in the lagoon and the intertidal–supratidal flats are only poorly developed (Figure 14.8). Most of the shoreline is vegetated by mangroves. Some of the shoreline consists of an antecedent topography formed of Pleistocene rocks and dead reefs that represent the former position of the distal edge of the shelf
- bioclastic sandbody facies with a barrier reef in the distal part of the lagoon
- patch reefs and an outer reef tract near the distal edge of the shelf
- sand- and mud-grade carbonate reworked from the reef and deposited beyond the shelf edge (Figure 14.25b).

Question 14.5

If the sea level were to fall, causing the carbonate rimmed shelf system shown in Figure 14.24a to prograde, complete a graphic log of the succession that you might expect to find, using the template provided in Figure 14.24b. Use the symbols provided on the key to Figure 14.24b. There is no need to colour in your log.

Activity 14.1 Carbonate platform environments

In this activity you will consolidate your understanding of carbonate platform environments.

Box 14.1 Cretaceous sedimentary rocks at Hunstanton, Norfolk, England

The spectacular cliffs near the old lighthouse at Hunstanton in Norfolk reveal some extraordinary rocks that have an extraordinary history. The cliffs face east towards The Wash and at low tide, modern-day channels, ripples, tube worms, trace fossils and mussels are all revealed. Back in the Cretaceous, around 100 Ma ago, conditions changed from warm and wet to very wet indeed, as oceanic islands were drowned and red, lateritic soils were gradually inundated by clear oceanic waters teeming with algae, all with calcareous skeletons that gradually accumulated as a thick layer of lime ooze (soupy mud). Figure 14.26 illustrates the succession exposed in the cliffs, from marginal marine sandstones of the Carstone, through shallow-marine sandy limestones of the Red Chalk, abruptly overlain by much purer limestones of the White Chalk.

Figure 14.26 A succession of mixed carbonate–siliciclastic sedimentary rocks exposed in cliffs at Hunstanton, Norfolk, England (Grid reference: TF 675420). Very shallow marine (upper shoreface) pebbly sandstones and slightly deeper marine fine sandstones of the Carstone Formation are overlain by red, highly bioturbated and often algal-coated chalk with abundant trace fossils, fossil fragments, including belemnites and brachiopods, rock fragments and phosphate. The junction with the overlying highly bioturbated White Chalk is sharp and encrusted by algal stromatolites. The succession represents a gradual deepening and it is likely that the iron-rich sediments typical of the Carstone and Red Chalk of the Hunstanton Formation were derived from an exposed, low-lying and arid, lateritic (highly oxidising, iron-rich) hinterland that was rapidly inundated as sea level rose. The White Chalk contains very little land-derived detritus, suggesting that the land was now fully submerged.

Chapter 14 Shallow-marine carbonate environments

14.7 Summary of Chapter 14

1. Shallow-marine carbonates differ from siliciclastics in the following ways: sediment is produced mainly by biological and chemical processes *in situ*; grain size is not necessarily related to the amount and/or type of transport; sediments are commonly both cemented and dissolved *in situ*; much carbonate mud is produced through bioerosion; precipitation and dissolution of different types of calcium carbonate have varied through geological time; and most shallow-water carbonates are deposited in warm, tropical waters.

2. Shallow-marine carbonates are deposited on carbonate platforms that vary in shape and extent; these include carbonate ramps and rimmed shelves.

3. Carbonate deposition is controlled by the type and number of organisms, the climate (which in turn influences the water temperature, circulation, oxygenation and salinity), siliciclastic sediment supply and tectonic setting.

4. The physical processes that affect shallow-marine carbonates include waves, tides and currents. The chemical processes include precipitation and dissolution. Biological processes are particularly important in shallow-marine carbonates and affect both the amount of deposition and erosion; both the type(s) and number of organisms are important.

5. Tidal flat deposition is dominated by the deposition of micritic limestones; these often include stromatolites, desiccation cracks and evaporite minerals.

6. Lagoonal successions are dominated by micrites, biomicrites and pelmicrites, with occasional evidence of tidal processes and storms.

7. The formation of carbonate sandbodies typically results in the deposition of cross-stratified oosparites and biosparites.

8. Carbonate buildups and biostromes are commonly cemented *in situ*, preserving their original features. Their form and make-up varies in time and space depending on the type and amount of organisms present, and their evolutionary stage.

9. Prograding carbonate ramp successions (such as those formed during the Carboniferous) ideally comprise, from the base, micrites and biomicrites deposited below fairweather wave-base, overlain by an oolitic and/or bioclastic sandbody, which in turn is overlain by micrites, biomicrites and pelmicrites of the back-barrier, peritidal environment.

10. A prograding rimmed shelf such as the modern Florida Shelf comprises, from the base, micrites and redeposited limestones, the remains of a carbonate buildup, an oolitic and/or bioclastic sandbody, and lagoonal micrites with evidence of periodic subaerial exposure.

14.8 Objectives for Chapter 14

Now you have completed this chapter, you should be able to:

14.1 Summarise the chemical, physical and biological processes that are important in shallow-marine carbonate environments.

14.2 Summarise the differences between shallow-marine carbonates and siliciclastics.

14.3 Describe the controls on carbonate deposition.

14.4 Recognise and describe the main facies in tidal flats, carbonate sandbodies, lagoons and carbonate buildups.

14.5 Summarise the variety of architecture and palaeoecology in buildups and biostromes.

14.6 Give examples of graphic logs summarising the main biological and sedimentological features of a rimmed-shelf carbonate platform and a ramp-type carbonate platform.

Now try the following questions to test your understanding of Chapter 14.

Question 14.6

Examine RS 21, RS 22 and RS 24 and explain whether each is more likely to have been deposited as part of a lagoonal succession or a carbonate sandbody.

Question 14.7

(a) Which way was the current flowing when the beds in Figure 14.6 were deposited?

(b) Explain whether the materials in Figures 14.9 and 14.10 have been deposited by biological, physical or chemical processes.

(c) Explain whether you would find the type of minerals illustrated in Figures 14.9 and 14.10 in the carbonate environments shown in Figures 14.8, 14.15 and 14.16.

Chapter 15 Deep-sea environments

15.1 Introduction

The 'deep sea' is used to include all of the oceans and seas that are on the oceanward side of the continental shelf. There is a change in gradient at the distal edge of the very gently dipping continental shelf, to a more steeply dipping zone called the **continental slope** (Figure 15.1).

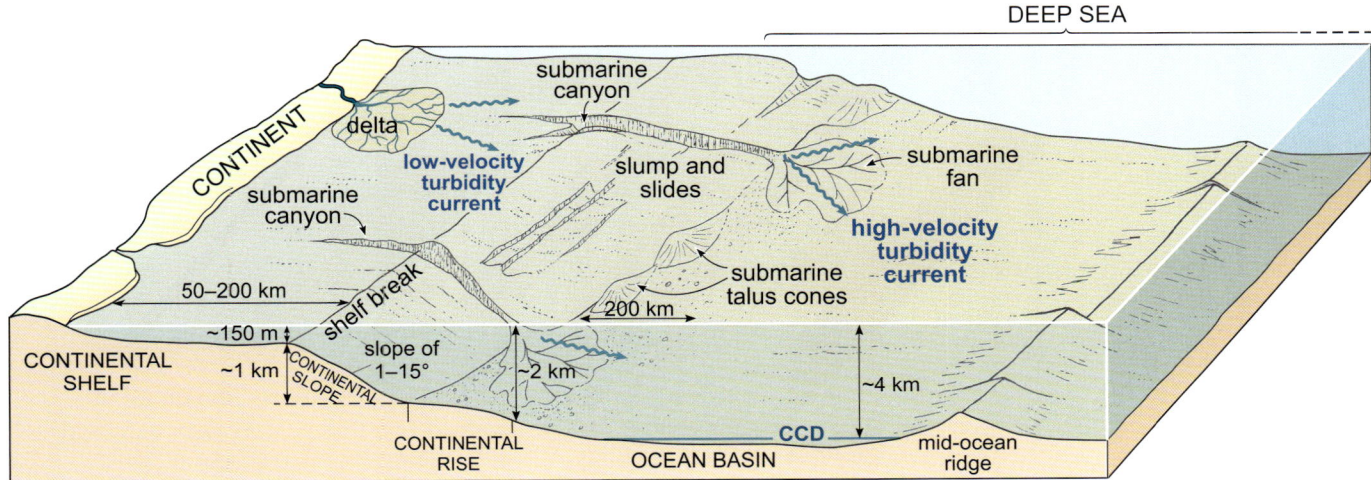

Figure 15.1 The main morphological features of the continental shelf and adjacent deep-sea areas. Note exaggerated vertical scale.

The continental shelf usually dips at less than 0.1° and is typically covered by seawater to a depth of about 150 m. By contrast, the continental slope dips at between 1° and 15° and has a vertical fall of about 1000 m. So water depths at the bottom of the continental slope are of the order of 1150 m – quite a change! The continental slope is often deeply cut by **submarine canyons** that are orientated roughly perpendicular to the continental slope and look a bit like steep-sided valleys (Figure 15.1). These canyons are often extensions of rivers on land, but some are too deep to be related to rivers, or to down-cutting during glacial times when sea level was much lower. Instead, the canyons are the product of faulting and/or submarine erosion by turbidity currents and debris flows (Section 7.6). At the base of the continental slope is the **continental rise**. This region is slightly elevated above the rest of the ocean floor and mostly comprises sediment that has been transported down the continental slope, often through submarine canyons. Beyond the continental rise is the abyssal plain, which lies beyond the usual reach of most **terrigenous** (continental-derived) sediment.

■ Thinking back to Book 2, what type(s) of crust would you expect to find underlying the continental shelf, continental slope, continental rise and abyssal plain?

☐ There is a lateral transition between continental and oceanic crust across a continental margin (Book 2, Section 2.2.1). The continental shelf rests on thinned and stretched continental crust whereas the abyssal plain must be on true oceanic crust. The intervening slope and rise are in the transitional region. However, it is probably best to regard the continental rise as oceanic crust buried by a fringe of sediment.

Today, and throughout much of the Phanerozoic, the oceans have covered about 75% of the Earth's surface making them, in terms of area, a very important site of sediment deposition. It is also worth noting that a significant proportion of deep-sea sediments are eventually destroyed during subduction of the oceanic lithosphere (Book 2, Section 2.3.1), so they are not as common as they might be in the geological record. However, during subduction, some of these sediments get scraped off and stuck to the edges of the overriding plate and then become preserved in the geological record.

15.2 Processes and sedimentation in the deep sea

A number of physical, chemical and biological processes affect sedimentation occurring in deep-sea areas. They can be grouped into three categories: sediment gravity flows (Section 7.6), deep-sea currents, and pelagic deposition from suspension.

Sediment is moved from the continental shelf down the continental slope into the ocean basin by slumping or sliding or by sediment gravity flows (mostly via submarine canyons). Movement is often triggered by earthquakes, faulting or by over-steepening of the sediment. The sediment gravity flows comprise a dense fluid composed of large amounts of sediment, i.e. 20–70% sediment mixed with seawater.

■ Will a marine sediment gravity flow be more or less dense than the seawater around it?

☐ A sediment gravity flow will be more dense because it is composed of seawater and grains of sediment that together are more dense than seawater.

The greater density of the sediment gravity flow compared to the surrounding seawater means that movement will be maintained through the water and that gravity is the driving force. Sediment gravity flows include slides and slumps, debris flows and turbidity currents. They result in deposition of large amounts of sediment at the base of the continental slope (Section 7.6).

■ What is the name of the characteristic succession of beds deposited by turbidity currents, and which mechanism is responsible for the deposition of progressively finer beds?

☐ A Bouma Sequence (Section 7.6.3), deposited during waning turbidity current flow.

The second category of processes in the deep sea involves deep-sea currents. Winds blowing across the ocean surface not only produce waves, but can also generate currents. Other currents are a consequence of the Earth's rotation, and slight differences in both temperature and salinity in the ocean water result in density variations in the water column and density-driven currents.

Deep-sea currents produced by density differences can be powerful enough to erode, transport, rework and deposit sediment in the oceans. There are also contour-hugging currents that affect both the continental shelf and the ocean floor, and they flow around rather than down the slope (Figure 15.2). These contour currents are driven by winds operating in shallow-water areas and by the thermohaline circulation (driven by density differences in oceanic water masses). The sediments deposited by contour currents are called **contourites** but their importance in the deep-sea geological record is a matter of debate. The currents transport only a small amount of sediment, which is likely to be reworked by bioturbation, winnowing, down-slope debris flows and turbidity currents. Figure 15.2 shows the different transport directions associated with contour currents compared to gravity-driven processes (turbidity currents, debris flows, slumps and slides).

The third category of processes that occur in the deep sea involves predominantly biological and chemical processes. These are discussed further in Section 15.4.

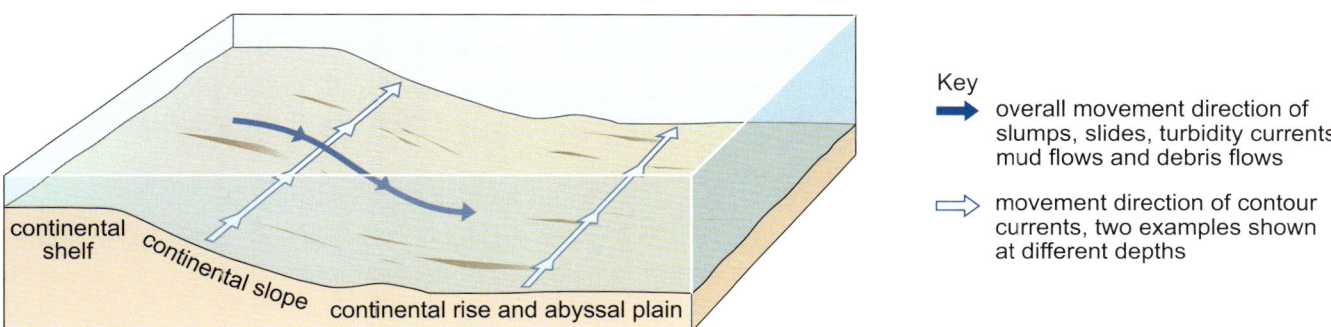

Figure 15.2 General movement direction of contour currents compared with slumps, slides, debris flows and turbidites. Note that contour currents can be present anywhere below storm wave-base.

15.3 Sediment gravity flows in the deep sea

15.3.1 Sediment gravity flow processes

The relatively steep gradients found on the continental slope, in submarine canyons, and along the steep distal edge of some carbonate platforms, mean that sediment gravity flows are the most common mechanism transporting sediment in the deep sea. Sediment gravity flows are driven downslope by gravity (often triggered by earthquakes, fault movement or by over-steepening of the sediment pile) and, once in motion, they can travel long distances – for instance, they may start by collapse of an unstable delta front, then travel across the continental shelf and down a submarine canyon before finally coming to rest on the continental rise or even extending out onto the abyssal plain.

As discussed in Section 7.6, there is a spectrum of sediment gravity flow types, depending on the amount of consolidation of the sediment at its starting point, and the concentration of sediment and degree of turbulence in the flow. This balance can change during transport, so that one type of flow can grade into another. Debris flows (Section 7.6.2) and turbidity currents (Section 7.6.3) are responsible for the most common sediment gravity flow deposits both today and in the sedimentary record, and they can form in both siliciclastic and carbonate sediments.

■ Why do some carbonate platforms have a steep distal edge from which turbidites and debris flows could be initiated?

□ The buildups that fringe carbonate platforms are almost vertical if they are built by organisms that produce a framework. This steep edge may be exaggerated by erosion during former times of lower sea-level (Figure 14.17a).

Sediment gravity flows can form where the depositional slope is greater than about 1° and where sufficient sediment is available. Indeed, both debris flows and turbidites may form in other environments besides the deep sea (e.g. wadis and lakes respectively), but it is in the deep sea that they are most common.

Question 15.1

Sediment gravity flows move either by fluid flow or by plastic flow. Which of these flow types do (a) turbidity currents and (b) debris flows represent?

Question 15.2

Is the sediment in (a) a turbidity current and (b) a debris flow supported by fluid turbulence or the strength and buoyancy of the matrix?

15.3.2 Turbidites and the Bouma Sequence

Turbidity currents produce turbidites – a characteristic fining-upwards succession of beds known as a Bouma Sequence – with a predictable pattern of clast sizes and sedimentary structures deposited as the current gradually slows down. Five divisions A–E have been recognised within a complete Bouma Sequence (Figures 7.42 and 7.43). The basal division (A) is the coarsest grained and the base often shows evidence of erosion (sharp contact with flute marks and tool marks) and rapid deposition (flame structures and ball-and-pillow structures – Section 7.7). Together with division B, which is planar-stratified, these two divisions represent deposition in the upper flow regime (Figure 7.22). Division C is cross-stratified, reflecting the formation and migration of ripples and division D is planar-stratified; both of these divisions represent deposition in the lower flow regime. Division E has been subdivided into two: division E(t) comprises the finest-grained sediments carried by the turbidity current, representing deposition in the final stage of the turbidity flow; division E(h) comprises hemipelagic, fine-grained sediment deposited after the turbidity current has stopped. Divisions A to D represent bedload deposition; division E represents suspension deposition.

If division E does not have time to solidify before another turbidite is deposited on top, then the bedding may become deformed or convoluted because as the overlying heavier sediment loads onto the unlithified beds below, the underlying beds rapidly expel water upwards due to the increased downward pressure. This leads to the formation of flame structures and ball-and-pillow structures.

Question 15.3

What are the sedimentary structures shown in Figure 15.3? Which part of the bed do they represent, and which way was the current flowing?

Figure 15.3 Sedimentary structures for use with Question 15.3.

Question 15.4

In relative terms, explain how you think the period of time taken to deposit divisions A–D of a Bouma Sequence might compare to the time taken to deposit division E.

Question 15.5

Which division of a Bouma Sequence do you think is most likely to be bioturbated, and why?

The sedimentary record usually comprises a whole series of Bouma Sequences stacked on top of each other, each representing a single turbidity current. An example of a succession of such turbidites is shown in Figure 15.4.

Figure 15.4 A succession of turbidites and one debris flow from Portugal. Hammer is about 30 cm long.

Figure 7.42 represents a complete Bouma Sequence with all of the divisions (units) present. However, in reality, incomplete Bouma Sequences are not unusual because of the changing conditions of the flow, the variety of grain sizes carried in different turbidity currents and the proximity of the flow to the source. Incomplete sequences showing repetitions of Bouma divisions BCDE, CDE and DE can all be found.

Siliciclastic turbidites tend to be immature, both texturally and compositionally.

■ Why do you think turbidites are generally texturally immature?

☐ Two factors contribute to this: (i) turbidites generally represent a complete mixture of material that has become unstable and has flowed in a turbidity current that has picked up everything in its path; (ii) rapid deposition from suspension means that there is not enough time to sort all of the sediment (remember that at first *all* grains are carried in suspension by fluid turbulence).

There is almost no time for grain rounding to take place during transport, but if the sediment grains in the source area are well rounded, then the grains in the resulting turbidites will also be well rounded.

■ If texturally immature sediments are common in turbidites, name a rock type in which they might frequently be found.

☐ Wacke (or greywacke, Section 9.6.1; see also Book 1, Section 7.1).

RS 12 is a muddy conglomerate and could have formed part of a Bouma Sequence, probably division A. RS 27 could represent the fine-grained division E of a siliciclastic Bouma Sequence. Carbonate turbidites are typically composed of clasts of various sizes of shallow-marine carbonates, mixed with carbonates that formed in deeper water.

The main factor controlling the composition of turbidites is the nature of the sediment source rather than the amount of transport. So if the sediment supply is derived from fine-grained sandstones and mudstones, then the turbidites will be composed entirely of these. However, if there are granules and pebbles present, these will also be incorporated.

15.3.3 Debris flows and mudflows

Debris flows and mudflows are the other common types of sediment gravity flow found today and in the geological record (see Section 7.6.2). Debris flows and mudflows behave plastically and comprise clasts up to several metres in diameter, enclosed in a finer-grained matrix of clay, silt or fine- to medium-grained sand that provides buoyancy during transport. Mudflows (Figure 7.40) are typically matrix-supported and the clasts are usually unsorted and ungraded. Debris flows are usually grain-supported and may display some sorting and/or stratification (Figure 15.5). The flows can form in both siliciclastic and carbonate environments.

Figure 15.5 A debris flow of Neogene age, Cyprus. The clast size varies from centimetres to large blocks several metres in diameter and there is some crude stratification picked out by the long axes of clasts (e.g. the large white block on the left). The exposure shown is about 12 m high.

15.3.4 Sediment gravity flow successions and facies models

Bouma Sequences are often repeated and stack up to form turbidite successions (Figure 15.4). They are often found in association with debris flows and mudflows at the base of submarine canyons where they form a fan-shaped feature called a **submarine fan** (Figure 15.6). Fan sediments are fed from the mouth of the submarine canyon, which effectively forms a point source. However, turbidites, mudflows and debris flows are also deposited from linear sources (Figure 15.6), such as along unstable slopes that are controlled by active faulting, or at the base of slopes produced by oversteepened organic growth (e.g. a carbonate buildup).

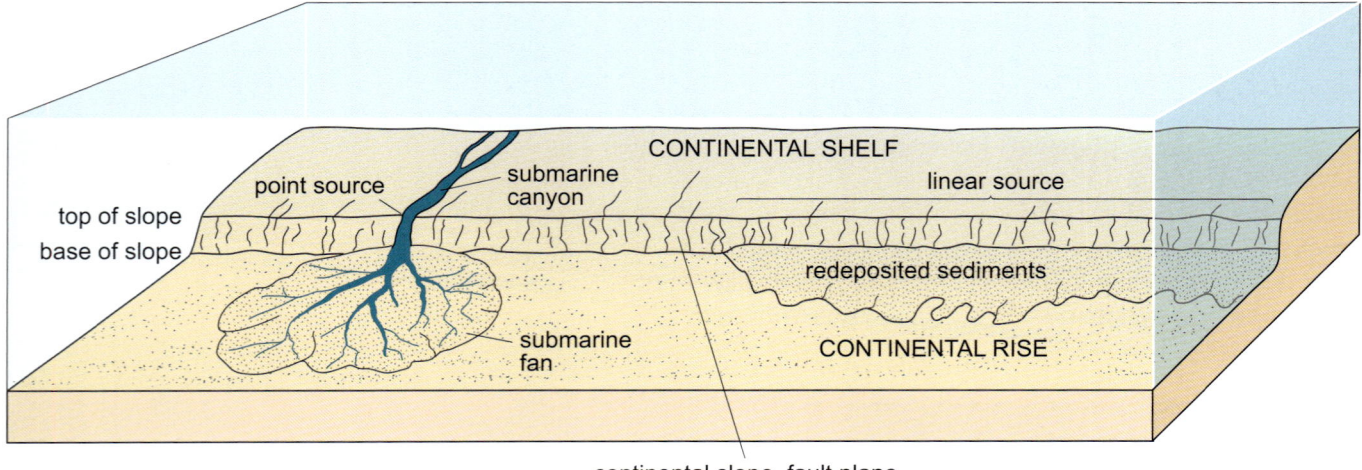

Figure 15.6 Turbidites and debris flows from a point source, forming a submarine fan, and a linear source such as a fault scarp.

Figure 15.7 Neogene carbonate turbidites derived from a carbonate shelf environment, Waiau Basin, New Zealand.

■ What would be the plan view shape of deposits found at the base of a submarine fault scarp or at the edge of a linear morphological feature such as a coral buildup?

□ For a linear source, the sediment would be deposited in a long belt roughly parallel to the base of the fault scarp or buildup, similar to that shown on the right-hand side of Figure 15.6.

Siliciclastic turbidite, mudflow and debris flow systems can be fed either by a point source or a linear source, but most *carbonate* turbidite, mudflow and debris flow systems are fed from linear sources because linear buildups at the shelf break provide the topographic feature from which they originate. An example of a carbonate turbidite succession derived from a fault-bounded, linear shelf source is illustrated in Figure 15.7.

The nature of the slope and the sediment supply are two important factors that influence the creation of turbidite, mudflow and debris flow successions. If the slope is stable, and there is no sediment supply, then there will be no deposition of sediment gravity flows. Submarine canyons form conduits that transport much of the sediment from the continental shelf where there is abundant terrigenous sediment supplied by subaerial erosion (see Figure 15.1). Sediment being fed to the continental shelf through deltas, strandplains and barrier islands will be carried down the submarine canyons by currents and end up being deposited in a submarine fan.

Submarine fans are similar to deltas in their overall shape and the way in which the sediment is fed along channels to the distal parts of the fan though, of course, submarine fans are *entirely* under water. The SE Poster illustrates a submarine fan in the deep-sea environment. At the end of the canyon, where the gradient lessens, the sediments are dispersed from the major channel into smaller channels, which serve to deposit the sediment as distal lobes. At different times, each of these channels may be active, abandoned, braided or meandering. The channels also have levées similar to those in alluvial channels (Figure 15.8).

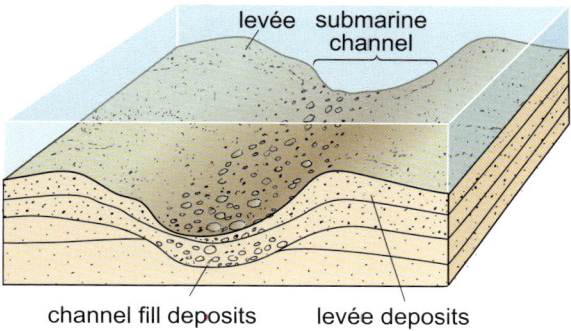

Figure 15.8 Block diagram to show the form of a channel and levées in a submarine fan.

■ Which factors will influence the grain size of sediment deposited in the different areas of the submarine fan, and what will be their distribution?

☐ The type of sediment supplied will influence the grain size distribution and this may change through time. However, assuming that there is a fairly constant supply of sediment with a constant grain size distribution, then – like all sedimentary systems deposited by currents – the coarsest-grained material will be deposited nearest the sediment source because the energy is not usually sufficient to carry it very far. As the current decreases away from the mouth of the canyon, the finer-grained material is deposited. There will also be a difference in grain size distribution between the channels and their levées. The channels contain the coarser-grained material and the levées the finer-grained material.

An *idealised* graphic log through a submarine fan assuming a constant grain size distribution in the sediment supply is shown in Figure 15.9.

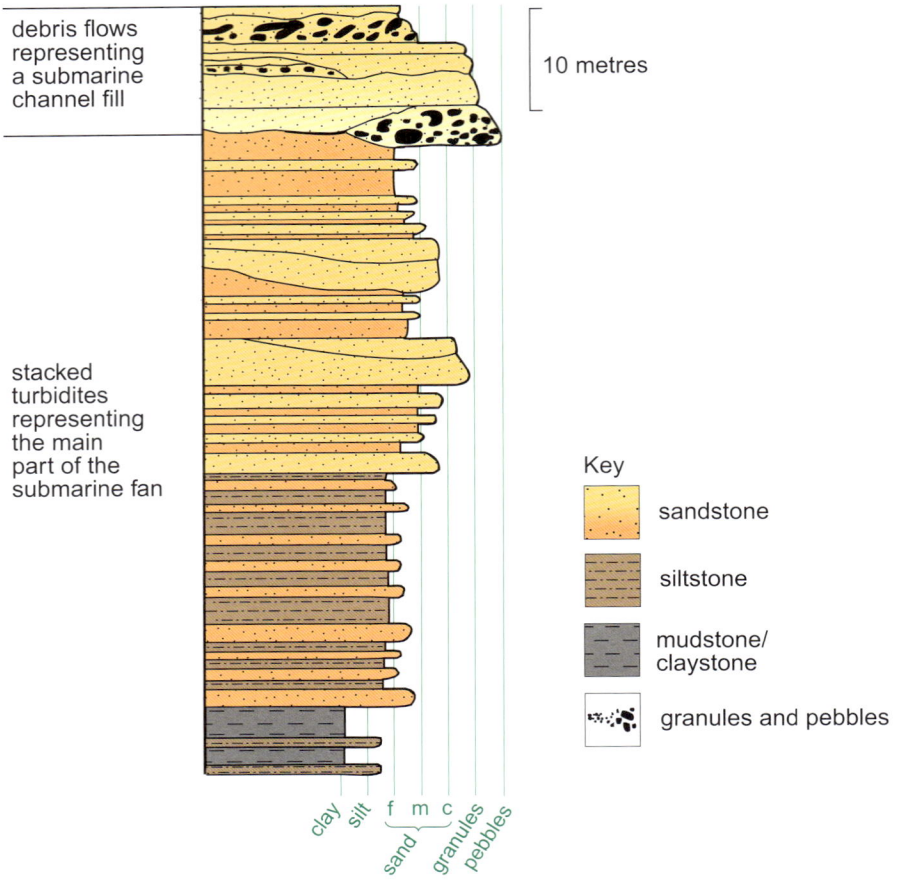

Figure 15.9 Idealised graphic log through a submarine fan, assuming that there is a constant supply of sediment with the same grain size distribution. A block diagram of a submarine fan is shown on the SE Poster.

15.4 Hemipelagic and pelagic sediments

In addition to the sediments deposited by turbidity currents and ocean currents, there is also constant deposition of fine-grained sediments, including weathered volcanic ash, wind-blown dust and clay minerals derived from the continent; these are collectively referred to as terrigenous sediment. Importantly, sediment is also produced biologically in the surface waters of the oceans, which then falls as 'biogenic rain' to the ocean floor. Some sediment may also be produced directly by chemical precipitation. Sediments that contain less than 25% terrigenous material, and which are therefore almost entirely made up of biogenic sediments produced and deposited within the ocean and seas, are termed **pelagic**. They include chalks and some bedded cherts. Some sediments deposited on the ocean floor will contain a mixture of pelagic and terrigenous material.

■ What is the name given to sediments containing a mixture of pelagic (up to 75%) and terrigenous material?

☐ Hemipelagic sediments (Section 7.6.3).

Pelagic and hemipelagic sediments are being deposited over more than 60% of the world's oceans today, but in general the sedimentation rate is very low, and so they rarely form thick deposits.

15.4.1 Hemipelagic and pelagic depositional and erosional processes

Hemipelagic and pelagic sediments are deposited from suspension, so they tend to appear massively bedded (i.e. the sediments show very little sign of bedding or lamination). However, slight variations in the composition of the sediment through time will give rise to lamination and bedding. The quantity of pelagic sediment produced depends on having favourable conditions to support biological activity in the surface waters of the ocean; these include light, warmth and nutrients.

■ In the deep sea, carbonate sediment preservation depends on another factor. What is this and how does it affect the sediment preserved?

☐ The carbonate compensation depth (CCD) is the other factor (Sections 4.1.6 and 8.1). The CCD is the depth at which the amount of carbonate grains being supplied to the ocean is balanced by the amount dissolving, so carbonate sediments will not accumulate below this depth.

The actual depth of the CCD varies with temperature, salinity, pressure and shape of the ocean basin. It is thus variable in both time and space.

Question 15.6

Assuming that there is a source of both siliceous and carbonate-secreting organisms living and dying in the surface waters, explain which one will provide remains able to accumulate below the CCD.

15.4.2 Pelagic successions

Pelagic sedimentary deposits include the remains of calcium carbonate-secreting micro-organisms such as coccolithophores (Figure 8.2) and foraminiferans, and the remains of siliceous micro-organisms such as diatoms and radiolarians (Figure 8.7). Carbonate-secreting organisms can result in the deposition of chalks and other deep-water carbonates (Figure 14.1), whereas siliceous organisms give rise to bedded cherts (Figure 8.8).

Figures 15.10 and 15.11 show two chalk successions. At first glance they might appear similar, but Figure 15.10 shows Cretaceous chalks mainly composed

Figure 15.10 Pelagic, coccolith-rich chalks of Cretaceous age near Dover, England. Height of cliff is about 12 m.

Figure 15.11 Interbedded pelagic, foraminiferan-rich chalks and marls of Neogene age, Cyprus. Height of cliff is about 10 m.

of coccoliths (coccolithophores), whereas Figure 15.11 shows Neogene chalks composed mainly of foraminiferans. This is because different organisms have dominated at different periods of geological time. High magnification microscope work is needed to recognise the type of microfossil of which the chalk is composed. Figures 15.10 and 15.11 also show some other features typical of pelagic chalk successions. The black nodules just visible in Figure 15.10 are chert nodules (Section 8.2). In this case, the silica originated from the skeletal parts of sponges and was remobilised during diagenesis. Many chert nodules are actually the site of large, horizontal burrow systems (*Thalassinoides*). During burial, silica from sponge spicules goes into solution and, as it is negatively charged, it is drawn towards the burrows because they contain small amounts of organic matter that are positively charged. The silica re-precipitates in the old burrow and, once it has started to solidify, this will encourage more silica to come out of solution and hence the nodules grow.

Figure 15.11 shows obvious bedding because it is a succession of pale-coloured chalks, composed of nearly 100% foraminiferans, interbedded with marls that consist of approximately 70% foraminiferans and 20–30% clay. The intermittent input of clay minerals might reflect climatic control on the rate of clay supplied from the land. Minerals may precipitate directly out of seawater onto the deep sea-floor, especially when there is very low sediment input. These include pyrite, manganese, glauconite and phosphate.

- What signs of life do you think may be found most abundantly in deep-sea sediments?

- Pelagic biota are commonly found in deep-sea sediments because, although they may live in the surface waters, on death they fall to the bottom where general lack of reworking by currents means there is good preservation potential.

- What factor will limit the preservation of calcitic remains in the deep sea?

- The position of the CCD (Section 15.4.1). If the calcitic fossil falls below the CCD, it will not be preserved.

Activity 15.1 Deep-sea environments

In this activity, you will consolidate your understanding of deep-sea environments.

15.5 Summary of Chapter 15

1   Physical processes in the deep sea include sediment gravity flows and deep-sea currents. Biological processes include the life and death of microorganisms in the surface waters. Direct chemical precipitation also occurs in the deep sea.

2   Decelerating turbidity currents deposit sediments with particular features. An idealised succession of units or divisions is called a Bouma Sequence; these fine upwards and can be divided into five divisions A to E (and E is subdivided into E(t) and E(h)), deposited during distinct stages of the waning turbidity current flow. The basal division A is the coarsest grained and often shows evidence for erosion (sharp base, flute marks and tool marks) and rapid deposition (flame or ball-and-pillow structures). Division B is planar-stratified, and division C is cross-stratified, reflecting the formation and migration of ripples. Division D is planar stratified, while division E(t) comprises the finest-grained sediments carried by the turbidity current and is deposited in the final stage of the turbidity flow. Division E(h) comprises hemipelagic, fine-grained sediments deposited after the turbidity current has stopped.

3   Sediment gravity flows are driven by gravity so they require a slope to start them in motion – a dip of 1° is sufficient. Both siliciclastic and carbonate sediment can be transported in this way.

4   Sediment gravity flows may come from a point source, such as a submarine canyon, and are deposited in a fan shape (submarine fan). The fan itself will be dissected by channels along which sediment is transported.

5   Sediment gravity flows derived from a linear source will form a linear mound at the base of the slope.

6   Large areas of the deep sea are the site of deposition of both hemipelagic and pelagic sediments. Hemipelagic sediments contain more than 25% terrigenous material (e.g. marls and division E(h) turbidites) whereas pelagic sediments (e.g. chalk) contain less than 25% terrigenous material. The pelagic material is either produced biologically in the surface waters (i.e. the remains of microorganisms such as diatoms and coccolithophores) or by direct chemical precipitation.

15.6 Objectives for Chapter 15

Now you have completed this chapter, you should be able to:

15.1   Describe the physical, chemical and biological processes that are important in deep-sea environments.

15.2   Describe and produce a summary graphic log showing all the divisions of a Bouma Sequence and explain how each division forms.

15.3   State the features of, and describe the difference between, point source and linear source sediment gravity flow successions deposited in marine environments.

15.4   Explain what is meant by hemipelagic and pelagic sediments and give examples of each.

Now try the following questions to test your understanding of Chapter 15.

Question 15.7

In the form of a table, summarise the four different depositional processes important in the deep sea and their corresponding sedimentary deposits.

Question 15.8

Explain whether submarine fans are more likely to form when sea level rises or when it falls.

Chapter 16 Glacial environments

16.1 Introduction

In 2009, approximately 10% of the Earth's total land area was covered by glacier ice, mostly within the Antarctic ice sheet that extends for 12.5×10^6 km^2 and reaches a thickness of more than 4 km. The Greenland ice sheet, by comparison, covers only 1.7×10^6 km^2, with a maximum thickness of 3.3 km. However, the West Antarctic and Greenland ice sheets are currently thinning at an accelerating pace, with the potential to cause significant sea-level rise (Figure 16.1). In more temperate latitudes, glaciers are confined to mountainous terrane; even in the tropics, glaciers occur in mountainous areas higher than 4 km. Glaciers and ice sheets develop wherever the rate of accumulation of winter snow exceeds the rate at which it melts in summer. Snow crystals are altered by a process known as sublimation – molecules of water vapour escape from snowflakes and re-freeze as they attach themselves to the central parts of other flakes to form crystalline granules called firn. These granules are progressively compacted beneath further snowfalls to produce glacier ice.

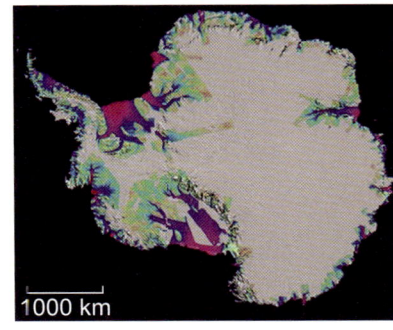

Figure 16.1 Antarctic ice loss between 1996 and 2006. The false colours reflect the relative speed of the ice loss: the dark colours (purple/red) denote fast melting; the pale colours (pale green and brown) denote slow melting.

During polar winters, the Antarctic continental margin and Arctic Ocean (Figure 16.2) are covered with pack ice (frozen seawater) and on land, intensely cold, but glacier-free areas are permanently frozen (permafrost), for example in Siberia and northern Canada. Freeze–thaw action leads to physical disintegration of rocks, mass movement of soil, frost heaving and the development of patterned ground. Frost heaving is the vertical movement of sediment due to pressure generated by ice crystallisation and gives rise to internal disturbance

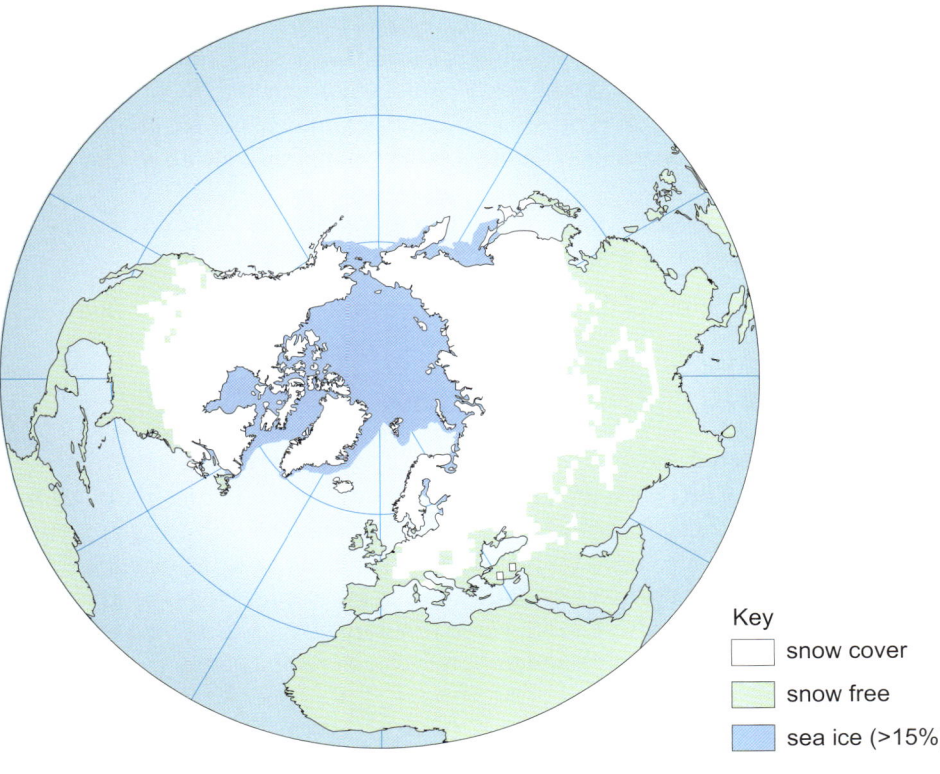

Figure 16.2 Extent of snow cover (including permafrost) and sea ice in the Northern Hemisphere during January 1995. The area of snow cover extends much further south in the continental interior of Asia, compared to the relatively mild coast of northwest Europe at similar latitudes. The area of permafrost is only slightly less extensive than the area of winter snow.

303

(cryoturbation). Patterned ground is characterised by distinctive polygonal shapes, caused either by the ground cracking at very low temperatures to produce ice-wedges (Figure 16.3a), or by sorting of sediment on gentle slopes (Figure 16.3b).

Figure 16.3 (a) Fossil ice wedge exposed in glacial sedimentary deposits, West Runton, Norfolk, England. Thermal contraction in winter opens vertical cracks in the permafrost, which subsequently freeze and enlarge. On melting, the ice is replaced by sediment. (b) Patterned ground, Alaska. Coarser-grained sediment is concentrated around the margins of the distinctive polygons.

Glacial processes have influenced the landscapes of much of North America and Europe during the past 2 Ma, giving rise to such features as ice-scoured U-shaped valleys (Figure 7.46a) and limestone pavements of upland Britain or the glaciers of the European Alps. Spectacular though these often are, such landscapes are less significant in the geological record because they are rarely preserved. Conversely, glacial sediments are often left behind, recording dramatic global climatic changes during several periods of Earth history, extending back to the Precambrian Eon. However, it is not always easy to recognise glacial facies, since few features are uniquely diagnostic. As with any depositional environment, all of the rocks, structures and contacts need to be carefully examined and the processes responsible thoroughly considered before a robust interpretation can be attempted.

■ Consider a poorly sorted, unstratified, pebbly sandstone. What environments could deposit just such a sedimentary rock?

□ Fluvial channel, alluvial fan, melting glacier, beach and submarine fan.

16.2 Erosive glacial landforms

You briefly considered glacial erosion and transport processes in Section 7.8. Moving ice armed with debris gouges out U-shaped valleys (Figure 7.46a) and leaves behind exposed rocks with glacial striations (Figure 7.46b). High up on the mountainside, the ice, aided by frost action, creates armchair-shaped hollows named **cirques** (often called corries or cwms) and knife-edge ridges or **arêtes** (Figure 16.4). **Hanging valleys** form where tributary glaciers flow into the main glaciated valley and are left high and dry after glacial retreat (Figure 16.5).

Chapter 16   Glacial environments

Figure 16.4 Arêtes, Mt Alexander, Joinville Island, off the northeastern tip of the Antarctic Peninsula.

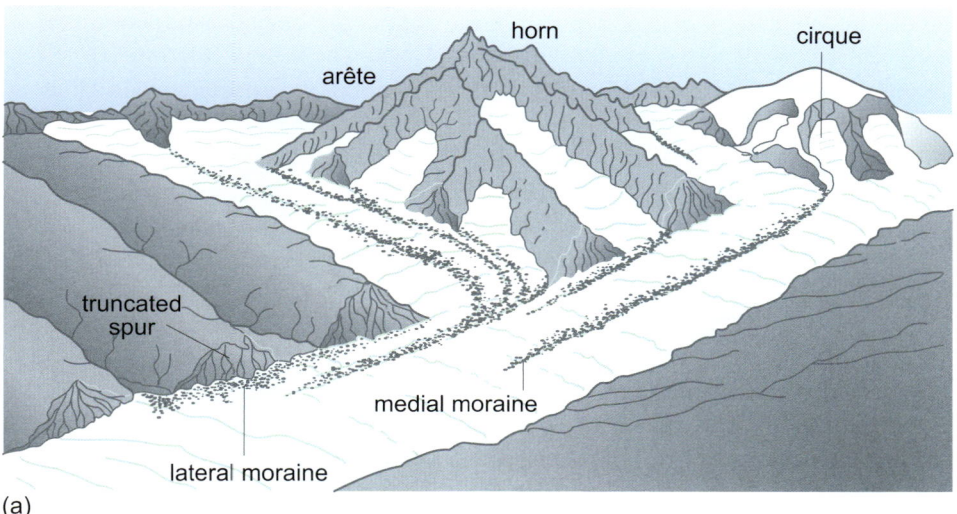

(a)

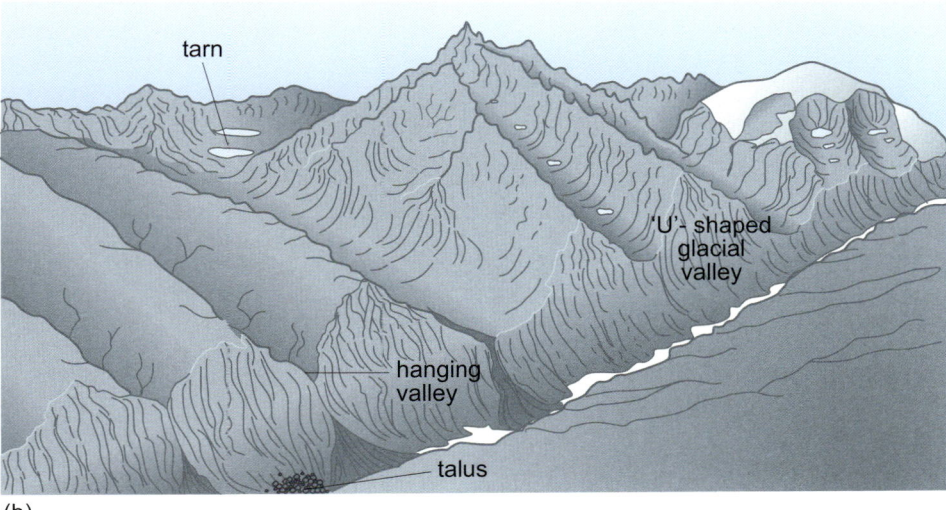

(b)

Figure 16.5 Landforms produced by valley glaciers (a) during and (b) after glaciation.

305

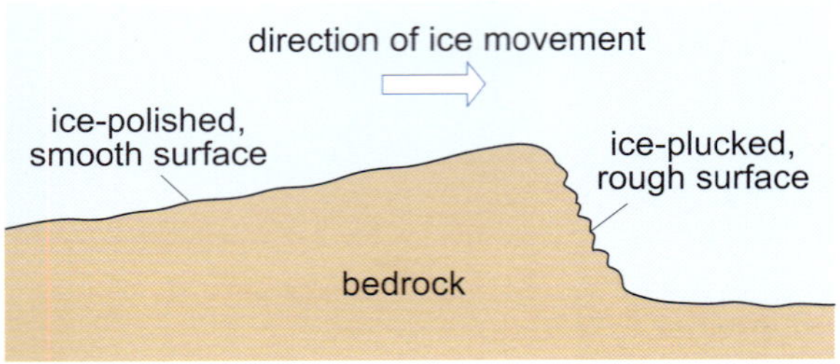

Figure 16.6 Roches moutonnées, which are polished outcrops of hard bedrock with a smooth up-glacier face, and a steep, ice-plucked down-glacier face.

Other features include **roches moutonnées** – 'sheep-like' mounds of bedrock with a smooth, abraded up-glacier face and a steep, ice-plucked, down-glacier face (Figure 16.6).

The preservation potential of erosive glacial landforms such as glaciated valleys or cirques is very low, but as the glacier melts and its front recedes, it deposits the sediments that it has carried along with it as a layer of melt-out till on the underlying scoured bedrock (see Section 7.8). This till protects the scoured surface from subaerial weathering and there are a surprising number of striated pavements and roches moutonnées in the geological record.

16.3 Depositional glacial landforms

Landforms created by glacial deposition are both complex and varied, and are more easily understood in modern glacial terranes, where their relationship to the parent glacier or ice margin is apparent. Poorly sorted and poorly consolidated glacial debris is deposited in two ways: either directly by advancing glaciers as *lodgement till*, where sediment is plastered onto the bed beneath the actively moving ice (Figure 7.46d), or by retreating glaciers as *melt-out till*, in which melting out of debris from the sides and top of the glacier (Figures 16.5a and Figure 16.7) forms irregular hummocks called moraines (Figure 7.46c). The tills – which when lithified are called **tillites** (Figure 16.8) – commonly overlie striated bedrock and they are frequently modified by **glaciotectonic processes** that deform the sediments beneath and in front of the glacier and even dislodge slabs of bedrock (Figure 16.9). Lodgement tills are often sheared and may contain striated pebbles. Most melt-out tills lack well-defined bedding or lamination, though they are often subsequently modified and reworked by fluvial processes from the action of glacial meltwaters. This creates **fluvioglacial**

Figure 16.7 Poorly sorted moraine on top of the Mer de Glace, Switzerland. Note the hammer (about 40 cm long) under the large boulder for scale.

Figure 16.8 Tillite comprising part of a lodgement till that formed at the base of an advancing glacier, West Runton, Norfolk, England.

Figure 16.9 Tillite deformed by glaciotectonic processes, West Runton, Norfolk, England. Note the raft of Chalk bedrock (just to the right of the centre).

sediments (outwash deposits) with characteristic features such as cross-stratification, imbrication and grading. The high meltwater discharge rates give rise to significant sediment transport and rapid erosion of exposed rocks. Figure 7.8 shows a meltwater channel cutting down into Carboniferous strata.

Sub-glacial streams carve out channels in the underlying sediment and tunnels in the overlying ice. Both the channels and tunnels are filled with gravel, the latter exposed as steep-sided ridges or **eskers** when the ice retreats. **Drumlins** are distinctive, generally ellipsoidal mounds of till or outwash sediment that form beneath moving glaciers and are oriented with their long axes parallel to the glacier flow direction (Figure 16.10). They are created by erosion of the underlying sediment and/or trapping of debris against subglacial obstacles (Figure 16.11). Drumlin fields are common in the northern parts of Britain, but the largest recorded field is in upstate New York, comprising around 10 000 individual drumlins.

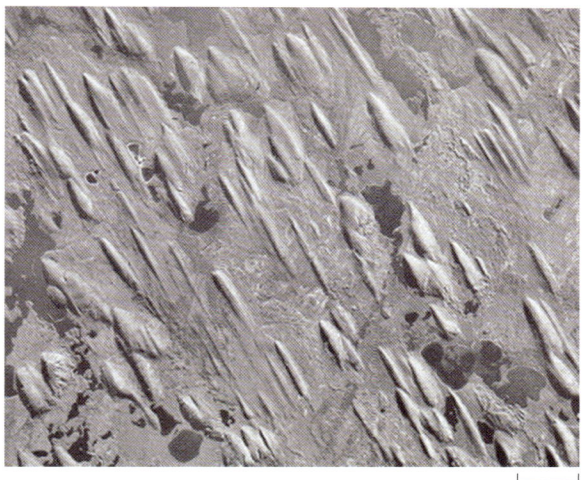

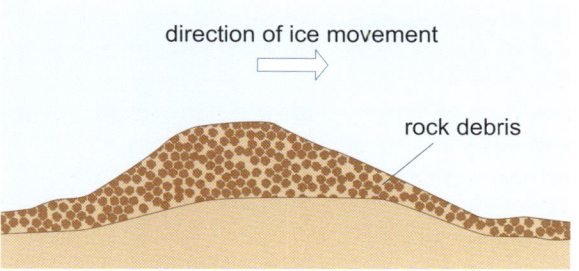

Figure 16.10 Aerial view of a drumlin field in Canada. Drumlins are oriented NW–SE and are typically 100–200 m in length.

Figure 16.11 Cartoon of a drumlin in cross-section. Each one forms a long, streamlined hill several tens of metres high that is composed of rock debris moulded into a mound by the advancing ice.

Isolated rocks up to boulder size, known as **erratics**, are left behind by the retreating ice sheet. They are distinctive because their exotic lithology reflects significant transport over tens or hundreds of kilometres (Figure 16.12); some may be striated.

Figure 16.12 Glacial erratic of Silurian grit perched on a plinth of younger Carboniferous Limestone at Norber Brow, Yorkshire, England.

16.4 Glacial lake sediments

Glacial lakes form in areas that have been deepened by glacial erosion, and/or dammed by ice or moraine (Figure 16.13). Though temporary, the lakes provide space for sediment accumulation, either as proximal, deltaic deposits, sediment gravity flows or as distal **varves**. Varves are annually-produced couplets of contrasting grain sizes resulting from summer melting (sand and silt) and winter freezing, when only the finest sediment (clay) settles out of suspension (Figure 16.14). Varves are distinctive and important glacial indicators, not to be confused with laminated mudstones that result from distal turbidity currents (Bouma divisions D and E; Figure 7.42). Varves provide a method of direct dating since each couplet represents one year. However, not all glacial lakes contain seasonally varved sediments, especially in mountainous terranes where avalanching and resedimentation processes operate all year round.

Figure 16.14 Varves from a glacial lake, Vermont, USA. Each dark and pale couplet represents one year. The thicker, paler-coloured sandy laminae were deposited from spring–summer meltwater, and the thinner, darker-coloured clay laminae were deposited from suspension during the winter.

Figure 16.13 A branch of a glacier has dammed a small, frozen lake in East Greenland and tabular icebergs have broken off from the floating glacier tongue. Ice-dammed lakes are a common feature of Arctic glaciers.

16.5 Wind-blown loess

Strong winds are common in glaciated regions, since large ice caps influence atmospheric circulation due to increased gradients of temperature and pressure. Significant quantities of fine-grained sediment (pulverised rock 'flour') are released by melting ice fronts and this is blown over vast areas, creating thick deposits of wind-blown, yellowish silt or **loess** (usually pronounced as 'lerss'). In northern and central China, loess from the last 2.5 Ma attains a thickness of 180 m and covers an extensive plateau. The loess grains, mostly quartz, range from silt to very fine sand-grade, averaging 30 µm in diameter, and are often angular and bladed in shape (Figure 16.15).

Question 16.1

How does wind-blown sand differ in shape and sphericity from loess grains, and how do you account for the differences?

Figure 16.15 Scanning electron microscope image of silt to very fine sand-grade loess grains.

16.6 Glaciomarine environments

Ice sheets enter the sea as ice cliffs or floating ice shelves and can extend to the edge of the continental shelf. Floating icebergs can extend beyond this zone to the continental slope and deep sea. The line between ice that is attached to bedrock (grounded) and floating ice is termed the *grounding line*. Proximity to an ice margin determines whether the environment is dominated by glacial processes (ice proximal) or marine processes (ice distal), and the climate dictates the volume of meltwater and sediment that are supplied to the shelf. Temperate oceanic environments receive large volumes of meltwater and mud, whereas deeply frozen polar areas receive very little sediment.

The glaciomarine environment shown on the block model of glacial environments (Figure 16.16) is the site of significant sediment deposition. It is also complex, reflecting not just glacial and marine processes, but also inputs from biogenic sources, rivers and wind. Two distinct geographical settings can be recognised:

- fjords, in which sedimentation is influenced by tidewater, floating glaciers and rivers
- continental shelves and the deep ocean, in which sedimentation is dominated by grounded ice margins, adjacent areas of marine ice called ice shelves and by open-marine processes.

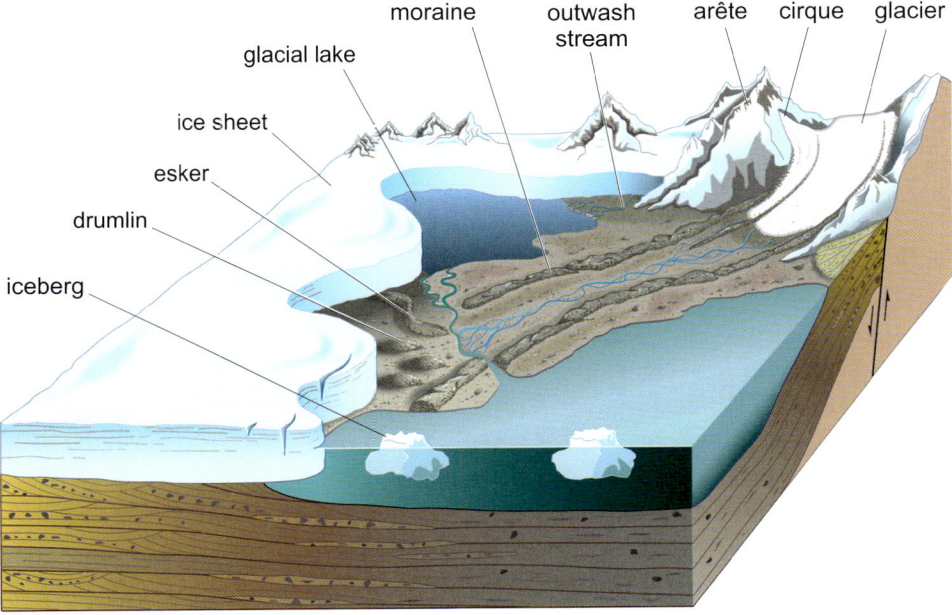

Figure 16.16 Generalised block diagram of the principal glacial environments.

In fjords, glacial sedimentation – mostly as muds and pebbly muds – is restricted to the end of the glacial cycle when ice retreats up fjord (when ice advances, any pre-existing glacial sediments are removed seawards). Sediment accumulating at the head of the fjord is prone to resedimentation by slumping and sediment gravity flows, similar to those occurring on continental slopes.

Figure 16.17 Dropstones in finely laminated mudstones, Dwyka Tillite Formation, Oorlogskloof, South Africa.

On continental shelves, ice margins construct large submarine morainal banks. The combination of ice rafting, deposition of fine-grained material from sediment plumes, and size sorting by currents, gives rise to a range of facies. **Dropstones** are particularly diagnostic, occurring as coarse-grained, usually isolated clasts that have been 'dropped' by melting ice into much finer-grained facies (Figure 16.17).

16.7 Glacial successions

Given that there are so many glacial-related sub-environments (Figure 16.16), it is difficult to encapsulate the range of facies that might be encountered in a single graphic log. Glacial melting leads to a range of glacial and glaciofluvial deposits on land, whereas transport into the marine realm produces glaciomarine facies that are often very similar to other marine facies, including bioturbated sediments, shelly horizons and turbidites. Terrestrial glaciated environments are prone to erosion, often during glacial retreat or advance, and so they are much less likely to be preserved than those accumulating in glaciomarine environments.

Terrestrial subglacial successions may consist of a series of lodgement tills that represent several phases of glacial advance and retreat, or changes of ice flow direction in a single glaciation.

- ■ What difference might you expect to see between two successive lodgement tills that represent a major change in flow direction?

- □ Since the composition of the lodgement till reflects the lithology of the bedrock over which the glacier moves, then a change in flow direction might be reflected in different clast compositions within the two tills.

A possible terrestrial succession is shown in Figure 16.18. Boundaries between successive tills are invariably irregular and erosive, and there is often evidence of deformation of the underlying till due to subglacial tectonism – likened to spreading peanut butter on very soft bread – and shearing within a particular till – comparable to spreading peanut butter on toast (where the toast surface acts as a shear plane). Gravel-filled channels result from subglacial stream flow, often with a planar upper surface created by erosion due to later ice movement.

Glacial lake successions often include coarse-grained, poorly sorted debris flows resulting from rapid deposition of suspended sediment and ice-rafted debris, coarsening-upward sand facies that reflect progradation of deltaic sediment into the lake, normally graded sands and muds deposited from turbidity currents, and rhythmically laminated varves – couplets of sand/silt and clay, representing spring/summer melt-out and winter freezing. These are commonly deformed by overlying glacial activity (Figure 16.18).

Chapter 16 Glacial environments

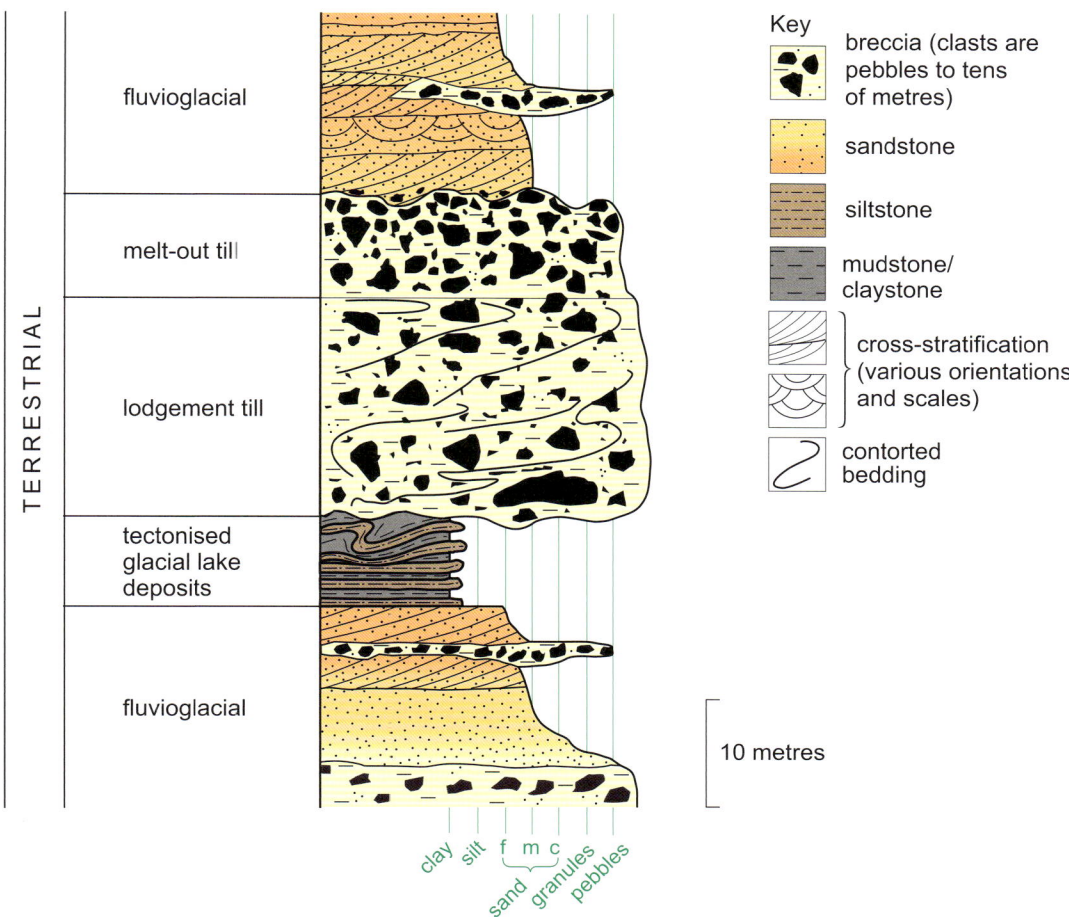

Figure 16.18 Graphic log of a typical terrestrial glacial succession.

Question 16.2

Considering features illustrated on the block diagram (Figure 16.16), what is the preservation potential of the following sedimentary features: glacial lake, esker, arête, drumlin, moraines and glaciomarine sediments? Hint: consider first whether they are erosional or depositional features.

Activity 16.1 Glacial environments

In this activity, you will consolidate your understanding of glacial environments.

16.8 Summary of Chapter 16

1. Approximately 10% of the Earth's total land area is covered with ice, mostly in polar regions, but also at high altitude in lower latitudes. Glaciers and ice sheets will grow wherever the rate of accumulation of winter snow exceeds the rate at which it melts in summer.

2. Erosive landforms are created by a combination of moving ice and freeze–thaw activity. U-shaped valleys, cirques and arêtes are the most spectacular erosional landforms, but they have little preservation potential. The most commonly preserved features are striated pavements, created by gouging of the underlying bedrock by debris-laden ice.

3. Depositional features on land consist essentially of subglacial lodgement till and melt-out till, creating irregular moraines. Many of these features are reworked by glacial meltwaters to produce fluvioglacial sediments. Gravel-filled channels (eskers) and ellipsoidal mounds fashioned from earlier outwash sediment (drumlins) are also produced beneath the ice.

4. Glacial lakes are formed by glacial erosion or dams of ice or moraine. The lakes are eventually filled with deltaic sediments, sediment gravity flows and distal varves (annually produced silt/sand and clay couplets).

5. Rock flour released by melting ice fronts is redistributed by strong winds and builds up thick loess deposits.

6. Ice cliffs or floating ice shelves release coarse debris into the marine environment to create submarine fans and distal muds that are influenced by normal marine processes. Glaciomarine sediments are complex but they have the greatest preservation potential of all glacial facies. Dropstones released by floating ice comprise isolated coarse clasts within much finer-grained material and they are highly diagnostic.

16.9 Objectives for Chapter 16

Now you have completed this chapter, you should be able to:

16.1 Describe the main erosional and depositional glacial landforms.

16.2 Evaluate the likely preservation potential of terrestrial versus marine glacial sediments.

16.3 Recognise and sketch simple graphic logs that depict the principal terrestrial glacial facies.

Chapter 17 A matter of time

In Book 1, Activity 1.1, the video sequence *Geological time* showed how geologists came to comprehend the immensity of geological time. By now, you should be used to thinking in terms of millions, tens of millions and even hundreds of millions of years when considering the span of history that can be read from rocks and their fossils. With this in mind, geologists often ask a fundamental question when standing in front of a quarry or cliff face exposing sedimentary rocks. This question is 'how long did that bed take to form?', rather than 'how long ago did it form?'

The 'how long did it take?' question usually cannot be answered by quoting a precise time measured in hours or years, but by applying the understanding of geological processes that you have gained in this book, you should be able to suggest a 'ball-park' estimate of the time it took for a given sedimentary feature to form.

Consider the two graphic logs shown in Figure 17.1. How quickly would the sandstone intervals have been deposited? In hours, days, months, years, decades, or centuries?

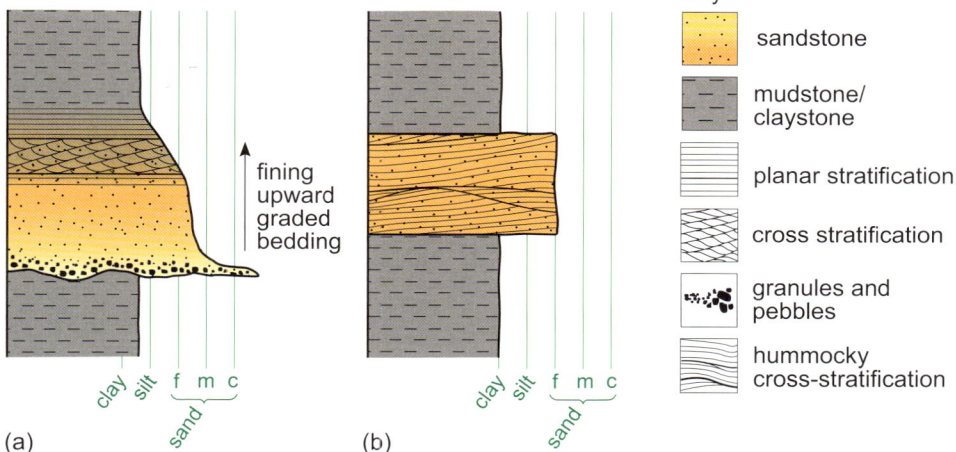

Figure 17.1 Graphic logs of sandstone units occurring between beds of mudstone. Each of these graphic logs may represent a thickness of about 1–10 m. Graphic log (a) depicts a Bouma Sequence deposited by a turbidity current. Graphic log (b) depicts a storm-generated sandstone with characteristic hummocky cross-stratification.

The turbidite succession in Figure 17.1a was deposited by a gravity-driven density current – a mixture of sediment and water. Once the turbulent part of the current had passed – and remember that turbidity currents may reach speeds in excess of 50 km h^{-1} (Section 7.6.3) – most of the sand held in suspension would have been deposited in a few hours as Bouma division A (Figure 7.42). The top part of the sand unit showing planar stratification and cross-stratification (Bouma divisions B, C and D; Figure 7.42) would probably have taken longer to be deposited, as the lower speed tail of the turbidity current may have continued to flow for a day or more. But the mud left in suspension after the flow had long gone may have taken weeks or even months to fall to the sea floor to form the Bouma division E(t). Moreover, the mud deposited from suspension as normal deep-sea background sedimentation would be deposited even more slowly, possibly over as much as hundreds or thousands of years (division E(h)).

- Consider the entire succession shown in the log in Figure 17.1b; which parts of it represent the longest periods of time?

□ By far the greatest amount of time – 99% or more – is recorded by the mudstones. This is because they were deposited very slowly from suspension over hundreds or thousands of years, whereas the sandstone with HCS is a storm bed, resulting from a single event that lasted for hours or at most a few days.

You also need to consider the time significance of the contacts between beds. If the base of a sandstone bed is erosional, as is the case with the turbidite shown in Figure 17.1a, then the contact between the sandstone and underlying mudstone may represent a long period of time during which fine-grained sediment was deposited and later removed as the turbulent head of the turbidity current passed by. Likewise, the contact of the HCS sandstone and underlying mudstone in Figure 17.1b will represent a long period of time if storm-generated currents eroded muddy sediments before the deposition of the sandstone.

Figure 17.2 shows the time duration of some of the sedimentary features described in this book, ranging from individual laminae, to successions of sediment deposited in depositional environments such as river systems, deltas and submarine fans.

- Do successions of sediment deposited over long periods in major sedimentary environments (such as those on the SE Poster) contain a continuous record of the past?

□ They definitely do not! There are many, often very large gaps in the record due to non-deposition or erosion.

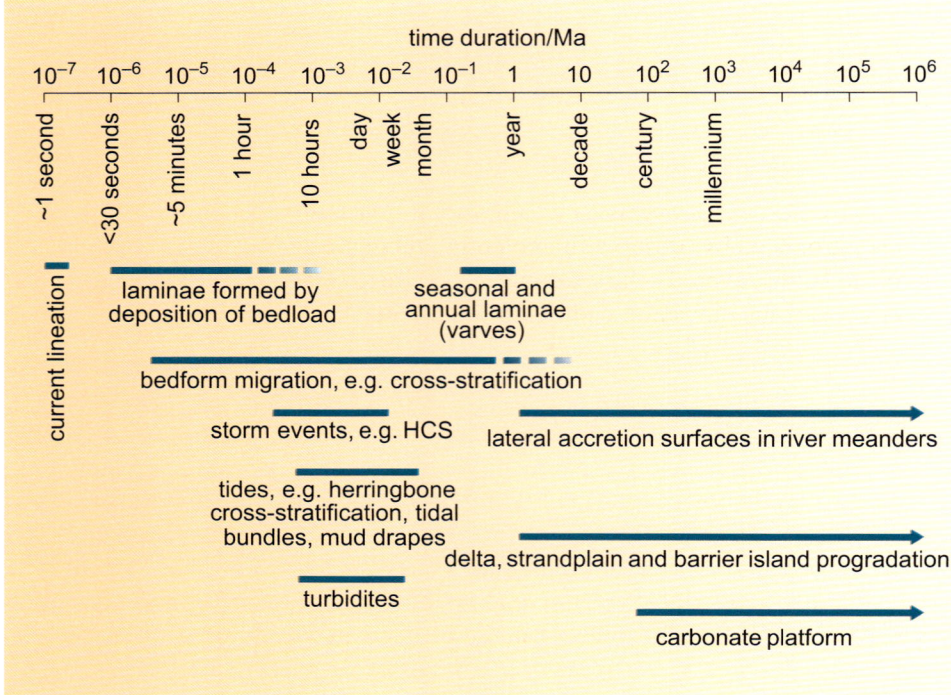

Figure 17.2 The length of time taken to form a variety of sedimentary structures and successions discussed in this book. The timescale is in years plotted with successive divisions representing a tenfold increase in time (a logarithmic scale).

As well as thinking about the time significance of beds and their bed contacts, you should also take into account the preservational potential of sediments accumulating in the different environments.

■ Where are sediments more likely to be preserved – on a modern-day beach or on a rapidly subsiding delta?

☐ On a rapidly subsiding delta, since here there is much less likelihood of erosion and removal of part of the succession, whereas on a beach, reworking and removal of sediment are commonplace.

Virtually all sedimentary successions represent the net accumulation resulting from periods of deposition, non-deposition and erosion. An environment in which steady, continuous deposition leaves behind a complete record is most unusual. One only has to think of familiar environments such as rivers and coasts. The former are subjected to seasonal changes in discharge and occasional floods, and the latter are affected by tidal currents that change daily, plus variable wave action, ranging from zero to storm force. Erosion and reworking of the underlying sediment is common in these environments. Even in the depths of lakes and oceans, where a continuous rain of suspended sediment might be expected, the resultant sedimentary records can be interrupted by deep-water currents such as turbidity flows.

Figure 17.3 Deer (centre) and bird footprints (bottom right) in Miocene siltstones of the Sorbas Basin, Spain. Coin is 2 cm in diameter.

As well as considering how long beds took to form, it is interesting to consider rates of deposition for a much thicker succession of beds. The rate of deposition for individual storm beds can be exceedingly high, perhaps as much as 10 000 metres per thousand years. However, this enormous rate is made up of short bursts of sedimentation during individual storms (often accompanied by erosion), interspersed by much longer, fairweather intervals when little sediment is deposited – so the net rate of sediment accumulation over long periods of time ($>10^4$–10^5 years) is likely to be only a few centimetres per year or even less. In fact, the longer the time period considered, the slower the overall rate of accumulation becomes, because the rapid, intermittent events add up to very much less time than the uneventful periods in between, when little sediment is deposited.

Despite this blurring of events as we go back in Earth history, and the bias introduced by preservation potential, there are still untold marvels to be discovered – for example the entire skeletons of extinct Cretaceous dinosaurs with feathers still attached found in China (Figure 2.4), to footprints left by birds and deer at a Miocene waterhole (Figure 17.3), to the unmistakable signs of a lightning strike in a Permian desert (Figure 12.8) or of a rain shower on a Jurassic beach (Figure 17.4).

Now that you have reached the end of this course, you should be able to interpret the rock record using the observation → process → environment approach. There is plenty of evidence to work on, but it is important to remember that the rock record is usually incomplete, so you need to keep in mind the consilience approach

Figure 17.4 Raindrop prints preserved on foreshore deposits during the Jurassic Period, Dorset, England. Coin is 2 cm in diameter.

(that is the gathering together of insights from different disciplines) implicit in the following quotation:

> The past is as biased as propaganda, as misleading as a fairy tale. The geological record contains a highly distorted picture of what has happened. Of shallow marine sedimentation we can learn endlessly, but of what took place in the high mountains, on the continents, of upland forests, we can find only the faintest shadows. And some things which we may suspect have happened in the past, we may find no trace of at all.
>
> As scientific optimists, we always tend to concentrate on what we can hope to know. Our paradigm is the successful detective, painstakingly searching for clues, to solve the crime. Try focusing on what we cannot know. Put yourself in the position of the criminal evading detection. How much of geological history is an irretrievable journey, a perfect crime? …
>
> The bias of history is, of course, not just a preserve of geology. We know far more about how the Romans built their drains than how they constructed their roofs. We know far more about prehistoric cemeteries than prehistoric nurseries. Far more of treasure hoarders than of treasure spenders, far more of the literate than illiterate. It is the job of geohistorians to try to make amends for these distortions; to understand, for example, how and where catastrophic events can freeze and preserve the most ephemeral, the most fragile the most vulnerable: like the buildings, possessions, even the fleeing inhabitants of Herculaneum, preserved by the eruption of Vesuvius.
>
> (Extract from 'Editorial', *Terra Nova*, vol. 8, pt. 5, p. 398, 1996)

Answers to questions

Question 1.1

(a) Global sea level was (i) highest in the late Ordovician and (ii) lowest around the Permian–Triassic boundary (with the late Triassic and late Neogene and Quaternary coming close). (b) Global temperature was (i) highest in the Devonian (with the late Cretaceous coming close) and (ii) lowest in the late Carboniferous and early Permian.

Question 1.2

(a) Observations

Property		Observations for RS 10
Composition	Grains	Quartz, rare white feldspar (altered at least in part to clay minerals)
	Matrix	Very little fine-grained intergranular material
	Cement	Red soft material, which is iron oxide coating the grains
		Quartz (note, however, the rock is poorly cemented, as it is easy to dislodge individual grains from the rock as you handle it)
Texture	Crystalline or fragmentary	Fragmentary
	Grain size	750 μm is the most abundant grain size (it is a coarse-grained sand)
	Grain sorting	Very well sorted
	Grain morphology:	
	shape or form	Equidimensional
	roundness	Rounded to subrounded
	sphericity	Spherical
	Grain surface texture	Frosted
	Grain fabric	No preferred orientation; grains closely packed; grain-supported
Fossils		None seen in specimen
Sedimentary structures		None seen in specimen

(b) Processes

(i) The sand-sized grains indicate deposition in moderately high energy conditions.

(ii) RS 10 is well sorted and the grains are rounded and spherical, with very little finer matrix material, indicating that either the processes responsible for its deposition sorted out the grains, leaving larger and smaller particles elsewhere, or that the rock represents sediment grains derived from a previous rock in which the grains were of equal size and well rounded. It is impossible to tell from this one specimen which of these scenarios may apply.

(iii) The grains were most likely to have been deposited by wind because the surface texture is frosted.

(iv) The lack of any fossils tends to support the fact that the sediment was deposited by wind.

[*Comment*: In general, fossils tend to be more abundant and better preserved in rocks that have been laid down under water. However, the specimen that you have is very small so the chances of finding a fossil are limited; thus, you cannot exclude the possibility that there may be fossils in these rocks, and you would need to examine the exposure to be sure.]

(v) Sedimentary structures are absent. This tells you that either the size of the sedimentary structures is greater than the specimen or that there are none at all.

[*Comment*: You would need to see a larger specimen or the exposure from which the specimen came to be sure. This demonstrates that it is important to make observations at a range of scales. The lack of sedimentary structures in RS 10 means that you cannot deduce anything further about processes in this case.]

(c) Environment

RS 10 has been transported and deposited by wind processes. The most likely environments of deposition are a desert or possibly the sand dunes at the back of a beach.

Question 2.1

(a) body fossil; (b) trace fossil (of an early human); (c) chemical fossils; (d) pseudofossil (i.e. not a fossil); (e) body fossil; (f) trace fossil (the skull itself is a body fossil); (g) body fossil; (h) body fossil; (i) trace fossil.

Question 2.2

Both ammonites and the pearly nautilus inherited a gas-filled shell, conferring buoyancy (c) and hence the ability to swim (a), from a common ancestor, so these features should be considered homologous. However, coiling of the shell (b), which ensured the maintenance of stability (d), evolved independently, and so should be considered analogous.

Question 3.1

You should have assigned the specimens as follows:

coral	FS E	bivalve	FS B
ammonite	FS D	trilobite	FS G
gastropod	FS A	echinoid	FS C
brachiopod	FS F	crinoid	FS H

Question 3.2

You should be able to see the pallial line about 8 mm within the ventral margin of the shell. Notice that the pallial line is parallel to much of the ventral margin of the shell, but that at the posterior end it has a sharp inward kink.

Answers to questions

Question 3.3

Circomphalus (FS I) lived within the sediment and fed on food particles suspended in seawater that were drawn into the inhalent siphon. It was thus an infaunal suspension feeder. Though stationary most of the time, it was capable of movement, so it could be called intermittently vagrant. FS B (*Crassatella*) lived similarly (though was more shallowly infaunal). Now enter FS I and FS B in the correct box of Table 3.2 (p. 28).

Question 3.4

The structure of the skeleton can be divided into two types of radial zone on the basis of the ornamentation: (a) wide zones, each with two prominent rows of large as well as many smaller knobs, alternating with (b) narrower and more subdued zones bearing smaller knobs, flanked on each side by paired pores, looking like double pinpricks.

Question 3.5

Pseudodiadema was epifaunal, moving across the sea floor using its spines and tube feet. It was essentially a grazer and probably fed on algae or sessile epifauna. *Pseudodiadema* should have been entered in Table 3.2 (p. 28) as a vagrant epifaunal grazer, though possibly also a predator.

Question 3.6

As it passed through the indentation, exhalent water (plus faeces) would have been directed obliquely back to the left and thus away from the head, instead of forwards, directly over it, as might have happened were no indentation present.

Question 3.7

The probable currents are indicated in Figure 18.1. Both specimens illustrate the importance of maintaining a clear separation of the clean inhalent water from the oxygen-deficient exhalent water, bearing faeces.

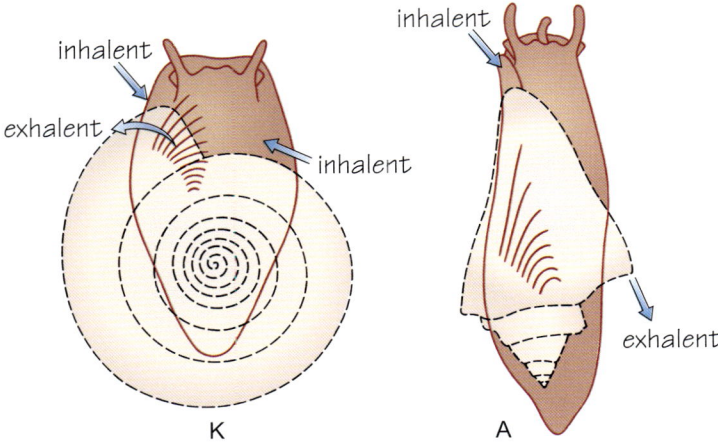

Figure 18.1 Inferred respiratory currents in FS K and A.

Question 3.8

FS K probably moved around over the sea floor. With two unelaborated inhalent current sites, it is unlikely that it could have located food accurately by smell and so it is improbable that it hunted vagrant animals. It is most likely to have fed on algae or detritus on the surface. FS A, in contrast, had a well-developed extensible inhalent siphon, so there is a good possibility that it might have been a predator or scavenger. Since it could have burrowed, it may have preyed on other burrowing organisms. (In fact, its modern relatives do burrow into sandy sediments and feed on infaunal prey such as bivalves, which they smother with the foot.) You could thus classify FS K as a vagrant epifaunal grazer (or possibly collector), and FS A as probably a vagrant infaunal predator (or possibly scavenger).

Question 3.9

You should have entered both FS F and L as sessile epifaunal suspension feeders, since both lived on the sea floor and strained off food particles from the water.

Question 3.10

You should have identified FS H as a sessile epifaunal suspension feeder.

Question 3.11

You should have entered both FS E and M as sessile epifaunal predators on microscopic prey, or even suspension feeders.

Question 3.12

Lenses face out to the front, the sides and the back of the eyes. The slight inclination of each bank of eyes also suggests that the lenses could have detected light changes above the animal. Thus, the trilobite seems to have had a field of view that went laterally all round it, and over it. The only zone not directly visible would have been beneath the animal.

Question 3.13

You should have identified FS G as vagrant, epifaunal and (probably) a predator, though it was probably also capable of scavenging when the opportunity arose. The completed version of Table 3.2 (p. 28) can be seen as Table 3.5 on p. 102.

Question 3.14

Echinoidea and Crinoidea have: (i) fivefold symmetry, (ii) a water vascular system, and (iii) an endoskeleton of many plates of calcite with sponge-like microstructure.

Question 3.15

In the Home Kit, the phylum Mollusca is represented by the classes Gastropoda and Bivalvia, as well as the Ammonoidea, which is a subclass (subdivision of class) belonging to the Class Cephalopoda. In gastropods and bivalves, the muscular foot is used for locomotion, though in ammonoids, such as ammonites, it was modified into tentacles used to grasp food and, together with the funnel, used in jet propulsion. Both gastropods and ammonoids have/had a radula, but this has been lost during the course of evolution in bivalves. Note that brachiopods are a separate phylum (see text for discussion).

Question 3.16

The slight bump on the slab represents the cast of a shallow scour hollow in the underlying mud. The scour can only just have reached the level of the network of galleries where their relief is greatest, since here the full thickness of the tubes was sand filled. The abrupt margin of this part of the trace fossil could indicate the limit of where the scour gained access to the network. The unexposed galleries beyond that limit remained unfilled by sand, and so were not preserved on the sole of the sandstone bed. Alternatively, the abrupt margin may simply represent the original edge of the network. At the other end of the trace, by contrast, the scouring action removed most of the network, such that the relief of the sand fills appears to fade gradually into the cast of the scoured surface.

Question 3.17

The trilobite first made the resting trace (*Rusophycus*) and then later left it, making a furrowing trace (*Cruziana*) across the sea floor. The sequence could not be the other way round because of the direction of the V-shaped scratch marks of the *Cruziana*.

Question 4.1

Groups with multiple skeletal elements (echinoids, crinoids and trilobites) would have been particularly prone to break-up, when the soft tissues holding them together decayed, unless they were rapidly buried (see discussion of these groups in Chapter 3). The paired valves of bivalves would likewise have been liable to separation, particularly since the ligament tends to open the shell on death as the muscles decay (Section 3.3.1). Brachiopods, except for some early Palaeozoic articulate brachiopods, generally break up less easily than bivalves (Section 3.3.5).

Question 4.2

Infaunal animals such as the bivalve (FS B) and the irregular echinoid (FS J) would have had relatively high intact fossilisation potential, provided they died in their burrows, and remained permanently entombed by sediment. In most shallow marine settings, however, the vast majority of individual infaunal organisms will not remain permanently in their burrows after death, as *net* sediment accumulation involves repeated episodes of erosion (which wash them out) as well as deposition. Occasional shallow burrowers, such as the trilobite (FS G) probably was, might sometimes have been preserved intact, though the chances of exhumation and break-up would have been greater than with fully infaunal organisms. The epifaunal forms would have tended to behave differently according to skeletal construction: the gastropod (FS K) has only a single skeletal element, and so has a relatively high probability of intact preservation, but the crinoid (FS H), like most trilobite carcasses, would have had a low intact preservation potential because of its multicomponent skeleton. Moulted trilobite exoskeletons are most unlikely to be preserved whole, in one piece.

Question 4.3

The internal mould of *Circomphalus* would show, in opposite relief to the shell interior (i.e. in positive relief), features such as the adductor and anterior retractor muscle scars, and the pallial line. You can verify this for yourself by gently pressing some softened modelling clay (or similar material) into the inside of the specimen and then withdrawing the mould thus made. (It may help to wet the shell first with detergent.)

Question 4.4

All these are taphonomic processes that might affect the fossilisation history of a brachiopod or similar organism with a skeleton made of more than one structural element.

Question 4.5

(a) Although the test is largely intact, the spines are all missing, as are also the small plates immediately around the anus (within the genital plates). The small plates around the mouth, as well as the teeth, also appear to be missing, allowing the test to be filled with sediment – although the oral surface, including the sedimentary fill, is partly obscured by encrusting worm tubes. These points were mentioned in Section 3.3.2.

(b) The most likely hypothesis is (ii). The few weeks' delay before initial burial could account for the loss of spines and teeth (Section 4.1.2), while the encrustation by worm tubes of the oral surface, including the sedimentary fill, points to subsequent re-exposure – a frequent occurrence in shallow marine environments (Section 4.1.7). In (i), an intact skeleton, with spines attached, would be expected. In (iii), by contrast, the plates of the test would also have fallen apart, while in (iv), the test would have sustained some damage, though spines could still have been attached.

Question 4.6

The durability and preservation quality of any potential fossil tends to *increase* as the grain size of the enclosing sediment decreases. Usually, the smaller the grain size, the weaker the currents involved in transport and deposition, and the less likely that the organism is damaged by fragmentation and abrasion. Small grains favour the preservation of delicate structures and fine details because the grains can fit closely around them (though compaction may occur). In addition, fine-grained, clay-rich sediments such as mudstones tend to inhibit the percolation of acidic waters that could dissolve calcium carbonate shells, and are more likely to be associated with anoxic conditions that inhibit decay and favour replacement of fine structures, for example by pyrite. Rapid, permanent burial by fine sediment in oxygen-poor environments is one of the most favourable situations for fossil preservation.

Question 5.1

The absence of any benthic fossils and of any bioturbation in the shale (shown by the preservation of the laminations) indicates that conditions were hostile at least to macroscopic organisms at the sea floor. The streamlined (nektonic) ammonites would have lived in the overlying water (Section 3.3.4), indicating that conditions there were suitable for life. The most likely explanation for the difference in conditions for life is that oxygen was depleted or absent at the sea floor but was available in the seawater. Further support for oxygen depletion at the sea floor is provided by pyrite in the sediment, which indicates anoxic conditions (Section 4.1.2).

Question 5.2

A bivalve must be aquatic, but could be freshwater (Section 3.3.1). A gastropod might be marine, non-marine or even indicate dry land (Section 3.3.3). Although most ostracods are marine, they also occupy a wide range of other environments (Section 3.5.3). Only a trilobite (Section 3.3.8), a crinoid (an echinoderm; Section 3.3.6) and a belemnite (a cephalopod; Section 3.5.1) unambiguously indicate a marine environment.

Question 5.3

With a thickness of 540 m deposited over 7.5 Ma:

$$\text{average net rate of deposition} = \frac{540 \text{ m}}{7.5 \times 10^6 \text{ y}} = 7.2 \times 10^{-5} \text{ m y}^{-1} = 7.2 \times 10^{-2} \text{ m per thousand years}$$

$$= 0.072 \text{ m per 1000 years} = 7.2 \text{ cm per 1000 years.}$$

Note that this is the average *net* rate; however, the presence of numerous gaps in the record means that any 7 cm layer taken at random may represent a shorter time (if no gaps are present in it) or a longer time (if gaps are present). In fact, the chance that such a layer represents close to a thousand years is really rather small.

Question 5.4

The valves are gaping open somewhat. Had the brachiopod been buried alive, the shell would be tightly closed. The shell of the dead animal therefore either lay freely at the surface or became temporarily re-exposed.

Question 6.1

Olivines are isolated groups; pyroxenes are single 1D chain structures; amphiboles are double 1D chain structures; micas are 2D sheet structures; feldspars and quartz are 3D framework structures.

Question 6.2

As K-feldspar is far less resistant to chemical weathering than quartz, but is present in the sandstone in a fresh, unweathered condition, then not much chemical weathering can have taken place. However, weathering must have been sufficient to decompose any other minerals that may have been present in the source rock. The presence of abundant quartz and K-feldspar would suggest a granitic (or possibly a metamorphic) source, so any micas present in the parent rock must have been removed from the site of deposition (probably because they are of very low density, and would easily wash away).

Question 6.3

The basalt and the granite are likely to show a surface coating of iron oxide, because they contain mafic minerals (albeit in very small quantities in the granite). Chemical weathering of these minerals releases Fe^{2+} ions, which immediately oxidise to Fe^{3+} ions and are deposited as ferric oxide. The quartzite would be composed almost entirely of quartz and therefore would not contain mafic minerals.

Question 6.4

(a) Physical weathering is by far the dominant process in Figure 6.1, which shows a mountain slope thickly covered by very large, shattered rock fragments. There is no sign of any fine-grained sediment, soil or vegetation. The evidence in Figure 7.19 suggests that chemical weathering is the dominant weathering process. The photograph records a well-vegetated, verdant landscape. There must be good soil cover, sufficient to support arable farming (note the fields in the right background) as well as small trees and shrubs (left background). There are no signs of rock exposures and the low hill profiles are rounded, not jagged.

(b) The rock exposures in Figure 6.1 are well bedded and/or jointed, allowing good access for percolating water, aiding frost shattering. The soil waters of the well-vegetated soils in Figure 7.19 are likely to be more acidic than rainwater because of the release of organic acids, which increase the rate of chemical weathering of rock fragments and mineral grains in the soil (as well as the bedrock beneath the soil).

Question 6.5

See Table 18.1.

Table 18.1 The end products of the chemical weathering of the main groups of igneous silicate minerals (completed Table 6.2).

Mineral group	Products in solution	New materials	Residual minerals
olivines	metallic ions, silica	ferric oxide, clays	none
pyroxenes	metallic ions, silica	ferric oxide, clays	none
amphiboles	metallic ions, silica	ferric oxide, clays	none
biotite mica	metallic ions, silica	ferric oxide, clays	none
muscovite mica	metallic ions, silica	clays	none
feldspars	metallic ions, silica	clays	none
quartz	a little silica	none	quartz

Note, however, that weathering does not always proceed to completion, so micas (especially muscovite) and feldspars may be found in some sedimentary rocks.

Question 6.6

(a) The gabbro (RS 19) contains a little olivine, but comprises mainly pyroxene and plagioclase feldspar. The quartzite (RS 5) is composed entirely of quartz.

(b) The olivine and pyroxene in RS 19 should chemically weather to produce metallic ions (including Fe^{2+}) and silica in solution, and new clay minerals. The Fe^{2+} would be oxidised immediately to form insoluble ferric oxide (new material). Chemical weathering of the plagioclase feldspar would produce metallic ions and silica in solution, and new clay minerals would also be formed. Chemical weathering of the quartzite (RS 5) can only lead to the release of quartz grains, by solution along grain boundaries, and the production of a little silica in solution. The quartz grains would not decompose and so would form an insoluble residue.

Question 7.1

Frosted grain surfaces would be the most important evidence of an aeolian origin, together with very well rounded and sorted grains and a lack of marine fossils. Marine sand waves might also be expected to possess good sorting, but grains should be glassy and less well rounded (due to the cushioning effect of water). It is important to bear in mind that frosting and roundness could be inherited from the source rock of the sediment, and that a lack of marine fossils could be due to post-depositional solution rather than to a terrestrial environment lacking marine fossils.

Question 7.2

(a) The lower flow regime is characterised by planar deposition, ripples and dunes; these will give rise to planar stratification and two scales of cross-stratification (Figures 7.21 and 7.22).

(b) As the current speed slows down, planar deposition (upper flow regime) would form first, and then current-formed ripples (Figure 7.22).

Question 7.3

As the current slows down, the coarsest-grained sediments will be deposited first because they are heavier. The succession from the base upwards would be pebbles, granules, coarse-grained sand, fine-grained sand, with clay at the top.

Question 7.4

As shown in Figure 18.2:

(a) The water current speed required is about 0.3 m s^{-1}.

(b) When the grain begins to move, it will be transported as bedload.

(c) The grain would be transported in suspension.

(d) The current speed must decrease to about 0.12 m s^{-1} before the grain can be deposited from the bedload.

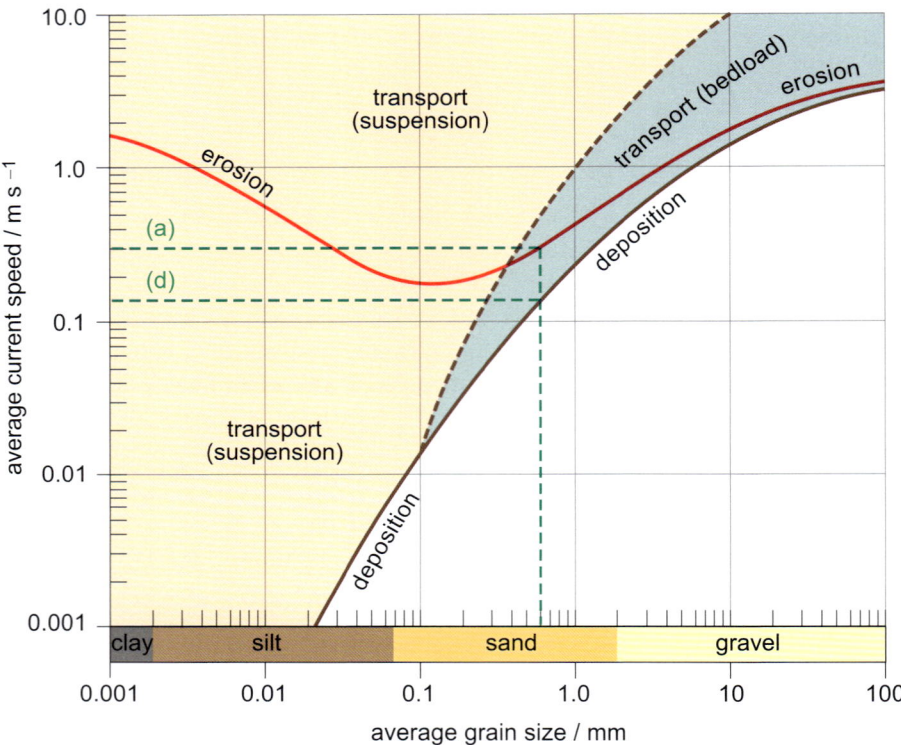

Figure 18.2 Graph for a water depth of 1 m showing the range of average current speeds at which sediment grains are set in motion, and at which they are transported and deposited. Answer to Question 7.4(a) and (d).

Question 7.5

According to Figure 7.22, initially there would be no sediment movement, but between 0.5 m s^{-1} and 0.6 m s^{-1} there would be a gradual development of lower flow regime planar bedforms, changing gradually to dunes (subaqueous), and then changing gradually to antidunes (Figure 18.3).

Answers to questions

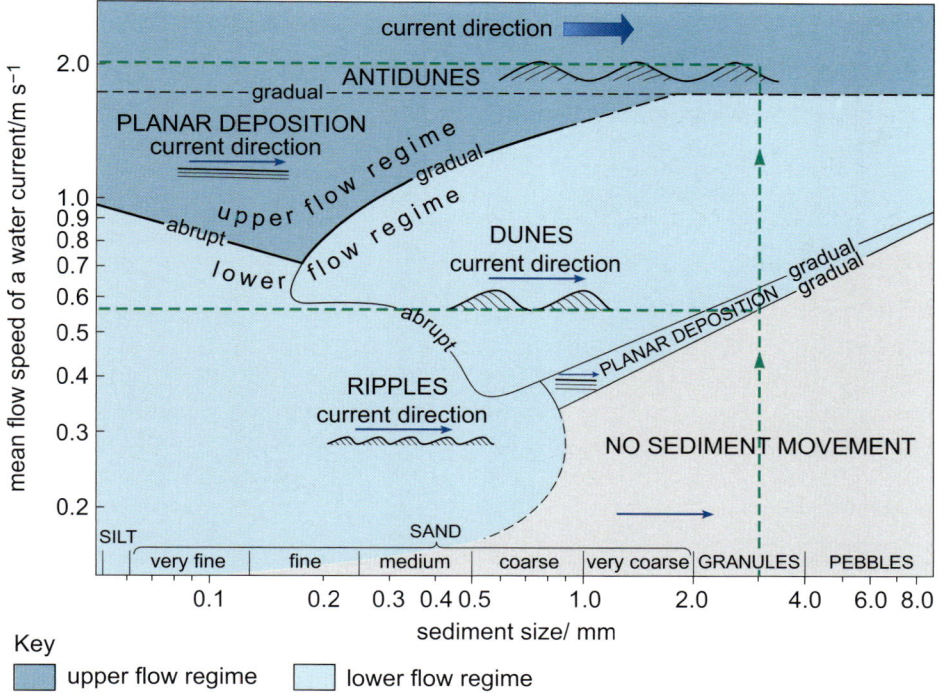

Figure 18.3 The relationship between current speed, grain size and subaqueous bedforms produced beneath a flow of around 0.25–0.4 m s⁻¹. Answer to Question 7.5.

Question 7.6

(a) As cross-stratification produced in the lower flow regime is inclined in the direction of current flow, the current must have flowed from right to left.

(b) The current speed must have varied because the cross-stratification (representing deposition as subaqueous dunes) is interbedded with planar stratification. Subaqueous dunes and planar deposition occur at different flow speeds (see Figure 7.22).

Question 7.7

(a) The ripple set that runs vertically in the photograph shows a symmetrical pattern and is therefore wave generated. The set that runs horizontally in the photograph appears to be asymmetrical in shape, with the steeper, downcurrent slopes in shadow. This second set is the product of unidirectional currents.

(b) The wave-generated, symmetrical ripples formed first, since the asymmetrical ripples have reworked (and cut across) some of the symmetrical ripples.

(c) The steeper downcurrent slopes of the asymmetrical ripples indicate a current direction from top to bottom of the photograph. The direction of wave motion will be perpendicular to the ridge crests of the symmetrical ripples, i.e. from right to left and left to right.

(d) There are boot marks (bottom left-hand corner, and left margin) and lugworm casts (scattered across the photograph). Both features are examples of bioturbation and they are both trace fossils – in the first case, the photographer, in the latter, burrowing worms.

Question 7.8

A Current ripples with an asymmetrical shape in cross-section.

B Current ripples or possibly climbing current ripples that build upwards as well as migrate laterally.

C Wave ripples with a symmetrical shape in cross-section.

Question 8.1

During this time interval in the Triassic, Cheshire must have been within some kind of marine basin that had restricted access to the open sea, so that evaporation and periodic replenishment of the water could take place. The climate must therefore have been very hot and arid, indicating a more tropical latitude than at the present day.

Question 8.2

Since $CaCO_3$ (and CO_2) is less soluble in warm water than in cold water, its precipitation would be favoured in lower latitudes.

Question 8.3

Ooids are precipitated in warm, shallow water. So you might infer that Britain was significantly warmer during the Jurassic than the present day. The presence of cross-stratification confirms that currents were active and water sufficiently shallow for such bedforms to be produced on the sea floor.

Question 9.1

(a) Muddy sandstone: clay and quartz, with quartz forming the highest proportion.

(b) Sandy limestone: quartz and calcite, with calcite forming the highest proportion.

(c) Calcareous mudstone: calcite and clays, with clays forming the highest proportion.

Question 9.2

Plotting the compositions on Figure 18.4 is tricky because of the distortions created by the 3D portrayal of the ternary plot. However, your answers should broadly correspond with those plotted on Figure 18.4.

(a) (i) subarkose (point A on Figure 18.4)

   (ii) lithic arenite (point B on Figure 18.4)

   (iii) arkosic arenite or arkose (point C on Figure 18.4).

(b) Of these three arenites, the subarkose would be the most compositionally mature, because it contains the most stable minerals (80% quartz) and the least amount of unstable components (20% feldspar + rock fragments).

(c) (i) quartz wacke (point D on Figure 18.4)

   (ii) feldspathic wacke (point E on Figure 18.4).

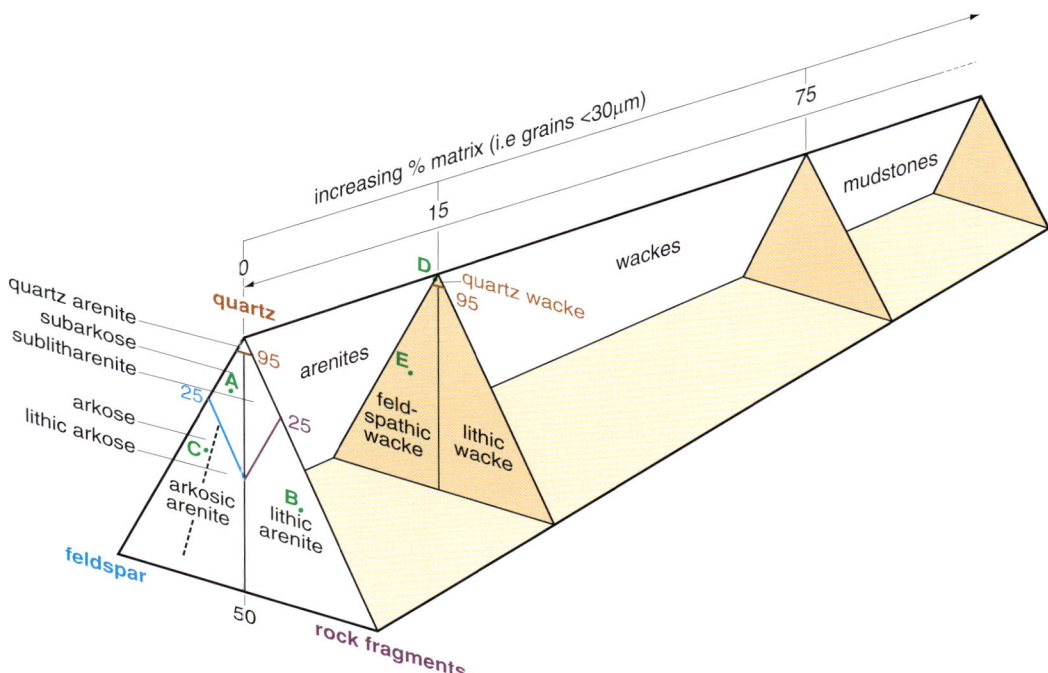

Figure 18.4 Sandstone classification using a QFR diagram. Answers to Question 9.2.

Question 9.3

(a) The main types of carbonate grains are bioclasts, peloids and ooids.

(b) The intergranular materials are micrite (either matrix or rare cement) and sparite (cement).

Question 9.4

Grains are typically in point contact (some outside the plane of the thin section), suggesting only minor compaction. It is likely that cementation took place before significant compaction, so that the cement prevented closer packing of grains during burial compaction.

Question 9.5

(a) *Claystone* or *mudstone*: the majority of the constituents are clay minerals, so that it is a claystone or mudstone (since the term mudstone incorporates clay-grade sediment; see Figure 9.17). The presence of bioclasts provides the qualifier bioclastic claystone or bioclastic mudstone. [The term shelly claystone/mudstone is sometimes used.]

(b) *Muddy limestone* or *muddy biomicrite*: the major constituents are calcitic (bioclasts + micrite), therefore it is a limestone, or biomicrite. It also contains significant quantities of clay, so this would provide the qualifier muddy limestone or muddy biomicrite.

(c) *Conglomerate*: the dominant component is rounded clasts of gravel grade.

(d) *Medium subarkose*. It is not an arkose because it contains <25% feldspar (Figure 9.17).

(e) *Pelmicrite* (see Table 9.1).

(f) *Quartz arenite*, since the grains consist almost exclusively (>95%) of quartz.

(g) *Arkosic arenite* or *arkose*, since the grains comprise >25% feldspar (stained yellowish-brown).

Question 10.1

They are both limestones, but their lithology and fossil content are distinctly different, so they represent different facies. RS 21 is a biomicrite with gastropods and bivalves, and RS 24 is a biosparite full of crinoid fragments.

Question 10.2

(a) The red dashed line X–X′ represents a time plane when all four facies co-existed. It marks the position of the beach and adjacent shallow subtidal surface.

(b) The lines labelled A–A′–A″, B–B′–B″, C–C′–C″ and D–D′–D″ represent individual facies boundaries.

(c) The red dashed lines W–W′, Y–Y′ and Z–Z′ represent the former positions of the top of the beach; like X–X′, they are also time planes, Z–Z′ being the oldest and W–W′ being the youngest.

Question 10.3

For RS 10 and RS 20, your specimen may have a slightly different grain size than ours, but this is what we found: RS 10 has a grain size of about 750 µm and, because it falls in the middle of the grain-size range for coarse sand grade on the graphic log, it would plot exactly on the coarse sand-grade line of the graphic log. RS 20 has a grain size of about 250 µm; it would therefore plot halfway between the fine sand- and medium sand-grade lines. RS 27 has no visible grains in it, and is below the resolution of the grain sizes shown on the grain-size scale. It is in fact a mudstone and would therefore plot to the right of the clay line on the graphic log (since mud includes both silt and clay).

Question 10.4

RS 21 is a biomicrite, since it contains bioclasts in a micritic matrix. Most of the bioclasts are in contact with each other, so it would plot on the line labelled 'micrite + grains, grain-supported'. RS 22 is an oosparite so it has no matrix; it would plot on the 'grains, no micrite' line.

Question 10.5

If an area of land is subsiding and sea level is rising, it is likely to produce less sediment because the rivers have less material to erode due to the decrease in the elevation of the river profile and the reduced land area.

Question 10.6

If the climate becomes cooler and wetter: (a) the area of evaporite deposition will become less and eventually diminish because evaporites require warm, dry conditions for their formation; (b) a siliciclastic coastline adjacent to a poorly vegetated land mass will receive more sediment because the increased run-off of water from the land will transport more sediment down to the coast.

Question 10.7

(a) A completed graphic log is shown in Figure 18.5.

(b) Beds A and B both contain marine faunas. This suggests that they were deposited in a marine environment. Bed C contains frosted sand grains, which are indicative of wind transport. The large-scale cross-stratification is consistent with an aeolian environment. Bed C is therefore most likely to represent a non-marine environment.

(c) The contact between Bed B and Bed C is sharp and could be termed an unconformity (see Book 1, Section 10.3.3).

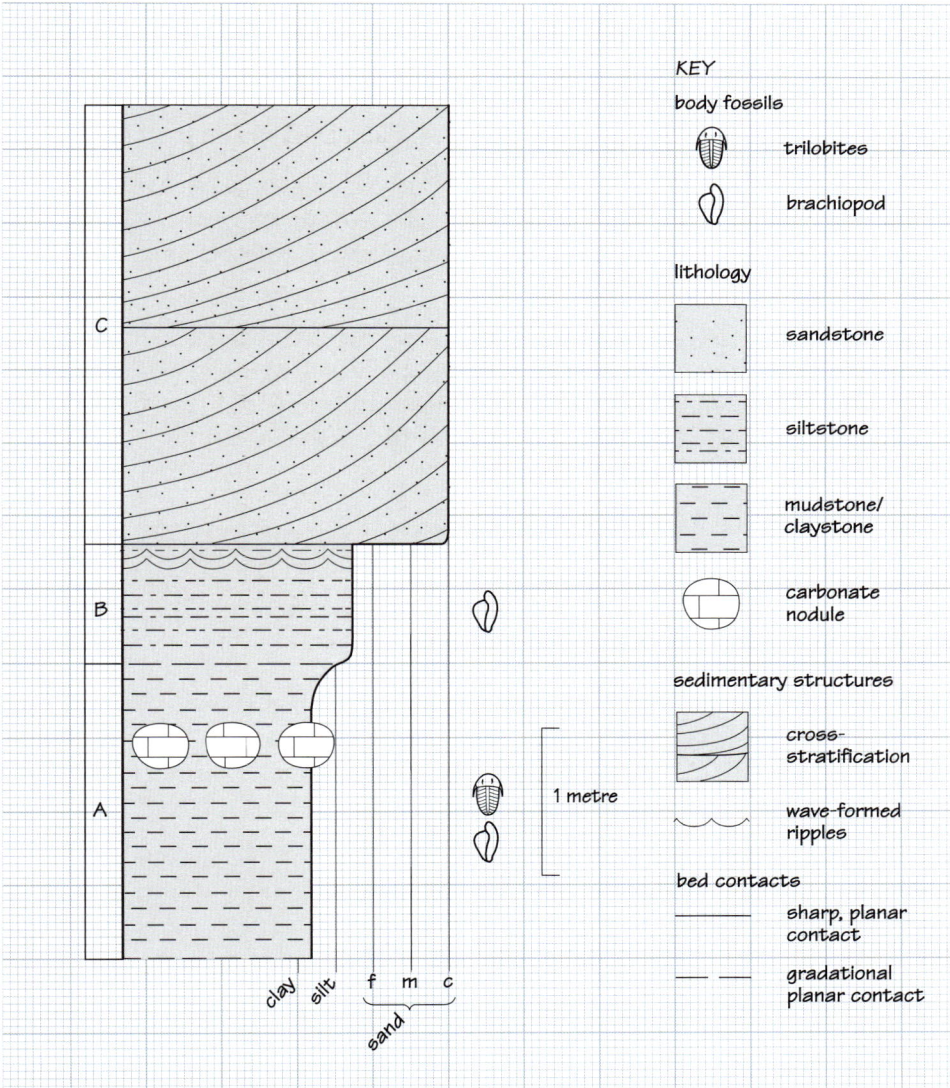

Figure 18.5 A completed graphic log for use with Question 10.7a.

Question 11.1

(a) If precipitation shows seasonal peaks, then the level of the water table will also show seasonal variations and would not remain constant throughout the year. When the seasonal variations are marked, then streams may only flow intermittently, or else start to flow higher up the valley in the wet season than in the dry season, leaving the upper part of the valley dry during the dry season.

(b) Rocks with a large amount of interconnected pores (highly permeable) will allow more infiltration than rocks with little interconnectivity (low permeability). The water table will thus lie closer to the surface when the drainage basin is underlain by such rocks as impermeable granite (which has very few pores) or clay (which has poorly connected pores), than when the underlying lithology has abundant connected pores, such as coarse-grained and poorly cemented sandstone. Hence the water table would not remain at a constant depth below the ground surface.

Question 11.2

(a) Thick flood plain suspension deposits are not very common because, at times of high discharge, further braiding or switching direction of the channels will occur rather than flooding, and whatever flood deposits are formed are liable to be eroded subsequently.

(b) Organic debris will be rare because sand bars and channels change location frequently, which prevents vegetation becoming established. Many of the locations that favour braided stream development (e.g. semi-arid regions and glaciated highlands) are climatically unsuited to extensive vegetation.

Question 11.3

(a) Meanders gradually migrate down river, so by tracing the successive positions of a meander, such as the one shown at the top of Figure 11.21, we can work out that channel D is the oldest (1765), channel B is 1820–30, channel C is 1881–1893 and channel A is 1930–1932.

(b) Feature X is an ox-bow lake. [The shift in meander position between 1830 and 1881 (51 years) is greater than during any other time periods – 1765 to 1820 (55 years); 1893 to 1930 (37 years). This suggests that channel C must have cut across the neck of the meander.]

(c) Point Y will be a point bar. The sedimentary structures will include lateral accretion surfaces, trough cross-stratification and small-scale cross-stratification formed by migrating ripples.

Question 12.1

Morphological feature	Sedimentary facies
sand dunes	cross-stratified sandstone
playa lake	mudstones with desiccation cracks; evaporites
wadis	coarsely stratified conglomerates, breccias and coarse-grained sandstones
alluvial fans	coarsely stratified conglomerates, breccias and coarse-grained sandstones

Question 12.2

The features to look for in large-scale cross-stratified sandstones to determine if they are deposited by aeolian rather than tidal processes are: frosted grains, very high-angle cross-stratification, lack of marine fossils and lack of features diagnostic of tides – such as mud-draped ripples and tidal bundles.

Question 13.1

	Fluvial	Tidal	Wave
oscillatory flow			✓
unidirectional flow	✓	✓	
bidirectional flow over ~12 hours		✓	

Question 13.2

	Fluvial	Tidal	Wave
herringbone cross-stratification		✓	
current-formed ripples	✓	✓	
wave-formed ripples			✓
cross-stratification	✓	✓	✓
tidal bundles		✓	
planar stratification	✓	✓	
trough cross-stratification	✓	✓	
planar cross-stratification	✓	✓	
mud-draped ripples		✓	
climbing ripples	✓	✓	✓
hummocky cross-stratification			✓

Question 13.3

To form a strandplain, it is necessary to have an abundant supply of sand or gravel, some tidal action and strong waves (Figure 13.6).

Question 13.4

Only the 'U'-shaped tube that descends below the surface of the sediment is likely to be preserved. The worm cast will probably be washed away by the next few waves that wash over the beach.

Question 13.5

HCS is almost invariably diagnostic of the offshore transition zone and this is defined as the zone below fairweather wave-base and above storm wave-base. It is rare that HCS is preserved anywhere other than in the offshore transition zone.

Question 13.6

(a) If the mouth bar sediments are reworked by waves, diagnostic features would include wave-formed ripples and cross-stratification, indicative of deposition on both sides of the ripple.

(b) If the mouth bar sediments are reworked by tides, the sedimentary structures may include tidal bundles, herringbone cross-stratification and other evidence of periodically changing current strength, including ripples with mud drapes.

Question 14.1

RS 22 and RS 10 are both coarse grained, but in RS 22 (an oosparite), the grains have been precipitated chemically and are therefore less closely related to the strength of the transporting current. Instead, the ooids grew successive coatings of aragonite as they were rolled around the sea floor. The coarse-grained nature of RS 10 (quartz arenite), together with the good sorting and rounding, and frosted grain surfaces, suggests fairly strong wind currents.

Question 14.2

RS 18 contains feldspar, which weathers rapidly, indicating that it has not travelled a great distance from the source area. RS 21 contains delicately preserved gastropods in a micritic matrix, which suggests that this rock too, has not moved far from its source.

Question 14.3

(a) It represents a shallowing-upward succession because shallower-water facies are shown near the top of the graphic log.

(b) Successively more proximal facies overlie more distal ones, which means that the carbonate ramp system must have been prograding seawards.

Question 14.4

Using the scales on the figure, the dip on the ramp is 20 m in 3.25 km, which is equivalent to a dip of approximately 0.35°. This is very shallow indeed!

Question 14.5

A graphic log typical of progradation of the carbonate-rimmed shelf and derived from the cross-section shown in Figure 14.24 is shown in Figure 18.6. We do not expect yours to be identical to ours, but it should be broadly similar.

Answers to questions

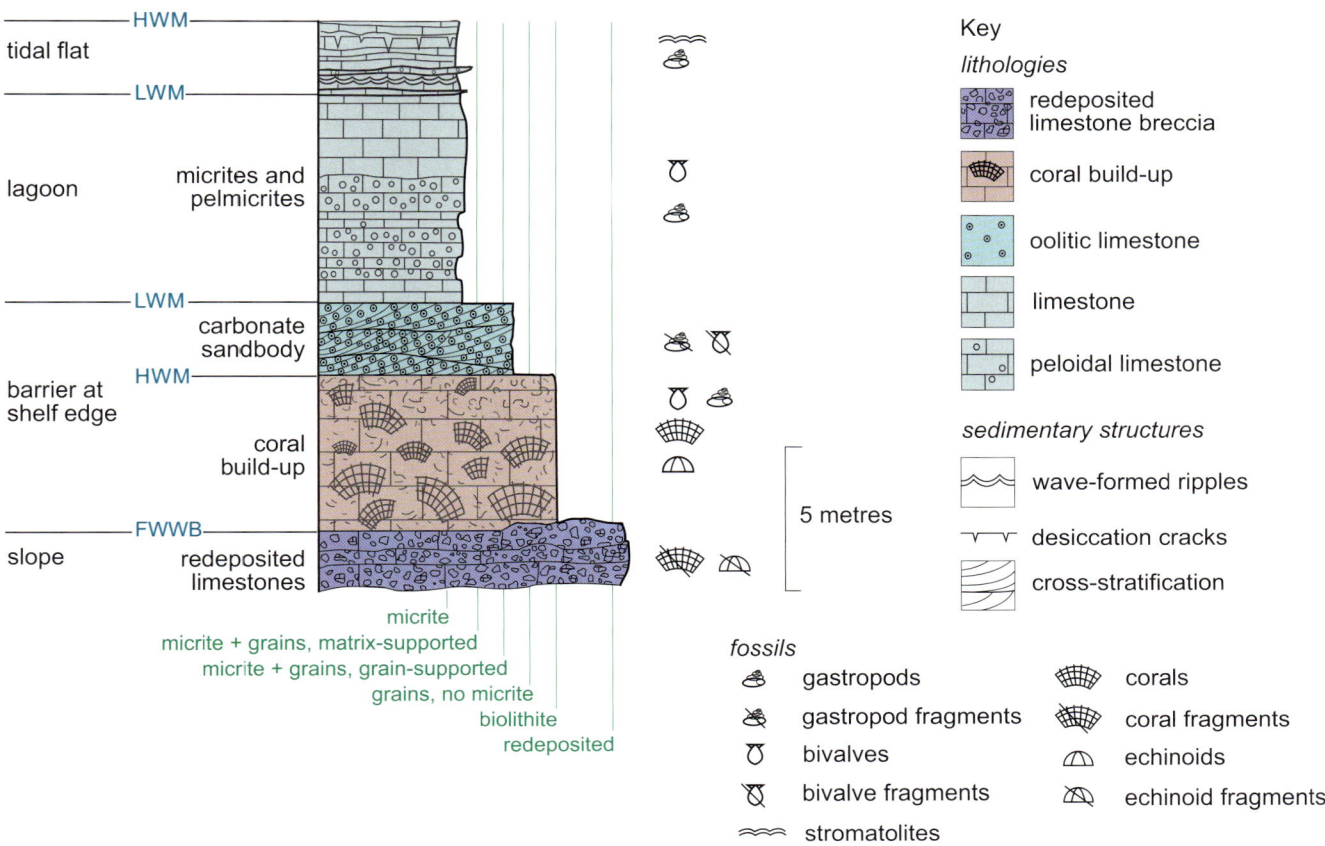

Figure 18.6 Completed graphic log of a typical prograding rimmed shelf carbonate platform.

Question 14.6

RS 21 is most likely to have been deposited in a lagoon since it contains only thin-shelled gastropods and has a micritic matrix, indicative of settling out of suspension in a low energy environment. The rock shows no evidence of strong wave or current activity because the bioclasts are not broken up. RS 22 and RS 24 are most likely to have been deposited as part of carbonate sandbodies, since they both contain grains in a sparite cement. In RS 24, the original bioclasts have been fragmented and transported, suggesting wave or current activity. Crinoids are unlikely to have lived in such high energy conditions, nor are they likely to have tolerated the abnormal salinity conditions in a lagoon.

Question 14.7

(a) The current was flowing from right to left, because the cross-stratification, representing the steep slope of the migrating bedform, is dipping to the left.

(b) Figures 14.9 and 14.10 are different types of evaporite minerals that have been formed by chemical processes.

(c) No, you would not find evaporites in the environments illustrated in Figures 14.8, 14.15 and 14.16, because all of these environments contain vegetation indicative of a humid climate. Evaporite deposition requires an arid climate.

Question 15.1

(a) Turbidity currents behave like fluid flows because all of the sediment is kept in suspension by turbulence (Section 7.6.3).

(b) Debris flows behave plastically because of the greater ratio of sediment to fluid (Section 7.6.2).

Question 15.2

(a) The sediment in a turbidity current is supported by the fluid turbulence (Figure 7.38).

(b) The sediment in a debris flow is supported by the strength and buoyancy of the matrix (Figure 7.38).

Question 15.3

The sedimentary structures shown in Figure 15.3 are flute casts (Section 7.6.3). They represent the underside of the base of a Bouma Sequence (probably division A, but it could be division B or C if the Bouma Sequence is incomplete). The current was flowing from right to left because the deepest and narrowest part of the flute (the bulbous end) is carved out first by the eddy.

Question 15.4

Divisions A–D represent a much shorter time than division E. Divisions A–D are deposited rapidly from a turbidity current, whereas division E settles out slowly. The subdivision E(t) is composed of fine-grained sediments in suspension in the turbidity current and E(h) is deposited by suspension settling from hemipelagic material that has not been carried along by the turbidity current, therefore representing the slowest deposition of all.

Question 15.5

Division E is the most likely to be bioturbated because: (i) it is the only part that is deposited slowly; any fauna around when the turbidity current was actually passing would have been swept up with it; (ii) division E is the part that is most likely to contain organic matter and, therefore, food for burrowing organisms; (iii) division E(h) represents a relative pause in deposition (see the answer to Question 15.4), so might allow sufficient time for organisms to re-colonise the sea floor, assuming there was sufficient oxygen in the bottom waters to sustain life.

Question 15.6

Only siliceous sediments will accumulate below the CCD because carbonate grains will dissolve.

Question 15.7

Depositional process	Sedimentary deposits
sediment gravity flows	turbidites, debris flows and mudflows
contour currents	contourites
pelagic deposition	fine-grained biogenic sediments
direct chemical precipitation from seawater	minerals, e.g. pyrite and manganese

Question 15.8

Submarine fans are more likely to form during sea-level fall, because the rivers on the continent and deltas will have to re-adjust their profile by eroding more sediment (Section 10.1). This will result in an increase in sediment supply to the continental shelf and hence to the base of the continental slope, to be deposited as submarine fans.

Question 16.1

Wind-blown sand is well rounded and spherical in shape and is the result of rapid grain collisions in air. Loess grains are subangular to subrounded and are often elongate in shape, i.e. they have low sphericity (Figure 16.15). Though they are transported by the wind, they are formed by frictional forces acting at the base of a glacier as it grinds the underlying rock into rock flour. During wind transport, the grains are too small to effect significant grain collisions.

Question 16.2

Glacial lake – depositional sediments such as varves, turbidites and avalanche deposits have good preservation potential, at least in the lower part, though the upper sediments may well be stripped off by later ice advances.

Esker – ribbon-like deposit of gravel filling a sub-glacial stream that may be removed during a later glacial advance.

Arête – erosional feature that will be destroyed during physical weathering of the exposed ridge.

Drumlin – depositional feature sculpted by ice that can be removed during subsequent glacial advance.

Moraines – depositional features left behind by retreating glaciers. They are likely to be reworked by meltwaters, generating fluvioglacial sediments.

Glaciomarine sediments – depositional environment with the greatest accommodation space and the best preservation potential, though reworking by tides and currents will mask the glacial origin of much of the sediment.

Acknowledgements

The authors would like to thank Bob Spicer for his contribution to the sections on fossil plants, and Helen Craggs for comments on the proofs.

Grateful acknowledgement is made to the following sources for permission to reproduce material in this book.

Figures

Cover: Andy Sutton.

Figure 1.1: www.mike-page.co.uk; Figure 1.2: AusAId/Robin Davies; Figure 2.1: Stuttgart Natural History Museum; Figures 2.2, 2.3 and 2.7c: Martin Lockley; Figure 2.4a: John Wyatt; Figure 2.4b: Xu, X. et al. (2003) 'Four-winged dinosaurs from China', *Nature*, vol. 421, pp. 335–340, 23 Jan 2003. Nature Publishing Group; Figures 2.5, 3.11a, 3.15, 3.27, 3.29b & c, 3.33b, 3.34b, 3.39a & b, 3.41a, 3.44, 3.48, 3.51b & c, 3.69, 4.3a, 4.7, 4.9 and 4.16: Peter Sheldon; Figure 2.6a: Alexei Tikhonov; Figure 2.6b: Adrian Lister; Figure 2.6c: Frédéric Plassard, Grotte de Rouffignac, France; Figures 2.7a & b, 2.8a & b, 3.10, 3.11b, 3.17, 3.25, 3.26b, 3.31, 3.34c, 3.39c, 3.41b, c & d, 3.51a, 4.2b, c & d, 4.3b, 4.8, 4.11, 5.5 and 7.5: Andy Tindle; Figure 2.8c: Professor Lee Berger; Figure 3.3: Raup, D.M. and Stanley, S.M. (1978) *Principles of Paleontology*, copyright © 1978, 1971 W.H. Freeman and Co; Figure 3.6: Wilbur, K. (1964) 'Shell formation and regeneration' in *Physiology of Mollusca*, 1, pp. 243–282, Academic Press; Figures 3.8, 3.20 and 3.22: Heather Angel/ Natural Visions; Figure 3.9: Trueman, E.R. (1975) *The Locomotion of Soft-Bodied Animals*, Edward Arnold; Figure 3.13: Smith, A.B. (1980) 'Structure, function and evolution of tube feet and ambulacral pores in irregular echinoids', *Palaeontology*, No. 23, The Palaeontological Association; Figures 3.14 and 3.35: Clarkson, E.N.K. (1980) *Invertebrate Palaeontology and Evolution*, Kluwer Academic Publishers B.V.; Figure 3.32: Kennedy, W.J. and Cobban, W.A. (1976) Special Papers in *Palaeontology*, 17, The Palaeontological Association; Figure 3.34a: Peter Wirtz; Figures 3.36, 4.10 and 8.2: © The Natural History Museum, London; Figure 3.38a: Charles G. Messing, Nova Southeastern University; Figure 3.38b: Brower, J.C. (1973) *Palaeontographica Americana*, Paleontological Research Institution, Ithaca, NY; Figure 3.42: Chip Clark, Smithsonian Institution; Figure 3.43: Sam Gon III, www.trilobites.info; Figure 3.52: Provided by Network Stratigraphic Consulting Ltd; Figure 3.54: David Siveter, University of Leicester; Figure 3.55a & b: Chris Hollis, GNS Science, New Zealand; Figure 3.55c: Mark Purnell, University of Leicester; Figure 3.56: Christa Hofmann and Reinhard Zetter; Figure 3.59, 4.13, 4.14, 11.6, 11.10 and 11.15: Bob Spicer; Figure 3.62: Jura-Museum, Eichstätt; Figure 3.63: Raup, D.M. and Seilacher, A. (1969) 'Fossil foraging behaviour: computer simulations', *Science*, 166, 21 Nov., pp. 994–995. American Association for the Advancement of Science; Figure 3.64b: Kent Chamberlain, C. (1971) 'Morphology and ethology of trace fossils', *Journal of Paleontology*, vol. 45, No. 2, The Society of Economic Paleontologists and Mineralogists and The Paleontological Society; Figure 3.66b: Basan, P.B. and Scott, R.W. (1979) 'Morphology of *Rhizocorallium* and associated traces …', in *Palaeogeography,*

Palaeoclimatology, Palaeoecology, Elsevier Scientific Publishing; Figure 3.68: Frey, R.W. and Seilacher, A. (1980) 'Uniformity in marine invertebrate ichnology', *Lethaia*, 13, Scandinavian University Press; Figure 4.1: Iowa Geological Survey Report of Investigations, No. 5, Iowa Department of Natural Resources, Geological Survey Bureau; Figures 4.2a, 5.2a & b and 14.21: Peter Skelton; Figure 4.4: Tucker, M.E. (1991) *Sedimentary Petrology: An Introduction to the Origin of Sedimentary Rocks*, Blackwell Science Ltd; Figure 4.6: Lawrence Witmer, Ohio University; Figure 4.12: Andrew Scott; Figure 4.15: Bill Crighton, supplied by Andrew Ross of National Museums Scotland; Figure 4.17: Philip Wilby, British Geological Survey; Figure 4.18: Reproduced with the permission of the British Geological Survey © NERC. All rights Reserved; Figure 5.1: McKerrow, W.S. (1978) *The Ecology of Fossils*, Duckworths; Figure 5.3: Ager, D.V. *Principles of Paleoecology*, McGraw-Hill; Figure 5.6: Fursich, F.T. and Palmer, T. (1984) 'Commissural asymmetry in brachiopods', *Lethaia*, 17, Scandinavian University Press; Figure 5.7: Margaret Collinson; see Collinson, M.E. and Hooker, J.J. (2000) 'Gnaw marks on Eocene seeds: evidence for early rodent behaviour', *Palaeogeography, Palaeoclimatology, Palaeoecology*, vol. 157, pp. 127–149; Figures 6.1, 6.8, 7.17, 7.18 and 7.19: Evelyn Brown; Figure 6.2: http://belmont.sd62.bc.ca/teacher/geology12/photos/erosion/exfoliation.jpg; Figure 6.3: Chinese International School, Hong Kong; Figures 6.4, 6.5, 6.6, 6.7, 7.9, 7.15, 7.20, 7.24, 7.25a, 7.26, 7.29f, 7.30b, 7.39, 7.40, 7.45a, b & d, 7.46a & b, 7.47, 7.48, 8.1, 8.9, 8.10, 8.11, 8.13, 9.7, 9.8, 9.9, 9.10, 9.11, 9.12, 9.13, 9.14, 9.15, 9.18, 11.3, 11.13, 11.20, 12.2, 12.8, 13.1, 13.4, 13.8, 13.10, 13.12, 13.14, 13.20, 13.27, 13.28, 13.31, 14.3, 14.5, 14.22, 14.26, 15.7, 16.7, 16.8, 16.9, 17.3 and 17.4: Fiona Hyden; Figure 7.2: From *SAND* by Siever © 1988 by W.H. Freeman and Company; Figures 7.8, 7.11, 7.14, 7.27b, 7.30a, 7.32, 7.34, 8.8, 10.5a, 11.17, 12.5, 12.7, 13.11, 13.16, 13.18, 13.19, 14.2, 14.6, 15.5, 15.10 and 15.11: Angela L. Coe; Figures 7.10, 7.35, 7.44b & c, 13.15, 13.22, 14.11, 15.3 and 15.4: Chris Wilson; Figure 7.12: The Canadian Sedimentology Research Group; Figures 7.37, 11.19, 14.8, 14.10, 14.12, 14.15 and 14.16: Jim Ogg, Purdue University, USA; Figure 7.41: Dr Jim Best, University of Hull; Figure 7.45c: Photo by Ted Nield © The Geological Society of London, www.geolsoc.org.uk; Figure 7.46c & d: Ólafur Ingólfsson, University of Iceland; Figures 8.3a and 8.4a: Roger Till; Figure 8.5: Green Electron Images; Figure 8.6: Professor Stan Awramik at University of California, Santa Barbara; Figure 8.7b: Professor Gordon T. Taylor, Stony Brook University/SF Polar Programs/National Oceanic and Atmospheric Administration/Department of Commerce; Figure 8.12: Cambridge Carbonates; Figure 9.5: Lysippos, used under a GNU Free Documentation License, Version 1.2; Figures 12.3 and 12.6: Dave Rothery; Figure 13.3: Stuart Wilding; Figure 13.5: NASA/GSFC/METI/ERSDAC/JAROS, and U.S./Japan ASTER Science Team; Figure 13.17: Glynda Easterbrook; Figures 13.24 and 16.1: NASA; Figure 14.1: Adapted from Jenkyns, H.C. (1986) 'Pelagic environments' in Reading, H.G. (ed.) *Sedimentary Environments and Facies*, Blackwell Science Ltd; Figure 14.4: Nhobgood, Used under a Creative Commons Attribution ShareAlike 3.0 License; Figure 14.9: Specimen loaned by Mike Henty; Figure 14.18: Haplochromis, used under the GNU documentation License 1.2; Figure 16.3a: The Quaternary Palaeoenvironments Group, University of Cambridge; Figure 16.3b: John Mather; Figure 16.4: Graham Hyden; Figure 16.10: Her Majesty the Queen in Right of Canada, from the collection of the National Air Photo Library with permission of Natural Resources of Canada; Figure 16.12: Walter Hunt; Figure 16.13: Michael Hambrey, Glaciers online; Figure 16.14: Jack Ridge, Tufts University; Figure 16.15: The Royal Society of Chemistry; Figure 16.17: Professor David L. Reid, University of Cape Town.

Every effort has been made to contact copyright holders. If any have been inadvertently overlooked the publishers will be pleased to make the necessary arrangements at the first opportunity.

Index

Entries in **bold** represent glossary terms. Page numbers in *italics* refer to figures and tables.

A

aboral surface 34, **36**, 37, 40, 41
abrasion 110, 111, 149, 177, 179, 186
abyssal plain *246*, 289–90, *291*
Acanthocardia 32
Acervularia 58
adaptation 13, 14, 15, 16, 34, 35, 41, 43, 53, 78, 209
adductor muscle scars 33, *34*, 35, *52*
adductor muscles 30, *31*, *32*, 33, 34, *52*, **53**, 98
aeolian bedforms 166–8, *169*, 180
aeolian dune 167–8, *169*, 171, 180, 212, 237, *239*, 253, *255*, 265
aeolian sediment 145, 148, 149, 151, 153, 237, *238*, 243
aerobic 105, 106, 121
Agnostus 63
algae 26, 27, 28, 37, *51*, 71, 73, 109, 127
    carbonate settings 184, 185, *188*, 189, 191, 210, 269, 276, 278, 279, 284
    earliest 91
    symbiotic 59, 62, 100
    toxicity 104
alluvial environments **207**, 221, 261
    alluvial fans *231*, 233, 234, 237, 242, 243, 304, *332*
    fluvial channels and sedimentary successions 223–32, 264, 266, 332
    river flow 221–2
    river valleys 221, 222–3
alluvial fans *231*, **233**, 234, 237, 242, 243, 304, *332*
amber 118, 120
ambulacra 37, 39, 41, 99
ambulacrum 37, 38, 40, 41
ammonites *18*, **21**, *24*, 67, 91, 318, 322–3
    anatomy of 46–8, 49, 59, 99
    extinctions 93, 94
    fossil formation 114, 132, 255,
    interpreting fossils of 46–51, 95, 101, 127
Ammonoidea 47, *66*, 99, 320
ammonoids 47, 67, 93, 320–1
Amoeboceras 46–51
amphibians 11, 14, 69, **70**, 92, 93, 94, 101
amphiboles 137, 184, 194, 323, 324
anaerobic 105, 106, 119, 121
analogy (in organisms) **15**–18, 48, 62, 318
Anchiornis 12
angiosperms 79
anhydrite 190, 191, 198, 199, 205, 273, *275*
annelid worms 69, 88, 89, 254
Annelida 69
Annularia 78
antennae 60, 61
anterior 30
Anthozoa 65, *66*
antidunes 157, *158*, 159, 254, 326, *327*
apatite *119*, 120, 121
aperture 42, 45, 48, 68, 99
Aptyxiella 46
Aquilapollenites **76**, 77
aragonite 36, 66
    carbon compensation depth 111
    stability of 30, 46, 47, 183, 186, 197, 205
Archaeopteryx 12
arenites 201–2, 205, 210, 328, *329*, 330
arêtes 304, *305*, *309*, 312, 337
arkose *201*, **202**, 205, 328, *329*, 330
arkosic arenite *201*, **202**, 205, 328, 330
Arthropoda 65, *66*, 74, 87, 88, 92, 100, 254, 255, 258
Articulata *66*
articulate brachiopods 52, 53, 100, 321
asexual reproduction 58
assemblage zones *96*, 97
Asteriacites 85, 89, 101
Asteroceras 18
Asteroidea 85
Athleta 42–6, 319–20
axis (of trilobite) **22**, 62

B

backshore zone **246**, *253*, 255, *257*, *259*, 265–6
backwash 253, 254
bacteria 11, 27, 62, 85, 90, 91, 116
    biological weathering 136
    cyanobacteria 75, 91, 100, 186
    decay of organisms 105–6, 118
ball-and-pillow structures *173*, **176**, 193, 292, 293, 301
barchan dunes 239, 243
barrier islands 207, 249, 257–9, 265, 297
bars 231–2, 234, 261, 263
basal apron 168, *169*
basalt 94, 324
beach
    cementation 196, 205
    facies 210, *211*, 212, 213
    fossils 68, 110
    sediment movement 143, 153, 167, 304
    siliciclastic coasts *245*, **246**, 251, 252–5, *257*, *259*, 265
beachrock 196, 205
bed contacts *214*, 215, 218, 315, 331
bed thickness 215
bedded chert 189, 191, 298, 299
bedding 154, *157*, 179, *285*, 299
bedforms 154
    aeolian bedforms 166–8, *169*, 180
    carbonate bedforms 188, 191
    tidal bedforms 161–2, 179, 258
    unidirectional current bedforms 155–60, *161*, 179, 232, 325, 326, *327*
    wave bedforms 163, *164*, 165–6
bedload 144, 145, *147*, 148, 149, 154, 166, 179
    alluvial environments 222, 234
    deltas 261, 262–3
belemnites 66, *67*, 100, *109*, 127, 284, 323

341

benthic 26, 53, 68, 72, 73, 74, 97, 100, 104, 126, 255, 263, 322
benthos 26, 71, 127
berm 246, *253*
Beyrichia 74
Big Five (mass extinctions) 92, **93**–4
bilateral symmetry 21, **22**, *24*, 29, 34, 35, 36, 41, 99, 100
bioclasts 183, 185, 190, 274, 277, 278, 329, 335
biodiversity 12, 16, 91, 92, 93, 101
bioerosion 187, 191, 268, *269*, *271*, 286
biogenic sediments 183, 190, 298, 337
   biogenic carbonate grains 184–5
biolithite 203, *204*, 206, *215*, 217, *282*, *335*
biological classification 14, 26, 28
   bodyplans and phyla 63–6
   classifying body fossils 19–23, *24*, 318, 319
   hierarchical 63–4
biological weathering 136, 139
biomicrite
   sedimentary rock classification 203, *204*, 206, 329, 330
   shallow-marine carbonate environments 273, 274, *275*, 276, 280, *281*–2, 284, 286
biosparite 203, *204*, 206, 274, *275*, 277, 286, 330
biostratigraphy 12, 64, 72, 73, 74, 75, 81, **95**–8, 101
biostromes 279, 286
biota 1, 6, 8, 19, 92
   coastal environments 254, 255, 276
   deep-sea environments 300
   sedimentary environments 209, 210, 218, 268, *269*, 270
biotite mica 137, 138, *324*
bioturbation 82, 127–8, 133, 177, 180, 255, 262, 263, 322, 336
biozones 72, **95**, 96–7, 101, 102
bipolar 161
birds 12, 14, 16, 69, **71**, 76, 92, 100, 108, 315
bivalves *21*, **22**, *24*, *25*, 64
   anatomy of *32*, *33*, 52, 53, 65, 74, 98–9, 320–1

carbonate coasts 274, 276
diversity 92
extinctions 93, 94
fossil formation 104, 105, 124, 125, *126*, 183, 321
interpreting fossils of 29–35, *45*, 95, 98–9, 129, *130*, 319, 323
siliciclastic coasts 254, 255, 258, 263
trace fossils of 85–6
Bivalvia 64, *66*, 320
body fossils 9, 11, 16, 318
   classifying 19–23, *24*
   see also individual groups
bodyplan 16, 41, 57, 63
   recognising 64–6, 85
boring organisms 35, *45*, 81, 82, *90*, *108*, 109, 110, 121, 122, *126*, 128, 197
Bouma Sequence 172, *173*, 174, 180, 181, 313–14
   deep-sea environments 290, 292–6, 301, 336
   glacial environments 308
brachial valve *52*, 53, 100, 130
Brachiopoda 65, *66*
brachiopods *21*, **22**, *24*, 26, 65, 127
   anatomy of *52*, 100
   carbonate coasts 277
   diversity 92
   extinctions 93, 94
   fossil formation 104, *109*, 112, *115*, 124, 255, 321, 322
   interpreting fossils of 51–4, 95, *113*, 130–2, 323
braided river systems **207**, 224, 230–2, 234, 332
braiding 223, 224, 230, 234, 332
breccias *200*, 217, *282*, *311*, *332*
bryozoans 68, 100, *109*, 184, 197, *204*, 269, *270*, 279
budding 58, 100
buildups 277–80, 286, 292, 296
burrowing (and burrows) 9, 14, *25*, 26, *80*–5, *88*–*90*, 101, 104, 105, 124, 127, 128, *176*, 180, *189*, 199, 226, 255, 300, 321, 327, 336
   bivalves *32*, 33, 34, 35, 98, *126*, 129
   brachiopods 52

echinoids 40, 41, 99
gastropods 45, 320
trilobites 62

C

Calamites 78
calcite 23, 30, 36, 46, 47, 52, 54, 60
   belemnites 66
   carbonate compensation depth 110
   cement 185, *187*, 195–8, *204*, 278
   coccoliths 73
   echinoderms 99
   nodules *198*, 199, 205
   permineralisation 115, 117
   stability of 138, 183, 205
   trilobites 100
   see also **micrite**
calcium carbonate 30, 68, 110, 111, 115, 183, 184, 190
calcium phosphate 75, 111, 115, *116*, *119*–*20*
calcrete 196
Calymene 62
Cambrian explosion 92, 101
carbon cycle 75
carbonaceous sediments 183, 189, 190, 191, 223
carbonate compensation depth 110–11, 183, 190, 299, 300, 336
carbonate platforms 207, 272, *276*, 277
   facies models 280–1, *282*–*3*, 284–5, 286
carbonate rocks
   cement 194, 195–6
   chemical weathering 138
   classifying 202–4, 206, 329
   graphic log *215*
carbonate sediments 183, 190, 328
   bedforms 188, 191
   biogenic carbonate grains 184–5
   carbonate mud 186–7, 191
   carbonate ooids 186, 190
   factors affecting deposition 268–71, 286
   siliciclastic sediments, compared to 267, *268*, 271–2, 286
   textural maturity 187, 191
   see also shallow-marine carbonate environments

carbonic acid 138, 140
cast (fossil) **112**–14, 121
catastrophic burial 104, 105, 124, 129, 133
Catenipora 58
CCD 110–11, 183, 299, 300, 336
cementation
    in limestones 197–8
    in sandstones 194–7, 205, 238
Cenoceras 47
cephalon 22, 62, *63*, 100
Cephalopoda *66*, 320
cephalopods 46–9, 51, 66, 93, 98, 99, 100, 120, 127, 255
Cerastoderma 32, *128*
Chalk 72, 74, 114, 184, *187*, 190, *284–5*
channel fills *228*, *232*, 234, 264, *265*, *297*, *298*
channels 150, 179
charcoal, fossil *116*, 121
chemical fossils 11, 16, 91, 318
chemical sediments 183, 186, 190
chemical weathering 136, 183, 184, 190, 324
    carbonate rocks 138
    rates of 139, 140
    silicate rocks 136–8, 201
chemostratigraphy 98
chert *see* **bedded chert; nodular chert**
Chicxulub Crater 94
Chondrites 80, *176*
chordates 67
cilia 27, 33, 43, 53, 55
Circomphalus 29–35, 36, 37, 53, 111, 319, 322
cirques 304, *305*, 306, *309*, 312
classes 63, 64, *66*, 69, 102
clay minerals 137, 146, 179, 194, 196, 201, 205, 300
claystones
    alluvial environments *229*
    compaction *193*
    graphic logs *214*, 215, *331*
    sedimentary rock classification *200*, 201
Climacograptus 67

climate change 8, 73, 75
climbing ripples 160, *161*, 173, *175*, *265*, 333
Cnidaria 65, *66*, 100, 125, 255
coal
    formation 189, 191, 223, 259, 260, 262, *264*
    fossils and 8, 12, 79
coastal erosion and deposition 1, 2, 3
coastal environment *see* siliciclastic coastal and continental shelf environments
coccolithophores *and* coccoliths 71, 72, 73–4, 94, 95, 100, 101, 184, *185*, *187*, 190, 210, 299–300, 301
coccospheres 184, *185*
cockles 32, *128*, *130*
Coelodonta 70
collagen 36, 56, 67, 110
collectors 27, *28*, 35, 49, 86, 99
colonial 20, 58, 59, 67, 68, 100
colony 58
compaction 120, 121, **193**–4, 204–5, 329
compositional maturity 137–8, 140, 153, 201
compound eyes 60, 65, 100
compression fossil 116–17, 121
computed tomography (CT) scanning 112, *113*
concretions 115, *116*, 198, 205
concurrent range zones *96*, 97
conglomerates 177, *200*, 201, 205, 217, 295, 329, *332*
Coniopteris 79
conodonts 75, 100
consecutive range zones *96*, 97
consilience 5, 14, 91, 315–16
continental rise 246, **289**–90, 291, *296*
continental shelf 12, 172, **245**, *246*, 289
    barrier islands 207, 257–9, 265, 297
    deltas 207, 248, *249*, 260–4, *265*, 266, 297
    fluvial, tidal and wave processes 247–52, 333
    strandplains 207, 218, 249–50, 252–7, 258, 265, 297, 333

    subdivision of coastal and continental shelf environment 246–7
continental slope 172, 246, 247, 263, **289**–90, 291, *296*, 308, 337
contourites 291, 337
convergent evolution 15
coprolite 81, **118**–19
corallites 20, 58
corals 20, *24*
    anatomy of *57*, 100
    classification 65, *66*
    diversity 92
    extinctions 93
    interpreting fossils of 57–9, 95, 127
    sediment formation 184, 268, *269*, 270
    symbiosis 59, 277
    see also reefs
Corollithion 72
crabs 25, 59, 65, *81*, 127, 254, 255
Crassatella 29, 319
crayfish 60
Cretalamna 17
crevasse splay 224, **226**, *228*, 229, 234, 261
crinoid 23, *24*, 66
    anatomy of 55–6, 64, 99
    extinctions 93
    fossil formation 104, 105, 320, 321
    interpreting fossils of 54–7, 95, 109, 127, 323
Crinoidea 64, *66*
crocodiles 92, 94
cross-sets 160, 161, 168, 180
cross-stratification 82, **154**, 156, *157*, 159, *169*, 179, 327
    alluvial environments 228–9
    coastal environments *252*, 255, 259, 263, 270, 334
    deep-sea environments 292, 301
    deserts 238, *239*, 240, 333
    facies *211*, 212, *213*
    see also **herringbone cross-stratification; hummocky cross-stratification; planar cross-stratification; swaley cross-stratification**
crustaceans 49, 60, 62, 65, 71, 74, 85, 87, 88, 89, 100, 106, 124, 185, 263, 274, 276

Cruziana 86–7, 88, 89, *90*, 101, 321
current-formed ripples 155, 156, 175, 228, 231, 254, 325, 333
current lineations 157, 158, 254
currents
    deep-sea 290, 291, 300
    longshore currents 156, 251, 265
    palaeocurrents 159, 160, 179
    rip currents 251, 265
    see also **turbidity currents**
cyanobacteria 75, 91, 100, 186
cyclostratigraphy 98
Cyrtograptus 67
cystoids *105*

D

debris flows
    deep-sea environments 289, 290, 291, 292, *294*, 295–6, *298*, 336, 337
    glacial successions 310
    sediment transport *170*, **171**, 180, 271
decay of dead organisms 9, 26, 27, 56, 103, 105–7, 115, 118, 120, 121, 189
deep-sea environments **207**, 289–90
    processes and sedimentation 290–1, 336
    sediment gravity flows 291–7, *298*, 300–1, 337
delta front *261*, **262**, 263, 264, *265*, 266, 291
delta plain 261–2, *263*, 264, *265*, 266
deltas 207, 248, *249*, 260, 297
    architecture and deltaic successions 261–4, *265*
    controls on delta building 261, 265
deposit feeders 27, 28, 35, 40, 62, 118
    trace fossils 82–5, 89–90, 101, 124, 321
deposition *147*, **151**–2
    alluvial sediment 227–9, 240
    bedding and lamination 154
    coastal 1–2, 3
    evaporites 330, 335
    factors controlling 208–10, 218, 268–71, 286, 330
    glacial 306–7, 312

grain orientation 154–5
planar deposition 158, 173, 325, 327
textural maturity 152–3
deserts 207, 237
    evaporites 240, 242, 243, *332*
    water in 237, 240, *241*, 242
    water table 238, 239, 243
    wind in 237–40
desiccation cracks 176, **177**, 180, 242, 274, *275*, 286, *332*
Desmoceras 48
detritus 27, 42, 49, 62, 82, 86, 99
diachronous 96
diagenesis 103, 111–20, 121, 193, 204, 205
diatoms *188*, 189, 299, 301
Dibunophyllum 57–9
Dicellograptus 67
diductor muscles 53
dinosaurs 10, 11, 12, 70, 92, 93, 94, *113*, 315
Diplocraterion 89, *90*, 101, 128
discharge 222–3, 234, 315
    braided river systems 230, 231–2, 332
    deltas *249*, 260, 261
    meandering river systems 226, 227, 230
    meltwater *150*, 307, 309
dissepiments 57
distal 208, 212, 263
distributaries 261–2, *263*
dolomite 115, 183, 195, 196, 198, 205
dormouse 132
dorsal 29, 30, 33, 42, 43, 53, 59, 65
dreikanters 240
dropstones 310, 312
drumlins 307, *309*, 312, 337
dunes
    subaqueous **156**, 159, 167, 179, 180, 327
    types of 239
    see also **aeolian dune**

E

Echinodermata 64, 65, *66*

echinoderms 23, 36, *39*, 42, 54, 64, 85, 93, 99, 105
    salinity and 127
echinoid 23, *24*, *25*, *285*
    anatomy of *37*, *38*, *39*, 55, 56, 64, 99
    diversity 92
    fossil formation 105, 106, 110, 124, 321
    interpreting fossils of 36–42, 95
Echinoidea 64, *66*
Echinus 39
ecology of marine animals 25–8
Ediacaran fauna 72, 91
edible sea urchin *39*
elephants 13, 14
endoskeleton 36, 54, 99
energy of the depositional environment 152
environment
    observation, process, environment concept 5–8, 315–16, 317–18
    palaeoecology and 11, 16, 123–33
epeiric carbonate platforms 272
epifauna 25–6, *28*, 34, 41, 99, 105, 109, 254, 319, 320
    shell damage 109
ergs 237
erosion
    bioerosion 187, 191, 268, *269*, 271, 286
    coastal erosion 1, 2, 3
    erosional structures 150–1, 179
    glacial 177, *178*, 304, *305*, 306, 307, 312
    mass mortality and 105
    river banks 225–6, 227, 230
    sediment transport 146, *147*, 148, 208–9
erratics 307
eskers 307, *309*, 312, 337
evaporites
    alluvial environments 223
    deposition 330, 335
    desert 240, 242, 243, *332*
    sediments from solution 183, 190, 191, 328
    shallow-marine carbonate environments 273, *275*, 286, 335
    supratidal 199, 205

Index

evolution 11–12, 13, 14, 15, 16, 91–2
exfoliation 136
exoskeleton 23
    arthropods 65, 100
    bivalves 30
    trilobites 59, 60, 65, 100, 321
external mould 112, *114*
extinctions 11, 12, 16, 87, 98
    mass extinctions 92, 93–4, 101
eyes
    arthropods 65, 100
    molluscs 42, 49
    trilobites 22, 60, 62, 100, 320

F

facies 210–11
    barrier islands 258
    beaches 210, *211*, 212, 213
    carbonate platform facies models 280–1, *282–3*, 284–5, 286
    environmental models 218
    graphic logs *211*, *213*, 214–17, 218, 330, *331*
    peritidal facies 273, 280, *281*, *282*, *283*, 284, 286
    sediment gravity flow facies models 296–7
    vertical successions of *211*, 212, *213*, 214–17, 218
    Walther's Law 212–13, 218, 264, *281*
fairweather wave-base
    aqueous bedforms *165*, 166, 180
    coastal and continental shelf environments *246*, *247*, 255, *257*, *259*, 265, 333
    shallow-marine carbonate environments 280, *281*, *282*, 286, *335*
family 63, 64
fan-worms 124, *125*
Favia 58
Favosites 58
feathered dinosaurs 11–12, 315
feeding types, marine animals 27–8, 102, 320
feldspars 137, 194, 201, 323, 324
feldspathic minerals 138
feldspathic wacke *201*, **202**, 205, 328, *329*

Fenestella 68
ferns 76, 77, 79
ferric oxide 138, 325
fish 11, 26, 49, **69**–*70*, 92, 94, 100, 111
    fossilisation *116*, *119*, 184, 255
fivefold symmetry 23, *24*, 36, 37, 54, 85, 99
fjords 309
flame structures 176, 193, 292, *293*, 301
flies 60, 118
flocculation 210, 256, 261, 265
floodplain 137, **221**, *224*, *228*
flute casts 150, *174*, 336
flute marks *173*, **174**, *175*, 179, 292, 301
fluvial channels 221, 223–4, 304
    braided river systems 230–2, 234, 332
    meandering river systems 207, 224–9, 234, 262, 332
fluvial processes 247–52, 333
fluvioglacial sediments 306–7, *311*, 312, 337
foot *32*, *33*, 34, **42**, *43*, 45, 98, 99
    see also **tube feet**
foraminifera 71, 72, 73, 74, 95, 100, 101, 299–300
foreshore zone **246**, 251, *252*, 253–4, *255*, *257*, *259*, 265–6
fossil 9
    see also **body fossils; macrofossil; microfossil; trace fossils**
fossil fuels 8, 12
    see also *specific fuels*
fossilisation 103–4
    biological attack and 108–9
    chemical influences on 110–11
    deaths of organisms 104–5, 121
    decay and disintegration 105–7, 121
    diagenesis 103, 111–20, 121
    final burial 103, 111, 121
    fossilisation potential 60, 92, 107–8, 116, 121, 124, 183, 315, 321, 322
    metamorphism 103, 120
    moulds and casts 111–14, 121, 322
    neomorphism 114, 121

    permineralisation and replacement 114–15, *116*, 117, 121
    physical influences on 109–10
    plants 116–17, 121
    sedimentary compaction 120, 121
fossilisation potential 60, 92, **107**–8, 116, 121, 124, 183, 315, 321, 322
frost shattering 135
fulgurites 242
funnel *46*, **49**, 99, 321
FWWB *see* **fairweather wave-base**

G

gabbro 325
gas, natural
    formation 71, 189, 191, 256, 260
    fossils and 8, 12, 71, 76, 116
Gastropoda *66*, 320
gastropods 21, *24*, *25*
    anatomy of *42*, *43*, 49, 99, 320–1
    carbonate coasts 274, 276
    diversity 92
    extinctions 93
    fossil formation 112, 114, 124, 183, 321
    interpreting fossils of 42–6, 51, 95, 99, 104, 113, 127, 129, *130*, 204, 323, 335
    siliciclastic coasts 254, 255, 258, 263
genera 63, 81
genus 63, 64, 77
geological time 313–16
gills *32*, *33*, 34, 35, 37, 42, 43–4, 45, *46*, 49, 61, 62, 65, 87, 98, 99, *119*
Ginkgo 79, *117*
glabella 62
glacial drift 177, 180
glacial environments **207**, 303–4
    depositional landforms 306–7, 312
    erosive landforms 304, *305*, 306, 307, 312
    glacial lake sediments 308, 312
    glacial successions 310, *311*, 312
    glaciomarine environments 308–10, 312
    sediment transport 177, *178*, 180
    wind-blown loess 308, 312, 337

glacial striations 177, *178*, 179, 180, 304
glaciotectonic processes 306
Glamys 132
global warming 1, 8
goethite 197, 198
Goniatites 47
goniatites 47, 93, 99
grain fall 168, *169*, 180, 238
grain flows 168, *169*, 170, **171**, 180, 238, 243
grain-supported fabric 152, *153*, 194
granite 135, 324
graphic log 214, 215–18
see also specific environments
graptolites 67, 93, 95, 100, 101
graptoloids 67
grazers 27, **28**, 45, 59, 62, 99, 109, 129, 209, 268, 278, 319, 320
'greenhouse' conditions 4
groove marks 174, 179
growth lines 30, 43, *57*, 114, 121, *130*
groynes 251
Gryphaea 35, *114*
guard 66, *67*, 100
gutter casts 151, 179, 251
gypsum 113, 183, 190, 191, 273

H

halite 183, 190, 191
Hallopora 68
hanging valleys 304, *305*
hardgrounds 90, *126*, **197**, 205
HCS *see* hummocky cross-stratification
head (mollusc) 42, *43*, 44, 49, 66
head-shield 22, 24, 60, 87, 100
hematite 113, 196–7, 205
hemipelagic sediments 299, 301, 336
gravity flows *173*, **174**, *175*, 293
herringbone cross-stratification 161, 179, 277, *314*, *333*, 334
high magnesian calcite 183, 190, 205
holdfast 54, *55*
homology 14, 16, 23, 29, 41, 48, 59, 64, 318
horseshoe crabs *81*, 127

hummocks 165, 166, *256*, 306
hummocky cross-stratification
aqueous bedforms **165**–6, 180
coastal and continental shelf environments *256*, *257*, *258*, *259*, 266, 333
sandstone deposition *313*, 314
shallow-marine carbonate environments 277
hydrogen sulfide 104, 106

I

ice sheets 177, 180, 303, 308–9, 312
icebergs 308, *309*
'icehouse' conditions 4
ichnology 80
ichthyosaurs 9, *70*, 94, 106, *115*
Illaenus 63
illite 137, 196, 198, 205
imbrication 155
impression fossil 117, 121
in situ (fossil) **125**
inarticulate brachiopods 52, 100, 127
index fossil 97
infauna 25, **26**, 28, 32, 41, 82, 85, 105, 319, 321
foreshore 254
siliciclastic coasts 254, 263
insects 16, 60, 65, 92, 93, 100, 118, 120
interambulacra 37, 38
interambulacrum 37
interdune areas 239–40, 242, 243
interference ripples 156, *182*
internal mould 20, 21, *46*, 47, 111, *112*, *119*
intertidal area
carbonate coasts 273, 274, *275*, 277, 284
fossils 75, 82
siliciclastic coasts **246**, *255*
inverse grading 152
invertebrates
ecology of marine invertebrates 25–8
see also individual invertebrates
iron oxide 138, *143*, 324
cement 194, 196–7, 205, 238

irregular echinoid 40, 41

J

jellyfish 25, 26, 65
jet propulsion 49, 99, 321

K

K-feldspar 137, 323
kaolinite 137, 196, 198, 205
keel 47
kerogen 189, 191
kingdoms 63, 64
Kosmoceras 50

L

lag deposit 109, 129, 148, 227, 231
lagoon
carbonate coasts 274, *275*, 276, *278*, *282*, 284, 286, 335
fossils 75, 90, 127
siliciclastic coasts 207, 249, **257**–9, 266
laminar flow 144, 178
lamination 82, **154**, 179, 322
lateral 29
lateral accretion surfaces 224, **227**, 228, 229, 262, *314*, 332
lee 155, **156**, 160, 168, *169*, 180
Lepidodendron 77–8, *117*
levées 224, **226**, 229, 234, 261, *263*, 297
life, a history of 91–4, 101
life position 124, 125, 126, 131, 133, 226, 276, 278
ligament 30, *31*, 33, 34, 53, 98
ligament groove 30
limestone
cementation in 197–8
chemical weathering 138
classifying 202–4, 206, 216–17, 329
formation 183, 184, 186, 190, 328
redeposited limestone 215, 282, 286, 335
limonite 138, **197**, 205, *223*
limpets 42

linear dunes 239
Lingula 52, 127
Lingulogavelinella 72
lithic arenite *201*, **202**, 205, 328, *329*
lithic arkose *201*, **202**, 205, *329*
lithic wacke *201*, **202**, 205, *329*
lithology
    carbonate platforms 276
    fossils 217
    glacial environments 307, 310
    graphic logs 214, 215, 216, 218
    sedimentary rocks 210, 218, 222, 330, *331*
Lithostrotion 57–9
lizards 92
load casts 176, 180
lobsters 59, *130*, 254, 255
local range zones 96
Lockeia 86, 89, 101
lodgement till 177, 178, 306, 310, *311*, 312
loess 308, 312, 337
longshore currents 156, **251**, 265
Lopha 35
lophophore *51*, **52**, 53, 65, *115*
low magnesian calcite 183, 190, 205
lower flow regime
    turbidity currents *173*, *175*, 292
    unidirectional current bedforms *157*, **158**, 325, 326, 327

M

macrofossil 11, 16, 19–20, 66
mafic minerals 138, 140, 197, 324
magnetic resonance imaging (MRI) 112
magnetostratigraphy 98
mammals 11, 14, 16, *64*, 69, **70**–1, 92, 94, 100
mammoths 13–14, *15*, 71, 120
manganese 17, *204*, 300, 337
mantle 31, **42**, 43, **52**, 65, *130*
mantle cavity *32*, **33**, 35, *42*, **43**, 44, *46*, 49, **52**, 53, 65, 98, 99
Marssonella 72
Marsupiocrinites 56
mass extinctions 92, **93**–4, 101

mass mortality 104, 105, 106
matrix-supported fabric 152, *153*
mean global temperature fluctuations 4, 317
meandering river systems **207**
    components of 224–6, 234, 262, 332
    vertical successions in 227–9, 234
meanders 224, 225, 226, 234, 332
melt-out till 177, 178, 306, 310, *311*, 312
Mesolimulus 81
meteorite impact 94
micas
    chemical weathering 137, 323, 324
    compaction 194
    deposition 151
mice 132
Michelinoceras 47
Micraster 40–1
micrite 186–7, 191, 197–8, 203, 216, 268–9, 329
    see also **biomicrite; pelmicrite**
microfossil 11, 16, 66, 71–5, 95, 100, 101, 174, 274, 300
Microraptor 12
Mollusca 65, *66*, 98–9, 255, 320–1
    see also **bivalves; gastropods; cephalopods**
Monograptus 67
moraine 177, *305*, 306, 308, *309*, 312, 337
morphology 6, **25**
    interpretation in fossils 13–16
mosaic evolution 78
mould 112
    see also **internal mould**; external mould
moulting 59–60, 65, 100, 104, 321
mouth bars 263, 264, 265, 266, 334
Mucrospirifer 51
mud drapes 161, 162, 179, 259, *314*, 333, 334
mudflows *170*, **171**, 295–6
mudmounds 278, *279*, *281*
mudstones 189, 191, 199
    alluvial environments *229*
    cementation 198, 205
    coastal environments 256

    compaction 120, 193, 205
    fossils in 76, 82, 114, 322
    graphic logs *214*, 215, *313*, *331*
    sediment movement 150, *154*, *176*
    sedimentary rock classification 200, 201, 205, 328, 329
muscovite mica 137, 324
mussels 34, 35, *108*, 124, 125, *128*, 184, 284
Mutilus 74
Mytilus 128

N

nannoplankton 71, 72
nautiloids 46–7, 48, 99
Nautilus 18, 46, 48, 49, 59, 99, *130*
neap tides 162, 179, 246, 253
nekton 25, **26**, **28**, 49, 50, 97
Neocrinus 55
neomorphism 114, 121
Nereites 83, 84, 85, 86, 89, *90*–1, 101
Neuropteris 79
Nilssonia 79
nodular chert 189, 191, 199, 205, 300
nodules 115, *116*, *136*, 190, 198–9, 205, 300
normal grading 152, 168, 310
Notelops 119

O

observation, process, environment concept 5–8, 315–16, 317–18
oceanic anoxic events 106
octopus 46, 65, *130*, 277
offshore transition zone
    carbonate coasts 277, *281*
    siliciclastic coasts *246*, **247**, 255–6, *257*, 258, *259*, 265, 266, 333
offshore zone *246*, **247**, 256, 265, 266
oil
    formation 71, 189, 191, 256, 260
    fossils and 8, 12, 71, 76, 116
Olenoides 60, 61, 62
olivines 137, 139, 194, 323, 324, 325
ooids 151, 186, 187, 190, 191, *203*–4, 277, 329

oomicrite 203, 273
oosparite 203, *204*, 206, 270, 277, 286, 330, 334
oral surface 36, 37
order 63, 64
ostracods 71, *74*, 100, 127, 323
overgrowths 195, 196, 205
ox-bow lake 224, **226**, 229, 234, 262, 332
oysters 22, 34, *35*, 114, 124, 132, 197, 258, 279

P

Pajaudina 51
palaeobiology 11, 16
palaeobotany 76
palaeocurrents 159, 160, 179
palaeoecology 11, 16, 123–33
palaeoenvironments 8, 16
Palaeoniscus 70
palaeontology 11
Paleodictyon 84, 85, 86, 89, *90*, 101
pallial line 31, *32*, 33, 318, 322
pallial sinus 31, 32, 33
palynologist 76
Paracidaris 37, 40
Paradoxides 63
parasites 28
partial range zones *96*, 97
Patella 42
patterned ground 303, 304
pearly nautilus *18*, 318
peat 189, 191, 223, 259, 262, 264, *265*
pedicle 51, *52*, 53–4, 100
pedicle valve *51*, *52*, 53, 54, 100
pelagic
 animals 25, **26**, *28*, 48, 67, 74, 97, 110
 sediments **298**–300, 301, 337
pelmicrite 203, 273, *282*, 286, 330, *335*
peloids 185, 190, 203, 268, 329
pelsparite 203, *275*
perforated plate *37*, **38**, 39, 40, 41, 55
peritidal facies **273**, 280, *281*, *282*, *283*, 284, 286

permafrost 14, 303, *304*
permeability 152
permineralised 115, *117*, 121, 122
petrifaction 115, *117*, 121
petroleum 17, 71, 73, 75–6, 116
photic zone 82
photosynthesis 59, 73–4, 75, 78, 82, 91, 94, 125, 277
phragmocone 66, *67*
phyla 63, 69, 91, 92, 98, 101
 recognising 64–6
phylum 63, 64
physical weathering 135–6, 139, 237, 324
phytoplankton 71, 73, 94, 184, *188*
pinna 79
planar cross-stratification 159, 166, 232, 240, 255, 259, 265, *333*
planar deposition 158, 173, 325, 327
planar stratification 154, 157–8, 179, 210, 254, 313, 327
 alluvial environments 229, 232
 deep-sea environments 292, 301
plankton 25, **26**, 27, *28*, 49, 57, *67*, 68, 71, 72, 73, *74*, 82, 90, 95, 97
 extinctions 93–4
 see also phytoplankton; zooplankton
plants 27, 28, 63, 75–9, 92, 100
 decay of 107
 extinctions 93, 94
 fossilisation of 116–17, 121
playa lakes 240, **242**, 243, *332*
plesiosaurs 70, 93, 106
Pleurotomaria 42–6, 319–20
point bar *224*, **225**, 226–9, 231, 234, *262*, 264, 332
pollen 71, 76, 77, 78, 95, 100, 101, 107
polychaetes 69
polyp 28, 37, **57**, 58, 100
pore pair *36*, *37*, **40**, 41, 99
porosity 152
posterior 30
Precambrian Era 72, 91
predators 27, *28*, 32, 34, 35, *45*, 49, 62, 80, 90, 92, 99, 104, 105, *113*, 320
 shell breakdown by 108, 129, *130*

preservation potential 60, 92, **107**–8, 116, 121, 124, 183, 315, 321, 322
pressure dissolution 194, 195, 205
prodelta *261*, **263**, 264, *265*, 266
progradation 212, 218, 263, *264*, 310, *314*, 334
protractor 33
proximal 208
Pseudodiadema 36–41, 122, 319, 322
pseudoextinction 97
pseudofossils 11, 318
pseudomorphs 190
Pseudotextulariella 72
pterosaurs 70, 93
pygidium 22, 62, *63*, 100
pyrite 106, 113, 115, 118, 120, 121, 127, 138, 217, 256, 300, 322, 323
pyroxenes 137, 194, 323, 324, 325

Q

quartz 115
 cement 194, 195, 205
 chemical weathering 137–8, 139, 323, 324, 325
 sand texture *146*
 sedimentary rock classification 201
quartz arenite *201*, **202**, 205, 210, *329*, 330, 334
quartz wacke *201*, **202**, 205, 328, *329*
quartzite
 composition 325
 weathering 324, 325
Quaternary Ice Age 3, *115*

R

radiolarians 74, *75*, 100, 188, 299
radiometric dating 95, 97
radula 45, *46*, 49, 65, 321
range charts 95
red beds 138, 196–7, 205
redeposited limestone *215*, 282, 286, 335
reefs 58, 59, 75, 92, 93, 127, 210, *270*, 274, *276*, 277–9, *283*, 284
regular echinoids 40, 41, 99, 105
replacement 115, 117, 119, 120, 121

reptiles *9*, 11, 66, 69, **70**, 93, 100, 106
retractor muscles 33
Rhacolepsis 116, *119*
rhinoceros *70*, 71
Rhizocorallium 88–9, *90*, 101
Rhizopterion 69
ribs 34, **47**, 50
rimmed shelf carbonate platforms *272*, 276, 281, *282–3*, 284, 286
rimmed shelves 272, 276–8, 281, *282*, 284, 286, 334–5
rip currents 251, 265
rivers
    alluvial fans *231*, 233, 234, 237, 242, 243, 304, *332*
    fluvial channels and sedimentary successions 223–32
    river flow 221–2
    river valleys *221*, 222–3
roches moutonnées 306
rock 'flour' 177, 308, 312
rudists 95, 126, 279
rugose (corals) **58**, 59, 93, 100, 127
Rusophycus 87, *90*, 101, 321

S

sabkhas 190
salinity 127, 209
saltation 144, 145, 146, 166, *167*, *174*, 222, 227
saltpan 242
sand waves 159, 169, 188, 191, *247*, 252, 325
sandbodies
    carbonate coasts *276*, 277, 286
    siliciclastic coasts 251, *257*, 258, 277
sandstones
    alluvial environments *229*
    cementation in 194–7, 205, 238
    classifying 200, 201–2, 210, 328
    deserts *238*
    graphic logs 216, *313*, *331*
Saturnalis 75
scallops 34, 124
Scaphites 50
scavengers 27, *28*, 45, 49, 62, 99, 103, 106, 125, 320

shell breakdown by 108
scleractinian (corals) **58**–9, 92, 100, 127
scours 150, *151*, 179, 251
SCS *see* **swaley cross-stratification**
sea anemones 57, 58, 59, 65, 255, 277
sea fans 65, 125
sea floor, ecology of 25–6
sea level fluctuations 3, *4*, 209, 317
sea slugs 42
sea snails 42
sea urchins 23, 36, *39*, 94, 184, 187; *see also* echinoid
seawhips 65
sediment gravity flows 170–4, *175*, 180, 191
    alluvial environments 222, 234
    deep-sea environments 291–7, *298*, 300–1, 337
sedimentary facies 210–11, 218, *332*
    see also facies
seed ferns 79
septa 20, **21**, *24*, *46*, 47–8, 49, 50, 51, 57, 58, 100
septarian nodules 199
septum 20, 21, *57*; *see also* septa
Serpula 69
sessile 25, *26*, *28*, 68, 104, 105, 114, 125, 131, 319, 320
sexual dimorphism 50
sexual reproduction 58, 91
shallow-marine carbonate environments 267
    carbonate buildups 277–9, 280, 286
    carbonate platform facies models 280–1, *282–3*, 284–5, 286
    carbonate platforms 207, 272, *276*
    factors affecting carbonate deposition 268–71, 286
    peritidal facies 273–7, 280, *281*, *282*, *283*, 284, 286
    tidal flats 273–4, *275*, *282*, 286, *335*
shell bed 109, 129
shells
    breakdown of 108–9, 110, 129, *130*, 254
    casts and moulds 111–14, 322
    fossilisation 107, *115*, 120
    growth of 30, 42, 49–50, 65, 130

shoreface
    carbonate coasts 277, 280, *281*
    siliciclastic coasts *246*, **247**, *251*, 255, *257*, *259*, 265
shrimps 65, 88, *119*, 185
siderite 115, **195**, 199, 205
silica 68, 73, 74, 100, 115, 188–9, 190, 300
silicate rocks
    chemical weathering 136–8, 201
    nodules 199, 300
siliceous sediments 188–9, 190, 336
siliciclastic coastal and continental shelf environments 245
    barrier islands 207, 257–9, 265, 297
    deltas 207, 248, *249*, 260–4, *265*, 266, 297
    fluvial, tidal and wave processes 247–52, 258–9, 333
    strandplains 207, 218, *249*–50, 252–7, 258, 265, 297, 333
    subdivision of 246–7
siliciclastic rocks 7, 135
    cementation 196–7
    classifying 199–202, 205, 216
    graphic log *214*, 216, *259*, 265
siliciclastic sediments 135, 136, 140, 183, 187, 188, 191, 245
    carbonate sediments, compared to 267, *268*, 271–2, 286
Sinornithosaurus 12
sinuosity 223–4
Siphonia 69
siphons 31–2, 33–4, *43*, **44**, 45, 98–9
siphuncle *46*, *47*, **48**
skeletal elements of fossils 20–3, *24*
slides 170, 180, *222*, 289, *290*, *291*
slipface 168, *169*
slugs 21, 42
slumps 170, 180, *222*, 289, *290*, *291*
smectite 196, 198, 205
snails 17, 21, 42
snakes 9, 92, 94
sockets 31, **52**
sole structures 175
solitary (coral) **20**, 58, 100
sparite 197, 203, 216, 329, 335

species 63, 64
    diversity 91, 92, 93, 101
    in biostratigraphy 95–8, 101
spicules 68, 71, 300
spines 26, *37*, **38**, 39, 40, 41, 62, 87, 99, 106, 124, 322
spiny cockle *32*
sponges *25*, 68, *69*, 71, 85, 87, 93, 100, 109, 127, 187, 188, 300
spores 71, 76, 77, 78, 79, 92, 100, 107
sporomorphs 76, 77, 100, 107, 116
sporopollenin 107, 121
spreite 88–9, 128
spring tides 162, 179, 246, 253, 255
squid *25*, 26, 46, 49, 66, 120
stalactites and stalagmites *138*, 197
star dunes 239
starfish 34, 85, 86, 130
Stenopterygius 9
Stigmaria 77, 78, *117*
storm wave-base
    aqueous bedforms **165**, 166, 180
    coastal and continental shelf environments 246, 247, 255, *257*, *259*, 265, 333
    deep-sea environments *291*
    shallow-marine carbonate environments *281*, *282*
stoss 155, 160, 168, *169*
strandplains 207, 218, 249–50, 252–7, 258, 265, 297, 333
stratification 154, 179
    see also **cross-stratification**
stratigraphic column 95, 101
Stratiotes 132
striated pavements 306, 312
stromatolites 72, 75, 91, 100, *188*, 203, *204*, 268, 274, *285*, 286
stylolites 194, 205
subarkose *201*, **202**, 205, 328, 329
subduction 3, 107, 290
sublitharenite *201*, **202**, 205, *329*
submarine canyons 289, 290, 291, 296, 297, 301
submarine fan 289, **296**, 297, *298*, 301, 304, 312, 314, 337
subtidal areas
    carbonate coasts *273*, *274*, *275*, 277

siliciclastic coasts **246**, 252
supratidal area
    carbonate coasts 273, 274, *275*, 284
    evaporites *190*, **199**, 205
    fossils 75
    siliciclastic coasts **246**
surface creep 145, 166, *167*
suspension feeders 27, 28, 90
    ammonites 49
    bivalves 34, 99, 319
    brachiopods 100, 320
    crinoids 99, 320
    graptolites 67
    sponges 68
    trilobites 62
    worms 24, 125, 131
suspension load 145, *147*, 148, 222, 261
suture 21, 47, *48*, 49–50, 194, 205
swales 165
swaley cross-stratification 165–6, 180, 255, 265–6
swallowers 27, *28*
swash 253, 254
SWB see **storm wave-base**
symbiosis 59, 62, 100, 277

T

tabulate (corals) **58**, 59, 93, 100, 127
tailpiece 22, *24*, 100
taphonomy 103, 106, 108, 110, 121, 133
taxa 64
taxonomic hierarchy 64
teeth 31, **52**
temperature, global *4*
tentacles 27, 38, 39, *42*, *46*, 49, 57, 67, 68, 100, 321
    lophophore *51*, *52*, 53, 65, 115
Terebratula 51
terrigenous sediment **289**, 297, 298–9, 301
test 36, 38, 99, 322
Tetragraptus 67
texturally immature 152, 187, 294
texturally mature 152, 153, 179, 187
thecae 67

theropods *10*, *17*
thorax (trilobite) **22**, 100
tidal bedforms 161–2, *179*, 258
tidal bundles 162, 179, *263*, *314*, 333, 334
tidal flats
    bedforms 155, *188*
    carbonate coasts 273–4, *275*, *282*, 286, *335*
    siliciclastic coasts 251, 258, *259*, 266, 274
tidal processes 247–52, 258–9, 333
tidal range 161–2, 247, 250, 258
till 177, 180, 307, 310
    see also **lodgement till**; **melt-out till**
tillites 306, *310*
time-averaged 129, 133
titanosaurs *10*
tool marks *173*, **174**, *175*, 292, 301
Torquirhynchia 130
torsion 43
total range zones 96
trace fossils 9, *10*, 11, 16, 17, 69, 80–2, 101, 126, 128, 129, *284*, *315*, 318
    coastal environments 254, 255, 277, 333
    current-swept environments 85–9, 131
    deposit feeders 82–5, 101, 124, 321
    water depth and 89–91, 101
transverse dunes 239, 243
trees 76, 77–9, 92, 94, 116–17
Trilobita 65, *66*
trilobites 22–3, *24*, 91
    anatomy of 59, 60–1, 100
    extinction 87, 93, 94
    interpreting fossils of 59–63, 323
    trace fossils of 86–7, 255, 321
Trinucleus 62, *63*
trophic pyramid 125
trough cross-stratification 159, 166, 228, 232, 255, 265, 332, *333*
tsunami 3, 5
tube feet 39, 40, 41, 55, 86, 99, *130*
tubercles 36, **37**, 38, 41, 99
turbidite 172, *173*, 292–5, 296

Index

turbidity currents
    deltas 263
    sediment gravity flows 150, *170*, **172**–4, *175*, 176, 179–80, 292–5, 301, 336
turbulent flow 144, 145, 146, 150, 155, 175, 178
Turricula 44
Turrilites 50, 51
turtles 92, 94
Tylocidaris 38
Tyrannosaurus 113

U

U-shaped valleys 177, *178*, 179, 180, 304, 312
umbo *29*, **30**, 35, *52*, 53, 54, 100
umbones 30, 130, 131
unidirectional current bedforms 155–60, *161*, 179, 232, 325, 326, 327
upper flow regime
    strandplains 254
    turbidity currents *173*, *175*, 292
    unidirectional current bedforms *157*, **158**, 159, 325, 327
    wave bedforms 165

V

vagrant *25*, 26, 27, *28*, 45, 319, 320
valves 21, 22, *24*, 29
    see also **brachiopods**; **bivalves**
varves 308, 310, 312, 337
ventral 29
vertebrates *64*
    body fossils 69–71
    earliest land 92, 101
    see also specific vertebrates
vertical successions
    in braided river systems 231–2, 234
    facies *211*, 212, *213*, 214–17, 218
    in meandering river systems 227–9, 234
visceral mass 42, 43, *46*, 49, *52*, 65

W

wackes 201, 202, 205, 294, 328, *329*
wadi 240, 241, 243, 292, *332*
Walther's Law 212–13, 218, 264, *281*
washover fan 259, 266
water
    aqueous bedforms 155–66, 179, 180, 327
    in deserts 237, 240, *241*, 242
    precipitation 221, 240, 243, 332
    sediment transport 143, 145–6, 148, 149, 152, 178–9, 222, 237, 326, *327*
water table
    alluvial environments 221, 335
    deserts 238, 239, 243
    sedimentary environments 209
    sedimentary rocks **193**, 198
water vascular system 38, 39, 40, 55, 99
wave-base 163
    see also **fairweather wave-base; storm wave-base**
wave bedforms 163, *164*, 165–6
wave processes 247–52, 253–5, 333
wave ripples 163, *164*, 165, 167, 181, 191, 270, 328
weathering
    biological weathering 136, 139
    physical weathering 135–6, 139, 237, 324
    process of 135, 139
    products of 143, 222
    see also chemical weathering
whelks 24, 42, 130
wind
    in deserts 237–40
    sediment transport 143, 145–6, 148, 178–9, 237, 242
    wind-blown loess 308, 312, 337
wind ripples 166, *167*, *168*, *169*, 180
winnowing 7, 109, 129, **148**, 166, 203, 291
wood 77, 78–9, 92, 107, 114, 115, 116, 117
wood mouse 132
worms *25*, 62, 69, 100, 131, 255, 263, 274, 284, 322
    trace fossils 81, 83, 88, 89

Z

zone fossils 67, **95**, 101
zooids 68
zooplankton 49, 67, 71, 73, 74, *75*, 94, 100, 184, 188
zooxanthellae 59